Origins of Mind

BIOSEMIOTICS

VOLUME 8

Aims and Scope of the Series

Combining research approaches from biology, philosophy and linguistics, the emerging field of biosemiotics proposes that animals, plants and single cells all engage insemiosis – the conversion of physical signals into conventional signs. This has important implications and applications for issues ranging from natural selection to animal behaviour and human psychology, leaving biosemiotics at the cutting edge of the research on the fundamentals of life.

The Springer book series *Biosemiotics* draws together contributions from leading players in international biosemiotics, producing an unparalleled series that will appeal to all those interested in the origins and evolution of life, including molecular and evolutionary biologists, ecologists, anthropologists, psychologists, philosophers and historians of science, linguists, semioticians and researchers in artificial life, information theory and communication technology.

For further volumes:
http://www.springer.com/series/7710

Liz Swan
Editor

Origins of Mind

Springer

Editor
Liz Swan
Longmont
CO, USA

ISSN 1875-4651 1875-466X (electronic)
ISBN 978-94-007-5418-8 ISBN 978-94-007-5419-5 (eBook)
DOI 10.1007/978-94-007-5419-5
Springer Dordrecht Heidelberg New York London

Library of Congress Control Number: 2012954274

Printed on acid-free paper

Springer is part of Springer Science+Business Media (www.springer.com)

For my husband Eric, who tells me to keep doing what I'm doing.

Acknowledgments

This volume is more than a collection of 21 chapters on the origins of mind. It is also a reflection of the people whose ideas and encouragement brought it into being. Lou Goldberg set me on a path of thinking about mind in a more scientific and realistic way. Joseph Seckbach kindly invited me to suggest a new Springer book on the topic of origins, and Marcello Barbieri and Jesper Hoffmeyer generously welcomed *Origins of Mind* into their book series on biosemotics. I am grateful to the wonderful Springer team who helped bring the book to fruition.

I'd like to thank all of the volume contributors for preparing such insightful chapters on mind, and also the reviewers who helped to make the content clearer and stronger. They include: Bernard Baars, Tom Barbalet, Marcello Barbieri, Megan Burke, Glenn Carruthers, Paul Cobley, Lou Goldberg, Justine Kingsbury, Curtis Metcalfe, Silvia Ouakinin, and David Skrbina.

And last but not least, I'd like to acknowledge Professor John R. Searle, on the occasion of his 80th birthday (in July 2012) for maintaining throughout his long career in philosophy that human consciousness is a biological phenomenon. Ab uno disce omnes.

Contents

Introduction: Exploring the Origins of Mindedness in Nature

Liz Swan

1 Mentis Naturalis

What is mind? This question is the single unifying force behind all efforts in philosophy of mind and cognitive science. When put in the context of biosemiotics and the broader life sciences, the question becomes, what is the nature of organic mindedness in the natural world? How did it evolve and why? Is it unique to humans or shared by other animals and even simpler forms of life? Is it peculiar to earthly life or is it part of the fundamental fabric of the universe?

A central underlying premise of this volume is that we will make more progress on understanding the phenomenon of mindedness if we conceptualize it as a natural process instead of as an object. The long tradition in the philosophy of mind and cognitive science of conceptualizing the mind as an object forces us to look for something that will fit our theoretical descriptions of it, even if this means forcing poor analogies between the mind and some object simply because we are in a better position to understand the object—only to discover that with this new knowledge we are no closer to a genuine understanding of organic mindedness.

By asking instead what the phenomenon of mindedness entails, we are already seeing it as a process instead of as a thing. The American pragmatists knew this and wrote exclusively from the perspective of this insight. Martin Heidegger too (in *Being and Time*) tried to circumvent the problem of mind by focusing instead on being—the experience or ongoing process we find ourselves in. Mindedness is a process that some organisms engage in, and each instance of mindedness will vary from one species to the next and even between individuals in a single species (as we know well in the human case).

L. Swan (✉)
Longmont, Colorado, U.S.A.
e-mail: lizstillwaggonswan@gmail.com

L. Swan (ed.), *Origins of Mind*, Biosemiotics 8, DOI 10.1007/978-94-007-5419-5_1,

Immanuel Kant's *transcendental idealism*, according to which we can infer a world of *noumena* (stuff) though the structure of the human mind limits us to experience only a world of *phenomena* (appearances), is one of the most robust attempts in the history of the philosophy of mind to conceptualize human mindedness as woven into the very fabric of the natural world. Though Kant didn't have the advantage of an articulated theory of evolution to draw on (Charles Darwin's *Origin of Species* would be published three-quarters of a century later), his notion that the human mind was in a sense determined to experience the world in particular ways in virtue of its intrinsic structure and function anticipated, by roughly 200 years, the application of evolutionary theory to the scientific study of the mind.[1]

Kant is singularly credited with having synthesized rationalism and empiricism in virtue of his progressive ideas about how the structure of the human mind fundamentally shapes how we construct our experience of, and thus come to know, the world. In essence, Kant's great contribution to philosophy of mind is the notion that knowledge comes neither from within nor from without. Rather, our knowledge of the world emerges from our particular human experience of it. This insight was particularly progressive, and reverberations of it can be seen in contemporary cognitive science.[2]

A philosophy of mind that followed Kant's lead, respecting a healthy balance of empiricist and rationalist intuitions, would be open to incorporating insights from the life sciences into its theories of mindedness, grounding abstract notions in hard fact. As it happened, however, the majority of twentieth century philosophy of mind, dominated as it was by the analytic tradition, enjoyed a robust existence completely insulated from discoveries and insights generated in the life sciences. It did, of course, engage with computer science in that functionalism—then the most popular philosophy of mind—was built on comparisons between machine functionality and human consciousness. The important point, however, is that discoveries in the biological sciences were for the most part not integrated into theories of mindedness in mainstream analytic philosophy of mind.

Focusing on what's wrong with contemporary philosophy of mind is an easy temptation to give in to. This volume resists this temptation and takes a different tack: It is about what mindedness is, from a naturalistic, scientifically informed perspective. Therefore, this book is intended to make a progressive contribution to our scientific and philosophical understanding of how organic mindedness emerged in the natural world.

Refreshingly, there is some crossover between the biosemiotics literature on mindedness and contemporary philosophy of mind literature. In particular, I've

[1] As Henry Plotkin (2004) explains, though one could identify the official beginning of psychology as a science with the establishment of Wilhelm Wundt's laboratory in Leipzig in 1879 and Darwin's 1859 publication of *The Origin of Species* as the first formal and popularized articulation of the theory of evolution by natural selection, the two branches of science were not synthesized in any real way until the emergence of the late twentieth century sciences of ethology and sociobiology and, later still, evolutionary psychology.

[2] See Brook (2004) for the full story.

been happy to discover the utilization of philosopher John Searle's work on mind and consciousness in the biosemiotics literature (e.g., see Brier 2012; Barbieri 2011; Kravchenko 2005; Hoffmeyer 1997). Searle's position that consciousness is a biological phenomenon, which he calls *biological naturalism* (1992), should, in my opinion, be the cornerstone of current and future work in the mind sciences; adopting this insight as a normative methodological principle would severely limit the creation and discussion of exceedingly abstract models of mind that are out of touch not only with the complex details of the brain but sometimes all of reality.

Mindedness is a biological phenomenon, thoroughly dependent upon a central nervous system in complex organisms such as humans and other primates, and a more diffuse kind of nervous system in less complex organisms. This simple observation implies that mindedness exists in degrees in the biological world, which entails that it certainly is not unique to humans, and that our particular kind of mindedness is just the most recent design in nature—its having existed in various forms long before hominins evolved.

1.1 Models in the Mind Sciences

Despite these simple insights that are fully supported by what we know about the natural world, it has been the tradition throughout much of the history of cognitive science to study nonliving, nonminded objects (in particular the computer) and draw inferences about the mind from such objects. Philosopher of physics and logician, R.I.G. Hughes, developed a metamodel of how models in science work—in particular how the results of models translate back to the phenomena in question (Hughes 1997). Figure 1 below captures the essence of Hughes' theory.

Certain elements of the natural phenomenon are *denoted* by certain elements of the model. The model is then used to *demonstrate* certain theoretical conclusions. And finally, those conclusions are *interpreted* in order to make predictions about the natural phenomenon. So, for example, physicists employ a ripple tank of a particular size and volume to model a certain stretch of coastline and use the tank to demonstrate some specifics of wave mechanics that are relevant to the coastline. The fact that the ripple tank bears no physical resemblance to the open body of

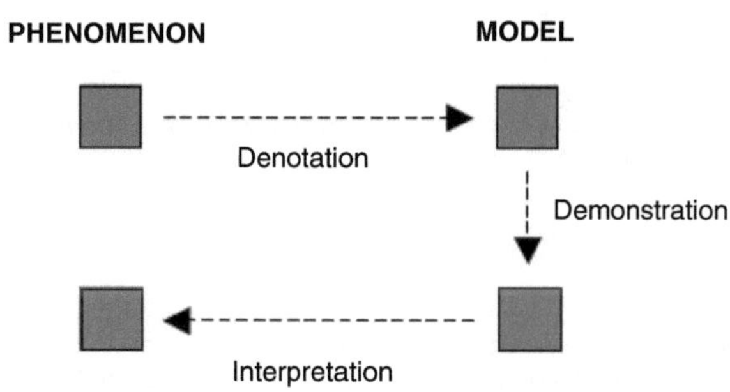

Fig. 1 Hughes' DDI metamodel of how models in science work

water makes no difference because the aim is an understanding of the behavior of water, and crucially, water is what is used in the model. And because of this crucial consistency of material composition between phenomenon and model, when experimental results are demonstrated in the ripple tank, experimenters are justified in using them to ground their predictions about various aspects of the behavior of naturally occurring bodies of water.

A further consideration makes the point of this exercise clear: It would be a mistake for physicists to infer that because the behavior of water in a ripple tank is sufficiently similar to that of open water, that therefore the ripple tank is an ocean or the ocean is a ripple tank. Physicists do not reason this way because to do so would be to confuse the model with the phenomenon. But this seems to be precisely what has happened in cognitive science. Applying the DDI model to the context of the mind sciences, the methodology expected is for biological systems to inspire models in nonbiological systems and for the results of those models to inform our understanding of biological systems (as demonstrated in Fig. 2 below).

The reigning view in cognitive science has been and still is functionalism. Functionalism, as a philosophy of mind, focuses only on the behavior of cognitive systems and not on their material instantiation. For this reason, cognitive scientists have systematically blurred the divide between biological and nonbiological systems. Discoveries made about the brain are believed to be implementable in hardware and software; likewise, discoveries made in hardware and software are presumed to translate to human brains. Contrary to the example from physics, however, where the material (water) is common to both the phenomenon and the model, in cognitive science, creations in silicon are used to model organic brains. It is philosophically irresponsible to talk as if the model were the same thing as the phenomenon and vice versa because the essential material is different in the model.

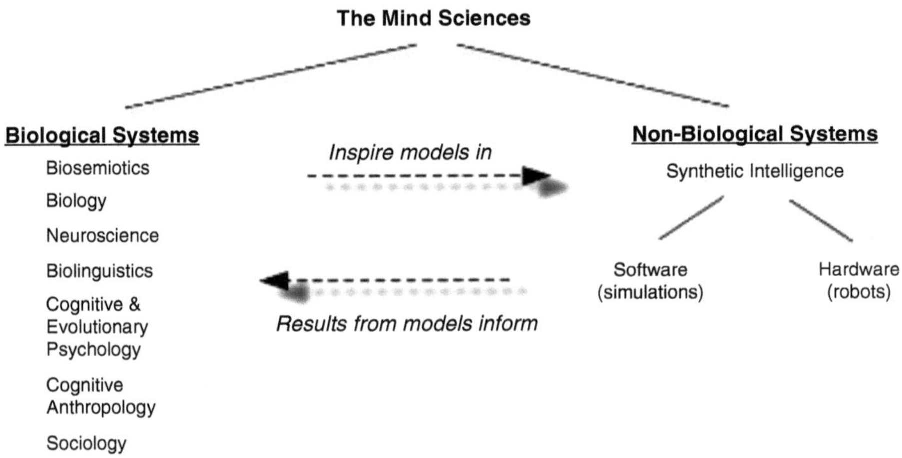

Fig. 2 The DDI metamodel as applied to the context of the mind sciences. This is a normative, rather than descriptive, picture of how models ought to function in cognitive science

Artificial intelligence as an enterprise has yielded considerable insight into the biological mind not because we have been able to reproduce in computer models exactly what is going on in the biological brain, but precisely because we have *not* been able to do so. What computer models of biological mindedness can do is reproduce certain elements of the brain's natural functioning, for example, calculation—and computers can perform this function much faster than human brains can. But it is wrong to infer from this example of successful modeling that therefore computers are brains (or conscious like brains) or, just as erroneously, that human brains are computers. Computationalism, when applied to anything but a computer, is useless until we can come up with good explanations for what it means for an organic brain to "compute," and if it computes symbols, what such symbols are like in the wet, gray matter of the brain.

In a paper illuminating how the pragmatists contributed to, and in some sense cleared a path toward, a naturalistic understanding of mindedness, philosopher Peter Godfrey-Smith outlines the principle of *methodological continuity* according to which: "Understanding mind requires understanding the role it plays within entire living systems. Cognition should be investigated within a 'whole organism' context" (Godfrey-Smith 1994). The reason this reasonable principle is not commonly followed is because those working on mind are typically not the same as those working on organisms and vice versa; in other words, philosophers employ abstract models and cognitive scientists employ software and hardware models to study mind, while life scientists who work with organisms typically are not working on problems of mind—with the obvious exception of discrete projects in experimental psychology that investigate various aspects of animal cognition.

The big question of how and why mindedness evolved necessitates collaborative, multidisciplinary investigation. Biosemiotics provides a new conceptual space that attracts a multitude of thinkers in the biological and cognitive sciences and the humanities who recognize continuity in the biosphere from the simplest to the most complex organisms and who are united in the project of trying to account for even language and consciousness in this comprehensive picture of life. The young interdiscipline of biosemiotics has so far by and large focused on codes, signs, and sign processes in the microworld—a fact that reflects the field's strong representation in microbiology and embryology. What philosophers of mind and cognitive scientists can contribute to the growing interdiscipline are insights into how the biosemiotic weltanschauung applies to complex organisms like humans where such signs and sign processes constitute human society and culture.

2 A Biosemiotic Theory of Mind

In this section, I outline a rough sketch of the beginnings of a biosemiotic theory of mind (BTM). I offer here the nutshell version of my view of mindedness: There is no such thing as "a mind" per se; rather, the term "mind" acts as a conceptual placeholder for a whole host of abilities that we and some other animals are able to do with our brains and bodies working in concert, such as communicate, show affection, imagine, satisfy our needs, learn, remember, hold beliefs, and plan. All living organisms have a host of abilities uniquely attuned to their particular environments, which in some cases, for example, the human case, we're inclined to conceptualize as "having a mind." Below, I explain how BTM differs from the picture of mind provided by other contemporary theories in analytic philosophy of mind and neurophilosophy.

2.1 BTM is Different from Analytic Philosophy of Mind

First, BTM is distinct from analytic philosophy of mind in that it is not concerned with abstract theories of mind that are developed and utilized exclusively within philosophy but rather with understanding mindedness as a natural phenomenon whose descriptions sit comfortably within the context of everything we know about the natural world (including brains and organisms) from the biological and cognitive sciences.[3]

An example concerning what philosophers call *qualia* (i.e., the qualitative aspects of experience) will help illuminate the distinction between analytic philosophy of mind and BTM. It has been argued in contemporary analytic philosophy of mind that if physicalism is right, then as physical beings, we should be able to perceive any color or sound or taste and respond to it appropriately without its being accompanied by any qualitative feel (Chalmers 1995). Important to note is the underlying presumption that human beings are physical things, and since physical things don't *experience* anything, neither should we. So, the argument goes, either we need something beyond mere physicalism to do the explanatory work of accounting for qualitative experience (which is philosopher David Chalmers' position) or we're not thinking about physicalism in the right way.

I believe that asking the question of why we have phenomenal experience shows that we are not thinking about physicalism in the right way—perhaps, for example, in virtue of lumping together animate and inanimate physical things and expecting them to behave the same, à la functionalism. Arguing that because bicycles and human beings are both physical things, we shouldn't have feelings because bicycles don't, shows poor reasoning in that it exemplifies the mistake of believing that one's ideas of the world somehow trump how the world really is.

Anything in the world that has meaning for us—a favorite song, a familiar face, a nagging headache, a green traffic light, a friend's embrace and the smell of coffee—is experienced qualitatively. Biosemiotics has boldly taken on the task of understanding

[3] BTM is similar to John Searle's *biological naturalism,* according to which human consciousness is a biological phenomenon like photosynthesis or digestion (his examples). My view differs from his, however, with regard to the nature of the relationship between mind and brain; specifically, he sees the relationship as *causal*, whereas I see it as *isomorphic*. We don't say that plants *cause* photosynthesis or that digestive tracts *cause* digestion and neither, I argue, should we say that brains *cause* minds; rather, the "mind" can be thought of as the brain as experienced by the agent. This mind-brain conceptualization is very similar to that presented in Fingelkurts et al. (2010).

how meaning emerges in biological systems. Since we know that we experience the world qualitatively with smells, sounds, sights, and feelings and that we're not unique among biological systems in doing so, the challenge is to explain how meaning emerges from matter, and here biosemiotics is more useful than a philosophical position that denies this is possible.

The question of qualitative experience is not a mystery in the context of BTM. To assume that qualitative experience is somehow superfluous to the mechanics of being begs the question of why and how beings would be motivated to do anything at all. It assumes, for instance, that we would know to take a drink of water without feeling thirst, or would pursue a particular academic subject without having a real passion for it, or be automatically driven to procreate without the excitement of sex and romance. Being attuned to the world through our senses offers obvious survival benefit in that it enables us to avoid drinking water that looks murky, eating food that smells rotten, or spending time with people or in places that make us feel unsafe. The fact that we experience the world qualitatively is what makes us different from robots and, moreover, makes us like all organisms that act in their environments in survival-enhancing ways.

Though the tradition in philosophy of mind has been to talk about the mind as if it were an atemporal and disembodied phenomenon, the gradual integration of insights from dynamic systems theory,[4] brain physics,[5] and neuroscience[6] into the philosophy of mind has forced us to think about how the brain actually works in a living organism, entailing a recognition of the mind as a necessarily embodied and thus spatial and temporal phenomenon.

2.2 *BTM Is Different from Neurophilosophy*

Much of the work in neurophilosophy (NP) is devoted to the effort of utilizing insights from neuroscience to inform questions in the philosophy of mind.[7] The research agenda therefore implicitly entails that in order to gain a deeper understanding of the mind, we must get a grip on how the brain functions. In what follows, I outline three significant problems with neurophilosophy which, taken as a whole, necessitate a more comprehensive biosemiotic theory of mind.

---

[4] For an excellent summary of the application of dynamic systems theory in cognitive science, see van Gelder and Port (1995).

[5] Fingelkurts et al. (2010) present a masterfully rich account of how the actual spatial-temporal structure of the physical world is presented to and experienced by the individual as phenomenal space-time in virtue of the brain's physiological operations, which are also spatial and temporal in nature. Their research breathes new life into Kant's theory that the particular structure of the human mind determines how we perceive the world.

[6] For an example of applying neuroscientific insights to the problem of mental representation, see Swan and Goldberg (2010).

[7] See, for example, Bechtel et al. (2001), Clark (2000), and Churchland (1989, 2002).

2.2.1 NP's Misguided Association with Eliminative Materialism

Neurophilosophy is unfortunately commonly conceptualized as a means to an unrealistic end, namely, the end of so-called folk psychology.[8] The idea is that as soon as we understand precisely how everything that we think and feel is just a result of particular neural events, we will no longer have the need for concepts such as thoughts, beliefs, and feelings. This strikes me as a particularly misguided goal for (at least) two reasons: (1) Folk psychology is familiar, useful, and integrated in our language and will thus be hard if not impossible to do away with, and (2) a thoroughly reductionistic approach to mind that overemphasizes the objective, third-person descriptive level to the neglect of the first-person level, is unsatisfactory because it does not account for meaning or the self or subjective phenomenal experience. As stressed by Marcello Barbieri, biology has traditionally shunned the problem of meaning, but biosemiotics provides a platform for grounding an account of meaning in biology.

2.2.2 NP's Implicit Disregard for the Brain's Evolutionary History

Because the discipline of neurophilosophy generally focuses on the human brain as it is now, it has carved out for itself a considerably narrow view of the mind, beyond which lay some of the most important and most interesting questions regarding the nature of mindedness, such as the following: In what ways is human mindedness similar to, and different from, (other forms of) animal mindedness? Was the emergence of human mindedness continuous with the emergence of earlier cognitive abilities in organisms with simpler nervous systems, or is human mindedness unique in the natural world? Given what we know about human evolution, why did mindedness evolve to be what it is now? What is it about organic minds that make them so difficult to reproduce synthetically?

2.2.3 NP's Scientifically Questionable, Overly Narrow Focus on the Brain

To its undeniable credit, what neurophilosophy does right is get philosophers of mind to think about the brain and encourage them to incorporate knowledge about the brain into their theories of mindedness. How useful or insightful could a philosophy of mind be if it's completely divorced from an understanding of the brain? And yet, the human brain doesn't do anything in isolation of a host living body that interacts with its environment, so a complete reduction of human mindedness to the

[8] This movement, known as *eliminative materialism*, is most closely associated with Patricia and Paul Churchland. For example, see Churchland (1999).

brain is as scientifically inaccurate as the thought experiments it was meant to replace (e.g., the brain in the vat).

In sum, a biosemiotic theory of mind, though it entails a thorough knowledge of the organismic brain, is distinct from neurophilosophy in that it (1) carves out a conceptual space for meaning understood in terms of beliefs, ideas, and other features of our "mental life," (2) embraces the biological origins and evolutionary development of mindedness as the necessary grounding for understanding human mindedness as it is now, and (3) focuses not just on the brain but on the entire living organism in its environment.

And finally, to illustrate the ways in which BTM differs from the other approaches to mind discussed in this chapter, we can invoke a jigsaw puzzle analogy. Analytic philosophy of mind can be a fun puzzle to play with when you have all the pieces—the terms, theories, concepts, and relations—but more and more it seems that the players have lost the cover to the puzzle box and are just endlessly rearranging the pieces without having any ultimate picture in mind. Neurophilosophy, on the other hand, has only some of the puzzle pieces and a part of the box cover design to work from. The major advantage that BTM has over these other theories is that it has the right cover to the puzzle box, and it has access to all the pieces (through interdisciplinary collaboration). The continual rearrangement and ultimate solution of the puzzle metaphorically represent the ideal, if not the reality, of how science works and thus how a scientific approach to mindedness ought to work (see Fig. 3).

3 The Volume's Contents

The purpose of this volume is to gather together a sampling of contemporary thinking on when, why, and how mindedness evolved in the natural world from researchers working in the biological, cognitive, and medical sciences. The question of the origin of mind is no longer the exclusive domain of philosophers; it has, in recent decades, become a respectable question for research scientists to work on as well.

The volume's contents are pluralistic. I've followed the tradition established by Marcello Barbieri in welcoming various viewpoints on mindedness to the table—some that thoroughly engage with the current biosemiotics literature, others that have less direct links to it, some that are consistent with my own views on mindedness, and others that are at odds with them. One element that most of the chapters in the volume have in common is in their adherence to the principle, endorsed by philosopher John Searle, and reflected in my own philosophical writings, that the phenomenon of mindedness, including the peculiarities of human mindedness, is a biological phenomenon. What I've actively sought out for this volume are thoughts, ideas, and theories that contribute to our naturalistic understanding of mindedness that address its biological origins and evolutionary development.

Key Term	Cognitive Science	Neuroscience	Biosemiotics
Meaning	Can we expect machines and robots of the future to find meaning in their computations and behavior?	Humans and other sufficiently complex organisms find certain things in their world meaningful. How does this happen?	What are the most basic components of meaning in biological systems? Is human meaning — e.g., in language — different in kind or only in degree from that of other beings?
Representation	Can robots exhibit intelligent behavior without utilizing stored representations? If they can, are they useful models of organisms, which do use stored representations?	Sufficiently complex brains represent features of the world and manipulate their representations mentally, for example when planning an action. How do they do it?	Are brain-objects a sufficient basis on which to build an account of mental representation or is something more needed (e.g., brain artifacts)?
Reductionism	Will reducing complex behavior exhibited in simulations (e.g., in artificial life) to its particulates illuminate how complexity emerges?	If we reduce human cognition and behavior to its neural correlates, do we lose or gain insight?	Are reductionistic explanations okay, and even useful, so long as they are not intended to *replace* more holistic explanations, for example of organic mindedness?
Mechanism	What kind of programs do we need to write to make the robot do x, y, or z?	In what ways and to what extent do humanoid robots illuminate human mindedness and behavior?	Can Cartesian mechanism be usefully updated with concepts from biosemiotics? Is natural selection the only mechanism of evolution or are there more?
Computationalism	Will computational power reach a critical threshold after which computers will be conscious and truly pass the Turing test?	What does it mean for biological brains to *compute?* What are they computing over? What are symbols in the wet, gray matter of the brain?	Does computationalism apply to simple organisms? Single cells? Or should the concept be sanctioned to computational models only?

Fig. 3 A comparison table of how questions concerning meaning, representation, reductionism, mechanism, and computationalism are typically formulated in the disciplines of cognitive science, neuroscience, and biosemiotics

The volume is divided into five parts devoted to the subtopics of biosemiotics, mental representation, consciousness, philosophy of mind, and synthetic intelligence. There is a chronological and hierarchical order to the chapters that might not at first be obvious to the reader. We begin with biosemiotics, which focuses on the most basic units of the biological world: codes, signs, and sign processes. Next is mental representation, which is ubiquitous in sufficiently complex organisms that can interact with their environments in survival-enhancing ways. Then we have consciousness, a level of awareness we believe to be characteristic only of some complex organisms. Next is philosophy of mind, understood in this context as an intellectual activity unique to humans. The last level in this system is synthetic intelligence, which emerges as a complex interactivity between humans and technology that can be utilized to investigate all of the levels below.

3.1 Biosemiotics

The first part of the book is titled *Biosemiotics* and contains four chapters that most closely engage with ideas currently circulating in the field on the topic of organic mindedness. The section opens with a chapter by Marcello Barbieri titled, "Organic Codes and the Natural History of Mind," who describes the idea that a neural code contributed to the origin of mind somehow like the genetic code contributed to the origin of life. More precisely, he suggests that mental objects are assembled from brain components according to coding rules, which means that they are no longer *brain objects* but *brain artifacts*. It also suggests that the parallel evolution of brain and mind was accompanied by the development of two new types of sign processes that gave origin first to *interpretive semiosis*, mostly in vertebrates, and then to *cultural semiosis*, in our species.

Next is a chapter by Angelo Recchia-Luciani, "The Descent of Humanity," that explores the notion of species-specific modeling which allows us to construct taxonomies of mental models. The taxonomic content is based on the models' differential capacity to adapt to behavior patterns controlled by neural networks. In humans, far more than any other primate, new cognitive devices are developed in fetalization that enables abstract thinking. Recchia-Luciani explains how fetalization and education are the two pillars that give rise to the human being's ability to accumulate a perceivable and collective knowledge, which is precluded to other animal cultures, and that the key to this evolutionary quantum leap is the advent of a new class of replicators—memes—defined as informational patterns of a signic nature with a metaphorical, relational organization.

Next, a chapter by Crystal L'Hote titled, "From Non-Minds to Minds: Biosemantics and the Tertium Quid," evaluates the prospects of the biosemantic program, understood as a philosophical attempt to explain the mind's origins by appealing to something that nonminded organisms and minded organisms have in common: representational capacity. She develops an analogy with ancient attempts to account for the origins of change, clarifies the biosemantic program's aims and methods, and then distinguishes two forms of objection, *a priori* and *a posteriori*. L'Hote offers reasons, by analogy with chemical combination and other everyday phenomena, to think that minded beings and their representational capacities might have their origin in nonminded beings. L'Hote concludes that an evolutionary explanation of mind is plausible.

Finally, the chapter "Cybersemiotics: A New Foundation for a Transdisciplinary Theory of Consciousness, Cognition, Meaning and Communication" by Søren Brier closes the section. In it, he explains why cybersemiotics shows that it is necessary to draw on our knowledge from the natural sciences and technologically founded information sciences, systems theory, and cybernetics to obtain a true transdisciplinary theory. He explains how the modern evolutionary paradigm combined with phenomenology forces us to view human consciousness as a product of evolution and accept humans as observers from "inside the universe." Brier explains how, therefore, the study of consciousness forces us to theoretically encompass the natural and social sciences as well as the humanities in one framework of absolute naturalism viewing the conscious life world with its intentionality as well as the intersubjectivity of culture as a part of nature.

3.2 Mental Representation

The second part of the volume is devoted to a subject close to my heart: mental representation. The subject has gotten a bad reputation in philosophy of mind due to its heavy baggage from Descartes' (understandably) scientifically naïve picture of how representation works in the human mind. I have high hopes for biosemiotics to provide a new and progressive discussion space for understanding how organisms with sufficiently complex nervous systems internally represent important features of their environments.

This section opens with a chapter by John Sarnecki, "The Emergence of Empathy in the Context of Cross-Species Mind-Reading," in which he explores how evolutionary accounts of the origins of mind-reading and empathy emphasize the selective pressures within human communities that contributed to our capacity to imagine ourselves in the spatiotemporal and cognitive place of other individuals. Sarnecki argues that these social accounts of empathy neglect the possible influence of mind-reading between humans and other species; for example, prehistoric hunting privileged the ability to take on the perspective of potential prey in tracking. A consequence of this view is that how we read the minds of other humans may have been conditioned by selective pressures for reading the minds of other animals, and thus, in our attempts to understand other humans, we may find echoes of the cognitive lives of animals.

Next is a chapter by Rob Arp titled "The Evolution of Scenario Visualization and the Early Hominin Mind" in which he argues that *scenario visualization*—namely, a mental activity whereby visual images are selected, integrated, and then transformed and projected into visual scenarios for the purposes of solving problems in the environments one inhabits—emerged in our hominin past and accounts for certain kinds of vision-related creativity. The kinds of problems with which our hominin ancestors were confronted most likely were of the spatial-relation and depth-relation types related to basic survival—such as judging the distance between an object and oneself, determining the size of an approaching object, etc., so the capacity to scenario visualize would have been useful for their survival. Arp concludes that scenario visualization has been and continues to be relevant for *vision-related* forms of creative problem solving.

A chapter by Michael Nair-Collins, "Representation in Biological Systems: Teleofunction, Etiology, and Structural Preservation," follows in which he proposes a novel thesis about the nature of representation in biological systems. He argues that what makes something a representation is distinct from what determines representational content and thus that it is useful to conceptualize *what it is to be* a representation in terms of fundamental concepts from biology, in particular teleofunction. By contrast, he explains, representational *content* is best understood as a structured relation involving two parts, and the explanation of how states of biological systems have content involves the preservation of internal structural relations and causal history. He explains how his theory provides a unifying theoretical framework within which a variety of neurophysiological mechanisms involved in a sensory discrimination task can be explained and interpreted as representational.

Concluding this section is a chapter by Isabel Barahona da Fonseca et al., "Beyond Embodiment: From Internal Representation of Action to Symbolic Processes," who link symbol formation to efferent processes that occur in an organism capable of movement. Action planning and command involves an anticipatory stance in which symbolic meanings are created and referred to the agent in the internal model that binds perceptual past, present, and future desirable states. They argue that symbols are abstract when projected beyond immediate instantiations and thus lie beyond embodiment.

3.3 Consciousness

This part begins with a chapter by Ellen Fridland, "Imitation, Learning, and Conceptual Thought: An Embodied, Developmental Approach," that offers a strategy for moving from imitation to conceptual thought. She argues that imitation plays a vital role in accounting for the facility with which human beings acquire abilities, but that successful task performance is not identical to intelligent action. In order to move beyond first-order behavioral success, she suggests that the orientation humans have toward the means of intentional actions also drives us to perfect our skills in ways that produce fertile ground for florid thought.

The next chapter "Evolving Consciousness: The Very Idea!" is by Jim Fetzer who explains that discovering an adequate explanation for the evolution of consciousness is an outstanding problem. He further explains that this difficulty is compounded by the introduction of notions like the unconscious and the preconscious and that an evaluation of the prospects for unconscious factors as exerting causal influence on human behavior depends upon understanding both the nature of evolution and the nature of consciousness. Fundamentally, this chapter advances a conceptual framework for understanding the evolutionary function of consciousness in genetic and cultural contexts. It becomes increasingly apparent that, given a suitable theoretical and semiotic perspective, an adequate explanation for the evolution of consciousness may be possible.

Teed Rockwell's chapter which is titled "Mind or Mechanism: Which Came First?" constitutes an anomaly in the volume in not assuming that human mindedness is an evolutionary phenomenon with biological origins. His chapter questions the reductionist assumption that bits of lifeless matter must have grouped themselves into complex patterns that eventually became living conscious beings. He argues that there is no reason to question Charles Sanders Peirce's suggestion that mind came first and that mechanical causality emerges when regions of a fundamentally conscious universe settle into deterministic habits. Rockwell reasons that if we define consciousness in a way that ignores clearly accidental properties such as looking and behaving like us, some form of panpsychism is not only possible but plausible and concludes that ignoring this possibility could cause us to subconsciously exclude legitimate avenues of research.

This part concludes with a chapter by Jonathan Tsou titled "Origins of Qualitative Aspects of Consciousness: Evolutionary Answers to Chalmers' Hard Problem" in which he analyzes philosopher David Chalmers' formulation of the hard problem of consciousness in terms of various "why-questions": Why does the physical processing of the brain give rise to a rich inner life? Why is the performance of brain functions accompanied by experience? Tsou explains that Chalmers suggests these questions are mysterious and that materialist explanations of mentality fail to adequately address them. Tsou argues that either Chalmers' why-questions do not fall within the proper purview of science, or there are evolutionary answers to them. With respect to the latter, he discusses evolutionary explanations for the qualitative aspects of various conscious states including pain, color vision, and orgasms.

3.4 Philosophy of Mind

This part begins with a chapter by Tibor Solymosi titled "Neuropragmatism on the Origins of Conscious Minding" who argues that the philosophy of pragmatism has much to offer mind and life scientists in thinking about the origins and nature of experience. He provides an introduction to neurophilosophical pragmatism by reviewing how classical pragmatists like John Dewey reconceived concepts such as experience, mind, and consciousness in light of Darwinism. He explores a recent debate in cognitive science and neurophilosophy over how to think about conscious mental activity and, in so doing, draws on and modifies the pragmatist framework sketched in the first part of the chapter.

Next is a chapter by Andrew Winters and Alex Levine, "Not So Exceptional: Away from Chomskian Salationism and Towards a Naturally Gradual Account of Mindfulness," who argue that a chief obstacle to a naturalistic explanation of the origins of mind is the position of human exceptionalism, as exemplified in the seventeenth century by René Descartes and in the twentieth century by Noam Chomsky. As an antidote to human exceptionalism, the authors turn to the account of aesthetic judgment in Darwin's *Descent of Man*, according to which the mental capacities of humans differ from those of lower animals only in degree, not in kind. They explain why a naturalistic explanation of these capacities is attainable by shifting away from the substance-metaphysical implications of the search for an account of *mind*, toward an account of the origins of *mindfulness*.

Tom Ray's chapter, which is titled "Mental Organs and the Origins of Mind," introduces a new hypothesis of the origins of complex mindedness through the emergence of "mental organs," defined as populations of neurons that bear a specific G-protein coupled receptor (GPCR) on their surface. He explains how mental organs provide a direct connection between mental properties (compassion, comfort, awe, joy, reason, consciousness) and the genes and regulatory elements associated with GPCR, and that mental properties associated with mental organs have heritable genetic variation and are thus evolvable. His chapter provides a comprehensive account of how the genetic and regulatory systems that control the more than 300 different GPCR expressed in the human brain allows evolution to richly sculpt the mind.

This part concludes with a chapter by Frank Scalambrino titled, "Mnemo-Psychography: The Origin of Mind and the Problem of Biological Memory Storage" in which he argues that the internal logic of a semiotic view of life seems to point to either the "brain-object" thesis or the "mnemo-psychography" thesis as being the solution to the problem of the origin of mind. By providing a biosemiotic reading of the results of contemporary memory research, specifically the work of Kandel, Schacter, and Nicolelis et al., Scalambrino argues for the thesis of mnemo-psychography over the brain-object thesis, which he takes to be a variety of the identity theory of mind. He advocates for the biosemiotic view that the mind writes itself out of memory, that is, "mnemo-psychography."

3.5 *Synthetic Intelligence*

This part opens with a chapter titled "Minimal Mind" by Alexei Sharov who explores the features of this theoretical minimal mind, which is defined as a tool for classifying and modeling objects. The emergence of minimal mind marks an evolutionary transition from protosemiotic agents that use signs to directly control their actions to eusemiotic agents that can associate signs with ideal (mental) objects. Sharov argues that the hallmark of mind is a holistic perception of objects, which is not reducible to individual features or signals, and that epigenetic mechanisms likely play a crucial role in the origin and function of mind because chromatin states serve as rewritable memory signs. He allows that primitive forms of mind may exist at the cellular level where the nucleus plays the role of a brain and thus that a multicellular brain in animals is a community of cellular minds of individual neurons.

In their chapter, "Concept Combination and the Origins of Complex Cognition," Liane Gabora and Kirsty Kitto present theoretical and computational arguments to address the question of how advanced human cognitive abilities arose. They propose cognitive mechanisms that were likely underlying both the earliest indications of cultural sophistication such as tool use around the time of the arrival of Homo erectus and the cultural explosion following the arrival of modern humans in the

Middle-Upper Paleolithic. The first shift, they propose, involved the onset of the ability to recursively reprocess one thought in terms of the previous thought, and the second shift, they propose, involved the onset of the ability to shift between analytic (convergent) and associative (divergent) modes of thought.

Tom Barbalet's chapter, "The Mind of the Noble Ape in Three Simulations," offers an account of the applied mind through computer simulation. He demonstrates three prominent simulation methods that can be used together to create a coherent view of the human mind as a raw survival device, tuned to social hierarchical interaction with a strong undercurrent of programmed language. The simulations show three potential origins of mind through the movement of humans from raw survival into primitive social groups and finally to fully conversing (both with others and with themselves) entities. The use of computer simulation in the endeavor to understand mindedness implicitly comprises a critique of nonapplied philosophy of mind.

This part, and this volume, concludes with a chapter by Massimo Negrotti, "From the Natural Brain to the Artificial Mind," who notes that in discussing the mind we face a clear asymmetry: While the brain can be scientifically observed, the mind cannot. He notes that in order to reproduce something we need to observe it, yet argues that the artificial reproduction of mental activities is not helpful in understanding the mind. In fact, what any school of AI tries to reproduce is not the mind but a model of it coming from a specific psychological paradigm. Therefore, the "eradication" of the mind from the brain's evolution and activity adds a further degree of arbitrariness to the unavoidable bias and transfiguration that characterizes every attempt to reproduce natural objects, that is, to design *naturoids*. The chapter will discuss the methodological steps that any designer of naturoids has to follow, namely, the selection of an *observation level*, the boundaries of the natural *exemplar*, and its *essential performance*.

This volume marks a new beginning for the mind sciences—a formal entrance of biosemiotics into the discussion on the origins of organic mindedness in the natural world. Biosemioticians have of course already contributed to our understanding of mind; however, this volume initiates a more directed engagement on the topic from a multidisciplinary group of researchers attracted to the field of biosemiotics for what it has to offer in explaining organic mindedness. The reader is invited to join the discussion, and I hope will be inspired to do so after having read the chapters in this volume.

References

Barbieri, M. (2011). Origin and evolution of the brain. *Biosemiotics, 4*(3), 369–399.

Bechtel, W., Mandik, P., & Mundale, J. (2001). Philosophy meets the neurosciences. In W. Bechtel, P. Mandik, J. Mundale, & R. S. Stufflebeam (Eds.), *Philosophy and the neurosciences*. Malden: Blackwell.

Brier, S. (2012). Cybersemiotics: A new foundation for a transdisciplinary theory of consciousness, cognition, meaning and communication. In L. Swan (Ed.), *Orgins of mind*. Dordrecht: Springer.

Brook, A. (2004). Kant, cognitive science, and contemporary neo-Kantianism. *Journal of Consciousness Studies, 11*, 1–25.

Chalmers, D. J. (1995). Facing up to the problem of consciousness. *Journal of Consciousness Studies, 2*(3), 200–219.

Churchland, P. S. (1989). *Neurophilosophy: Toward a unified science of the mind-brain*. Cambridge, MA: The MIT Press.

Churchland, P. M. (1999). Eliminative materialism and the propositional attitudes. In W. G. Lycan (Ed.), *Mind and cognition: An anthology* (2nd ed.). Malden: Blackwell Publishers, Inc.

Churchland, P. S. (2002). *Brain-wise: Studies in neurophilosophy*. Cambridge, MA: The MIT Press.

Clark, A. (2000). *Mindware: An introduction to the philosophy of cognitive science*. New York: Oxford University Press.

Fingelkurts, A., Fingelkurts, A., & Neves, C. (2010). Natural world physical, brain operational, and mind phenomenal space-time. *Physics of Life Reviews, 7*(2), 195–249.

Godfrey-Smith, P. (1994). Spencer and Dewey on life and mind. In R. A. Brooks & P. Maes (Eds.), *Artificial life IV* (pp. 80–89). Cambridge, MA: The MIT Press (A Bradford Book).

Hoffmeyer, J. (1997). Biosemiotics: Towards a new synthesis in biology. *European Journal for Semiotic Studies, 9*(2), 355–376.

Hughes, R. I. G. (1997, December). Models and representations. *Philosophy of Science, 64*(Supplement). Proceedings of the 1996 biennial meetings of the Philosophy of Science Association. Part II: Symposia papers (pp. S325–S336).

Kravchenko, A. (2005). Cognitive linguistics, biology of cognition and biosemiotics: Bridging the gaps. *Language Sciences, 28*(1), 51–75.

Plotkin, H. (2004). *Evolutionary thought in psychology: A brief history*. Malden: Blackwell Publishers, Inc.

Searle, J. R. (1992). *The rediscovery of the mind*. Cambridge, MA: The MIT Press.

Swan, L. S., & Goldberg, L. J. (2010). How is meaning grounded in the organism? *Biosemiotics, 3*(2), 131–146.

Van Gelder, T., & Port, R. (1995). It's about time: An overview of the dynamical approach to cognition. In *Mind as motion: Explorations in the dynamics of cognition*. Cambridge, MA: The MIT Press.

Part I
Biosemiotics

Organic Codes and the Natural History of Mind

Marcello Barbieri

Abstract The purpose of this chapter is to show that organic codes played a key role in the origin and the evolution of mind as they had in all other great events of macroevolution. The presence of molecular adaptors has shown that the genetic code was only the first of a long series of codes in the history of life, and it is possible therefore that the origin of mind was associated with the appearance of new organic codes. This would cast a new light on mind and would give us a new theoretical framework for studying it. The scientific models that have been proposed so far on the nature of mind can be divided into three major groups that here are referred to as the *computational theory*, the *connectionist theory* and the *emergence theory*. The new approach is based on the idea that a neural code contributed to the origin of mind somehow like the genetic code contributed to the origin of life. This is the *code model of mind*, the idea that mental objects are assembled from brain components according to coding rules, which means that they are no longer *brain objects* but *brain artefacts*. The model implies that feelings and perceptions are not side effects of neural networks (as in connectionism), that they do not come into existence spontaneously by emergence and that they are not the result of computations, but of real manufacturing processes. In the framework of the code model, in short, feelings and perceptions are *manufactured artefacts*, whereas according to the other theories, they are *spontaneous products* of brain processes. This is relevant to the mind-body problem because if the mind were made of spontaneous products, it could not have *rules of its own*. Artefacts, on the other hand, can have such autonomous properties for two different reasons. One is that the rules of a code are conventions, and these are not dictated by physical necessity. The second is that a world of artefacts can have *epigenetic* properties that add unexpected features to the coding rules. The autonomy of the mind, in short, is something that spontaneous brain products cannot achieve whereas brain artefacts can.

M. Barbieri (✉)
Dipartimento di Morfologia ed Embriologia, Università degli Studi di Ferrara, Via Fossato di Mortara 64a, 44121, Ferrara, Italy
e-mail: brr@unife.it

L. Swan (ed.), *Origins of Mind*, Biosemiotics 8, DOI 10.1007/978-94-007-5419-5_2,

Keywords Organic codes • Macroevolution • Origin of brain • Origin of mind • Semiosis • Modelling systems • First-person experiences

1 Introduction

Mind is defined by its actions. An organism has mind when it has feelings, sensations and instincts—more generally, when it has *first-person* experience. The origin of mind was the origin of *subjective* experience, the event that transformed some living systems into living *subjects*. There is a large consensus today that mind is a natural phenomenon and that mental events are produced by brain events. More precisely, it is widely accepted that mind is made of higher-level brain processes, such as feelings and instincts, which are produced by lower-level brain processes such as neuron firings and synaptic interactions (Searle 2002). We need therefore to understand *how* the brain *produces* the mind and *what the difference is* between them.

This chapter describes a new idea about these problems. The idea is that there has been a (nearly) universal neural code at the origin of mind as there has been a (nearly) universal genetic code at the origin of life. The parallel between neural code and genetic code, in turn, is part of a wider framework according to which the genetic code was only the first of a long series of organic codes in the history of life. This framework—which is referred to as *the code view of life*—is based on the fact that we can actually *prove* the existence of many organic codes in nature with the very same procedure with which we have proved the existence of the genetic code (Barbieri 2003, 2008).

Any code is a set of rules of correspondence between two independent worlds and is necessarily implemented by structures, called *adaptors*, that perform two independent recognition processes (the adaptors are required because there is no necessary link between the two worlds, and a set of rules is required in order to guarantee the specificity of the correspondence). The genetic code, for example, is a set of rules that link the world of nucleotides to the word of amino acids, and its adaptors are the transfer RNAs. In signal transduction, the receptors of the cell membrane create a correspondence between first and second messengers, and have all the defining characteristics of true adaptors because any first messenger can be coupled with any second messenger. This means that signal transduction takes place according to the rules of a code that has been referred to as the *signal-transduction code* (Barbieri 1998, 2003).

Molecular adaptors have also been found in many other biological processes, thus bringing to light the existence of *splicing codes*, *cell compartment codes* and *cytoskeleton codes* (Barbieri 2003, 2008). Other organic codes have been discovered with different criteria. Among them, the *metabolic code* (Tomkins 1975), the *sequence codes* (Trifonov 1987, 1989, 1996, 1999), the *adhesive code* (Readies and Takeichi 1996; Shapiro and Colman 1999), the *sugar code* (Gabius 2000; Gabius et al. 2002), the *histone code* (Strahl and Allis 2000; Turner 2000, 2002; Gamble and Freedman 2002), the *transcriptional codes* (Jessell 2000; Marquardt and Pfaff 2001; Perissi and

Rosenfeld 2005; Flames et al. 2007), a *chromosome folding code* (Boutanaev et al. 2005; Segal et al. 2006), an *acetylation code* (Knights et al. 2006), the *tubulin code* (Verhey and Gaertig 2007), and the *splicing code* (Pertea et al. 2007; Barash et al. 2010; Dhir et al. 2010).

The living world, in short, is literally teeming with organic codes, and we simply cannot understand the history of life without them. This paper is an attempt to reconstruct the natural history of mind by taking the organic codes into account, and to this purpose it is divided into two parts. The first is about the events that culminated in the origin of mind and the second is dedicated to its evolution.

2 Part 1: The Origin of Mind

2.1 Organic Codes and Macroevolution

The existence of many organic codes in nature is an experimental fact—let us never forget this—but also more than that. It is one of those facts that have extraordinary theoretical implications. It suggests that the great events of macroevolution were associated with new organic codes, and this idea—the *code view of life*—gives us a totally new understanding of history. It is a view that paleontologists have never considered before and yet we have at least one outstanding example before our eyes. We know that the very first event of macroevolution—the origin of life itself—was associated with the genetic code, because it was that code that brought biological specificity into existence. But let us examine a few other examples of the deep link that exists between organic codes and macroevolution.

1. *The Three Domains of Life*
   The data from molecular biology have revealed that all known cells belong to three distinct primary kingdoms, or domains, that have been referred to as Archaea, Bacteria and Eukarya (Woese 1987, 2000). The fact that virtually all cells have the same genetic code suggests that this code appeared in precellular systems that had not yet developed a modern cell design. According to Woese, those systems were not proper cells because they had not yet crossed what he called the 'Darwinian threshold', an unspecified critical point after which a full cell organization could come into being (Woese 2002). According to the code view, the ancestral systems that developed the genetic code were not modern cells simply because they did not have a signal-transduction code. It is this code that gives context-dependent behaviour to a cell because it allows it to regulate protein synthesis according to the signals from the environment. A signal-transduction code was therefore of paramount importance to the ancestral systems, which explains why there have been various independent attempts to develop it. It is an experimental fact, at any rate, that Archaea, Bacteria and Eukarya have three distinct signalling systems, and this does suggest that each domain arose by the combination of the universal genetic code with three distinct signal-transduction codes.

2. *The Difference Between Prokaryotes and Eukaryotes*
   According to the code view, the ancestral cells of the three primary kingdoms adopted strategies that channelled them into two very different evolutionary directions. Archaea and Bacteria chose a *streamlining* strategy that prevented the acquisition of new organic codes, and for that reason, they have remained substantially the same ever since. The Eukarya, on the contrary, continued to explore the 'coding space' and evolved new organic codes (splicing codes, compartment codes, the histone code, etc.) throughout the whole 3,000 million years of cellular evolution. In this theoretical framework, the key event that gave origin to the eukaryotes was the appearance of the splicing codes, because splicing requires a separation *in time* between transcription and translation, and this was the precondition for their separation *in space*, a separation that eventually became physically implemented by the nuclear membrane.

3. *The Origin of Multicellular Life*
   Any new organic code brings into existence an absolute novelty, something that has never existed before, because the adaptors of a code create associations that are not determined by physical necessity. Any new organic code was therefore a true macroevolution, a genuine increase in complexity, to the point that the best measure of the complexity of a living system is probably the number of its organic codes. This means that the evolution of the eukaryotes was due to a large extent to the addition of new organic codes, a process that turned the eukaryotic cells into increasingly more complex systems. Eventually, however, the complexity of the cell reached a limit, and new organic codes broke the cellular barrier and gave origin to three completely new forms of life, the great kingdoms of plants, fungi and animals (Barbieri 1985, 2003).

2.2 The Codes of the Body Plan

The origin of animals was a true macroevolution and gives us the same problem that we face in all major transitions: how did real novelties come into existence? In the case of the first animals, the starting point was a population of cells that could organize themselves in space in countless different ways, so how did they manage to generate those particular three-dimensional structures that we call animals?

The solution was obtained by three types of experiments. More precisely, by the attempts to form multicellular structures with one, two or three different types of cells (the *germ layers*). The experiment with one cell type produced bodies which have no symmetry (the sponges), two cell types built bodies with one axis of symmetry (the *radiata* or diploblasts, i.e. hydra, corals and medusae), and three cell types gave origin to bodies with three axes of symmetry (the *bilateria* or triploblasts, i.e. vertebrates and invertebrates) (Tudge 2000). In principle, the number of three-dimensional patterns that the first animal cells could form in space was unlimited, so it was imperative to make choices. These choices, or constraints, turned out to be sets of instructions that specify a body plan. More precisely, the cells are instructed that their position is anterior or posterior, dorsal or ventral and proximal or distal *in respect to the surrounding cells*. These instructions are carried by genes and consist of molecules that are referred to as the *molecular determinants* of the body axes (Gilbert 2006).

The crucial point is that there are countless types of molecular determinants, and yet all triploblastic animals have the same axes (top-to-bottom, back-to-front and left-to-right). This shows that there is no necessary link between molecular determinants and body axes, and that in turn means that the actual links that we find in nature are based on conventional rules, that is, on the rules of organic codes that can be referred to as the *codes of the body axes*.

It must be underlined that the relationships of the body axes are between *cells*, and this means that they do not determine only the axes of the body but also those of all its constituent parts. In the hand, for example, the proximo-distal axis is the direction from wrist to fingers, the anteroposterior axis is from thumb to little finger and the dorsal-ventral axis is from the outer surface to the palm of the hand. Right and left hands have different symmetries because their axes are mirror images of each other. There is therefore a multitude of axes in the animal body, and it turns out that many of them have the same molecular determinants. The products of the gene *Sonic hedgehog* (*Shh*), for example, determine the dorsoventral axis of the forebrain as well as the anteroposterior axis of the hand, which again shows that molecular determinants are mere labels and represent the conventional rules of a code.

The anteroposterior axis of the body (the head-to-tail direction) is determined by two small depressions that are formed very early on the outer surface of the embryo and that mark the signposts of mouth and anus. Between those two points, a third depression is produced by the movements of a colony of migrating cells that invade the space between the first two germ layers (ectoderm and endoderm) to form the middle germ layer (the mesoderm). The invagination point (the blastopore) can be set either near the mouth signpost (the *stomodeum*) or near the anus signpost (the *proctodeum*) and that choice determines the future organization of all organs in the body. The animals wherein the blastopore is formed near the signpost of the mouth (*stoma*) are invertebrates (technically *protostomes*): they have an outside skeleton, a dorsal heart and a ventral nervous system. The animals wherein the blastopore is formed away from the mouth signpost are vertebrates (more precisely *deuterostomes*): they have an inside skeleton, a ventral heart and a dorsal nervous system.

The whole organization of the body, in other words, is a consequence of a few parameters that determine the migrations of the mesoderm in respect to the body axes. The crucial point is that these migrations (the *gastrulation* movements) take place in countless different ways in both vertebrates and invertebrates, and this shows that they are not due to physical necessity but to the conventional rules of a *gastrulation code*. We realize in this way that the three-dimensional organization of the animal body is determined by a variety of organic codes that together can be referred to as the *codes of the body plan*.

2.3 Cell Fate and Cell Memory

All free-living cells, from bacteria to protozoa, react swiftly to environmental changes, but the cells of multicellular animals exhibit more sophisticated behaviour. Their reactions do not take into account only their present conditions but also their history. This is because in embryonic development, the cells learn not only to become different but also to *remain* different. They acquire, in short, a *cell memory*. In technical terms, they go through embryonic processes that fix their *histological* fate for the rest of their life.

This great discovery was made by Hans Spemann, in 1901, by studying what happens when small pieces of tissue are transplanted from one part of an embryo to another. Spemann found that embryonic cells can change their histological fate (e.g. skin cells can become nerve cells) if they are transplanted *before* a critical period, but are totally unable to do so if the transplant takes place *after* that period. This means that for every cell type, there is a crucial period of development in which *something* happens that decides what the cell's destiny is going to be, and that something was called *cell determination.*

Other experiments proved that determination does not normally take place in a single step but in stages, and that the number and duration of these stages vary from one tissue to another. The most impressive property of determination is the extraordinary stability of its consequences. The process takes only a few hours to complete but leaves permanent effects in every generation of daughter cells for years to come. The state of determination, furthermore, is conserved even when cells are grown in vitro and perform many division cycles outside the body. When brought back in vivo, they express again the properties of the determination state as if they had never 'forgotten' that experience (Alberts et al. 1994).

The determination of cell fate, in short, amounts to the acquisition of a *cell memory* that is maintained for life and is transmitted to all descendant cells. The various steps of determination are controlled by molecules, known as *molecular determinants*, which can be passed on by the mother upon fertilization or produced by the embryo at various stages of development. The crucial point is that the basic histological tissues are the same in all animals, but their molecular determinants are of countless different types, which shows that the link between determinants and histological fate is not dictated by physical necessity but by the rules of codes that have been referred to as *histological codes* or *transcriptional codes* (Jessell 2000; Marquardt and Pfaff 2001; Perissi and Rosenfeld 2005; Flames et al. 2007).

This is dramatically illustrated by the most fundamental of all cell distinctions that between somatic and sexual cells. In *Drosophila*, for example, that distinction is determined by the *pole plasm*, a substance that is deposited by the mother at the posterior end of the egg. All cells that receive molecules from the pole plasm become sexual cells and are potentially immortal, whereas all the others become somatic cells and are destined to die with the body. The distinction between somatic and sexual cells takes place in all animals but is produced by a wide variety of molecules, in some cases produced by the mother and in other cases by the embryo, all of which show that it is an outstanding example of histological code.

During embryonic development, in conclusion, the cells undergo two distinct processes of determination: one for their three-dimensional pattern and the other for their histological fate. Both processes are totally absent in free-living cells, which again show that the origin of animals was a true macroevolution. Both processes, furthermore, are based on conventional rules of correspondence between molecular determinants and cell states because the determinants can be of countless different physical types. In all animals, in other words, the body plan and the histological fate of tissues and organs are based on the rules of organic codes.

2.4 Evolving the Neuron

The organs of an animal are not larger versions of the cell organelles, but there is nonetheless a parallel between them because there is a similar division of labour at the two levels of organization. The same basic proteins, for example, are expressed in the muscles of an animal and in the contracting region of a cell, so it is likely that the evolution of the animal organs took advantage of the molecular mechanisms that had been developed in the organelles and compartments of the ancestral protozoa.

This makes sense from an evolutionary point of view and suggests that the first animals already had the potential to express an internal division of labour. Some of their cells, for example, could preferentially express the genes of locomotion, thus becoming the precursors of the future motor organs. Other cells could preferentially express the genes of signal transduction and thus become the precursors of the future sense organs. A third type of cell could establish a link between them and prefigure in this way the future *nervous system* because this system is, by definition, a bridge between sense organs and motor organs. Whatever happened, at any rate, we know that the cells of the nervous system have two key characteristics, both of which could be obtained by modifying pre-existing protozoan structures.

The first major feature of the neuron is the ability to communicate with other cells by chemicals that are released from vesicles at points of close contact between their cell membranes (the synapses). It is those vesicles that provide the components of the brain signalling system, but they did not have to be invented from scratch. They are very similar to the standard vesicles that exist in all eukaryotic cells and are routinely used for transporting molecules across membranes.

The second great feature of the neuron is the ability to transmit electrical signals, and this too can be explained with a modification of pre-existing structures. The cell is constantly exchanging molecules with the environment, and most of these molecules are electrically charged, so there is a constant flux of positive and negative ions across the cell membrane. These ions can travel only through channels provided by specialized proteins, and their movements take place either by active transport or by passive diffusion. In the first case, they are called 'ion pumps' and in the second case 'ion channels'. Most channels, furthermore, are opened only by specific stimuli (electrical, mechanical, chemical, etc.). The *voltage-gated sodium channels*, for example, are protein systems that let sodium in only when they are stimulated by electrical signals.

The transport of all ions across the cell membrane is influenced by the fact that the interior of the cell is always electrically negative in respect to the outside because most of the great molecules that are trapped inside carry negative charges. The combination of this structural electrical asymmetry with the currents produced by ion pumps and ion channels leads to a stationary state characterized by an electrical difference across the cell membrane that is referred to as the *membrane potential.*

This potential is the result of a dynamic equilibrium of forces, and any perturbation of it produces an electric pulse known as *action potential.* An electrical stimulus, for example, can open a sodium channel and let in a flux of positive ions that rapidly change the local value of the membrane potential. Such a change, however, is confined to a very small region under the cell membrane and can be propagated to other regions only if the membrane contains many other sodium channels at a close distance from each other. All cells, in short, have ion pumps and ion channels, but only an uninterrupted distribution of sodium channels can propagate an action potential. That was the novelty that allowed a cell to transmit electrical signals.

Chemical-releasing vesicles, ion pumps and ion channels, in conclusion, had all been invented by free-living cells during the first 3,000 million years of evolution and did not have to be redesigned. All that was required for the origin of the neuron was a new way of arranging them in space.

2.5 *The Intermediate Brain*

The nervous system is made of three types of neurons: (1) the *sensory neurons* transmit the electrical signals produced by the sense organs, (2) the *motor neurons* deliver electrical signals to the motor organs (muscles and glands), and (3) the *intermediate neurons* provide a bridge between them. In some cases, the sensory neurons are directly connected to the motor neurons, thus forming a *reflex arch*, a system that provides a quick stimulus-response reaction known as a *reflex*. Intermediate neurons, therefore, can be dispensed with, and a few animals do manage without them. It is a fact, however, that most animals do have intermediate neurons, and what we observe in evolution is that brains increased their size primarily by increasing the number of their intermediate neurons. The evolution of the brain, in other words, has largely been the evolution of the 'intermediate brain'.

It is well known, today, that most brain processing is totally unconscious, and we can say therefore that the intermediate brain is divided into a conscious part and an unconscious one. But when did this split occur? When did consciousness appear in the history of life? Here, unfortunately, we come up against the difficulty that consciousness is too large a category. It is associated with feelings, sensations, emotions, instincts, thinking, free will, ethics, aesthetics and so on. Some of these entities appeared late in evolution and only in a restricted number of species, so we can regard them as special evolutionary developments. The origin of consciousness, in other words, can be restricted to its most essential features—to the origin of something primitive and universal, something that even simple animals could have.

Feelings and instincts are probably the most universal of all conscious processes, and here it is assumed that consciousness came into existence when the primitive brain managed to produce them. Let us see how that could have happened.

The first nervous systems were probably little more than a collection of reflex arches, and it is likely that the first intermediate neurons came into being as a physical extension of those arches. Their proliferation was favoured simply because they provided a useful *trait-de-union* between sensory neurons and motor neurons. Once in existence, however, they could start exploring other possibilities.

Their first contribution was probably the development of a multi-gated reflex-arch system. The behaviour of an animal must take into account a variety of clues from the environment, and to this purpose, it is useful that a motor organ receives signals from many sense organs and that a sense organ delivers signals to many motor organs. This inevitably requires multi-gated connections between sensory inputs and motor outputs, and that probably explains why intermediate neurons had such great evolutionary success.

In addition to transmitting electrical signals, however, the intermediate neurons could do something else. They could start *processing* the signals, and that opened up a whole new world of possibilities. In practice, the processing evolved in two great directions and produced two very different outcomes. One was the formation of neural networks that give origin to feedback systems and provide a sort of 'automatic pilot' for any given physiological function. The other was the generation of feelings and instincts.

The first processing was totally unconscious and was carried out by a component of the intermediate brain that here is referred to as the *cybernetic brain*. The second processing was adopted by another major component of the intermediate brain that here is referred to as the *instinctive brain*. The intermediate brain, in short, evolved from a primitive reflex-arch system and developed two distinct types of neural processing, one completely unconscious and the second controlled by instincts. But why *two* types of processing? Why develop feelings and instincts if a cybernetic brain can work perfectly well without them?

2.6 The Instinctive Brain

A cybernetic brain can control all physiological functions and can cope with the vagaries of the environment, so there does not seem to be any need to also evolve feelings and instincts. We should not forget, however, that a cybernetic brain is an intermediary between sense organs and motor organs and can work only if there is a *continuous* chain of reactions between inputs and outputs. This means that all the operations of a cybernetic brain are linked together in a physically continuous sequence, and the initial input is inevitably a signal from the outside world. An animal with a fully cybernetic brain, in other words, is virtually a puppet in the hands of the environment. An instinctive brain, instead, is a system wherein the orders to act come from within the system, not from without. An animal with an instinctive brain makes decisions on the basis of its own instincts, of its own internal rules, and has therefore a certain autonomy from the environment. But does such autonomy have an evolutionary advantage?

In circumstances when there is no food and no sexual partner in the immediate surroundings, a cybernetic animal would simply stop eating and mating, whereas an instinctive animal would embark on a long journey of exploration well beyond its visible surroundings and even in the absence of positive external signals. An internal drive to act, irrespective of the circumstances, in short, can have a survival role, and that is probably why most animals evolved both a cybernetic brain and an instinctive brain.

It must be underlined, however, that an instinctive brain is not a system that can simply be 'added' to a cybernetic brain. An instinctive brain is a system that acts on the basis of internal drives, and that means that it has the ability to send its own orders to the motor organs, that is, to generate its own electrical signals. That in turn means that the signals delivered to the motor organs do not all come from the sense organs.

The evolution of the instinctive brain, in brief, required a major change in brain circuitry. The bridge between sense organs and motor organs provided by the cybernetic brain was *interrupted*, and the gap was filled by a new bridge made of feelings and instincts. The instinctive brain did not simply *add* feelings to a pre-existing system. It physically broke the continuity of the cybernetic bridge and introduced a new bridge in between. As a result, the intermediate brain acquired three distinct control systems, which are based respectively (1) on chemical signalling, (2) on neural networks and (3) on feelings and instincts. The first two make up the cybernetic brain, whereas the third system is the instinctive brain of an animal.

The origin of feelings and instinct, furthermore, can be associated with the origin of consciousness, but in order to appreciate this point, we need to discuss the concept of 'first-person' experience because it is this concept that is largely regarded, today, as the key component of consciousness.

2.7 The 'First-Person' Experiences

Feelings, sensations, emotions and instincts are often referred to as 'first-person' experiences because they are experienced directly, without intermediaries. They make us feel that we know our body, that we are in charge of its movements, that we are conscious beings and that we live a 'personal' life. Above all, they are quintessentially private internal states, and this makes it impossible to share them with other people.

The goal of science is to produce testable models of what exists in nature, and first-person experiences are undoubtedly part of nature, so we should be able to make models of them. Models, of course, are not reality ('the map is not the territory'), but they are ideas of reality and what really matters is that these ideas can be tested and improved indefinitely. In our case, the problem is to build a model that makes us understand, at least in principle, how first-person experiences can be produced.

Let's take, for example, the case in which a toe is injured. We know that electrical pulses are immediately sent to the central nervous system and that the intermediate brain processes them and delivers orders to the motor organs that spring the body into action. Here, we have two distinct players: an observer system (the intermediate brain) and an observed part (the injured toe). It is the observer that gets the information and transforms it into the feeling of pain, but then something extraordinary happens. We do not feel the pain in the intermediate brain, where the feeling is created, but in the toe, where the injury took place. Observer and observed have become one, and it is precisely this collapse into a single feeling unity that generates a 'first-person' experience.

Something similar takes place when we receive signals from the environment, for example, when we look at an outside object. In this case, an image is formed on the retina, and electrical signals are sent to the intermediate brain. Again, there is a separation between observer (the brain) and observed (the retina). What we see, however, is not an image on the retina, where the visual information is actually produced. The intermediate brain and the retina collapse into a single processing unity and what we see is an image in the outside world. This is again a first-person experience, and again it is generated by a physiological process that short-circuits the physical separation between sense organs and the intermediate brain.

What we call 'first-person' experiences, in brief, is nothing elementary, undifferentiated and indivisible. The exact opposite is true. They are the result of complex neural processes where many highly differentiated cells act in concert and create a physiological short circuit between observer and observed. First-person experiences, in other words, cannot exist in single cells. They could evolve only in multicellular systems, and their origin was a true macroevolution, an absolute novelty. Our problem, therefore, is to understand *how* it could have happened. What was the mechanism that brought them into existence?

2.8 The Difference Between Brain and Mind

Feelings, sensations, emotions and instincts are traditionally known as *mental* processes or products of the *mind*. There is a large consensus today that mind is a natural phenomenon, and that mental events are produced by brain events. At the same time, it is also widely acknowledged that there is a gulf between the physiological processes of the brain and the subjective experiences of the mind. Our problem, therefore, is to understand not only how the brain produces the mind but also what the *difference* between them is. Probably the best way to deal with this problem is by comparing it with the parallel problem that exists between matter and life. It is largely accepted, today, that life evolved from matter but also that life is fundamentally different from matter, because entities like natural selection and the genetic code, to name but a few, simply do not exist in the inanimate world.

How can we explain that? How can something give origin to something fundamentally different from itself? How could matter produce life if there is a fundamental difference between matter and life? Many have decided that no such difference can exist and therefore that '*life is chemistry*', a conclusion that goes in parallel with the idea that '*mind is brain*'.

The chemical view of life is still popular today, and it would be perfectly plausible if primitive genes and primitive proteins could have evolved all the way up to the first cells by spontaneous chemical reactions. But that is precisely what molecular biology has ruled out, because genes and proteins are never formed spontaneously in living systems. Instead, they are manufactured by molecular machines that physically stick their components together in the order provided by a template. Primitive genes and primitive proteins did appear spontaneously on the primitive Earth, but they could not give origin to the first cells because they did not have biological specificity. They gave origin instead to molecular machines, and it was these machines and their products that evolved into the first cells.

Genes and proteins, in short, are assembled by molecular robots on the basis of linear *information*, and this makes them as different from spontaneous molecules as *artificial* objects are from natural ones. Genes and proteins are *molecular artefacts*, that is, *artefacts made by molecular machines* (Barbieri 2003, 2008). They came from inanimate matter because their components were formed spontaneously, but they are different from inanimate matter because they need entities, like information and coding rules, that do not exist in spontaneous reactions. Only molecular machines can bring these entities into existence, and when they do, they produce artefacts, but above all, they produce *absolute novelties*, objects that are completely different from whatever is formed spontaneously in the universe.

This is the logic that explains, in principle, how genuine novelties appeared in evolution. Any biological system that makes objects according to the rules of a code is generating biological artefacts, and a world of artefacts is fundamentally different from the world where it came from. This makes us understand why life arose from matter and yet it is fundamentally different from it, as well as why mind is produced by the brain and yet it is fundamentally different from it. There is the same logic, the same underlying principle behind the origin of life and the origin of mind. This is the *code model of mind*, the idea that there was a *neural code* at the origin of mind as there was a genetic code at the origin of life (Barbieri 2006, 2010).

2.9 The Code Model of Mind

The parallel between the origin of life and the origin of mind can become a scientific model only if it takes the form of a coherent set of hypotheses, so let us see how this can be done.

In the origin of life, the key event was the appearance of *proteins*, and the genetic code played a crucial part in it precisely because it was instrumental to protein synthesis. In the origin of mind, the key event was the appearance of *feelings*, and our hypothesis is that a neural code was as instrumental to the production of feelings as the genetic code was to the production of proteins. The parallel, therefore, is between feelings and proteins, and this immediately tells us that there are both similarities and differences between the two cases.

Proteins are *space objects* in the sense that they act in virtue of their three-dimensional organization in space, whereas feelings are *time objects* because they are 'processes', entities that consist of flowing sequences of states. The same is true for their components. Proteins are assembled from smaller space objects like amino acids, and feelings are assembled from lower-level brain processes such as neuron firings and chemical signalling.

The idea of a deep parallel between life and mind leads in this way to a parallel between proteins and feelings and, in particular, to a parallel between the processes that produce them. We already know that the assembly of proteins does not take place *spontaneously* because no spontaneous process can produce an unlimited number of identical sequences of amino acids. The *code model of mind* is the idea that the same is true in the case of feelings, that is, that feelings are not the spontaneous result of lower-level brain processes. They can be generated only by a neural apparatus that assembles them from components according to the rules of a code. According to the code model, in short, *feelings are brain artefacts* and are manufactured by a codemaker according to the rules of the *neural code*.

In the case of proteins, the codemaker is the ribonucleoprotein system of the cell, the system that provides a bridge between genotype and phenotype. It receives information from the genotype in the form of messenger RNAs and assembles the building blocks of the phenotype according to the rules of the genetic code. It must be underlined, however, that the codemaking system has a logical and a historical priority over genotype and phenotype, and for this reason, it is a third category that has been referred to as the *ribotype* of the cell (Barbieri 1981, 1985).

In the case of feelings, the codemaker is the intermediate brain of an animal, the system that receives information from the sense organs and delivers orders to the motor organs. The sense organs provide all the information that an animal is ever going to have about the world and represent therefore in an animal what the genotype is in a cell. In a similar way, the motor organs allow a body to act in the world and have in an animal the role that the phenotype has in a cell. Finally, the intermediate brain is a processing and a manufacturing system, an apparatus that is in an animal what the ribotype is in a cell.

The parallel between life and mind, in conclusion, involves three distinct parallels: one between proteins and feelings, one between genetic code and neural code, and one between cell and animal codemaking systems. The categories that we find in the cell, in other words, are also found in animals, because at both levels, we have information, code and codemaker. The details are different, and yet there is the same *logic* at work, the same strategy of bringing absolute novelties into existence by organic coding.

2.10 The Neural Code

The term 'neural code' is used fairly often in the scientific literature and stands for the unknown mechanism by which the signals produced by the sense organs are transformed into subjective experiences such as feelings and sensations. It must be underlined that the term is potentially ambiguous, because it may indicate either a universal code or the code that an animal is using to create its own species-specific representations of the world. A similar ambiguity arises, for example, with the term 'language', which can mean either a universal human faculty or the specific language that is spoken in a particular place.

The parallel with the genetic code removes this ambiguity from the start and makes it clear that the code model of mind assumes the existence of a *universal* neural code. Our problem is therefore the scientific basis of that idea: on what grounds can we say that a (nearly) universal neural code exists in all animals as a (nearly) universal genetic code exists in all cells?

Let's consider, for example, the transformation of mechanical stimuli into tactile sensations. Rats have mechanoreceptors on the tip of their whiskers, while we have them on the tip of our fingers, and there is no doubt that our tactile exploration of the world is different from theirs, but does that mean that we use a different neural code? The evidence is that the physiological processes that transform the mechanical stimuli into tactile sensations are the same in all animals, and this does suggest that there is a universal mechanism at work (Nicolelis and Ribeiro 2006). As a matter of fact, the evidence in question comes from animals with three germ layers (the triploblasts), but they represent the vast majority of all animal taxa, so let us concentrate our attention on them. How can we generalize the experimental data and conclude that virtually all triploblastic animals have the same neural code?

We do know that the starting point of all neural processing is the electrical signals produced by the sense organs, but we also know that the sense organs arise from the basic histological tissues of the body and that these tissues (epithelial, connective, muscular and nervous tissues) are the same in all triploblastic animals. All signals that are sent to the brain, in other words, come from organs produced by a limited number of universal tissues, and that does make it plausible that they represent a limited number of universal inputs. But do we also have a limited number of universal outputs?

The neural correlates of the sense organs (feelings and perceptions) can be recognized by the *actions* that they produce, and there is ample evidence that all triploblastic animals have the same basic *instincts*. They all have the imperative to *survive* and to *reproduce*. They all seem to experience hunger and thirst, fear and aggression, and they are all capable of reacting to stimuli such as light, sound and smells. The neural correlates of the basic histological tissues, in short, are associated with the basic animal instincts, and these appear to be virtually the same in all triploblastic animals.

What we observe, in conclusion, is a universal set of basic histological tissues on one side, a universal set of basic animal instincts on the other side and a set of neural transformation processes in between. The most parsimonious explanation is that the neural processes in between are also a universal set of operations. And since there is no necessary physical link between sense organs and feelings, we can conclude that the bridge between them can only be the result of a virtually universal *neural code*.

3 Part 2: The Evolution of Mind

3.1 Two Universal Strategies

There are both unity and diversity in life. The unity comes from the presence of a universal genetic code in all living cells. The diversity comes from the existence of different organic codes in different groups of cells. The first cells, for example, were divided into three primary kingdoms (Archaea, Bacteria and Eukarya) by three distinct signal-transduction codes. After that original split, some cells (Archaea and Bacteria) adopted a streamlining strategy that prevented them from developing new organic codes, with the result that they have remained substantially the same ever since. The other cells (Eukarya) continued to explore the coding space and became increasingly more complex.

If we now look at the evolution of animals, we find again a split between a streamlining strategy and an exploring strategy. In this case, it was the split that divided invertebrates from vertebrates. The invertebrates adopted a streamlining strategy that reduced their brain development to the bare essentials, whereas the vertebrates appear to have explored almost without limits the potentialities of the brain space. In evolution, in other words, there seem to be two universal strategies at work, one that promotes streamlining and one that favours exploration. At the cellular level, these strategies divided prokaryotes from eukaryotes, and at the animal level, they divided invertebrates from vertebrates.

At the cellular level, furthermore, the exploring strategy of the eukaryotes was primarily based on the development of new organic codes, and this suggests that, at the animal level, the exploring strategy of the vertebrates could also have been based on organic codes. But can we prove it? Can we actually show that many organic codes appeared in vertebrate evolution?

Brains do not normally fossilize, but we can still obtain information on their ancestral organic codes. We can get such information from embryology, because the main driving forces of animal evolution were changes in embryonic development that have been passed on to their modern descendants. The embryonic brain, in short, is probably the best place where we can find information about the evolution of the brain and its organic codes.

3.2 Mechanisms of Brain Development

The embryonic development of the vertebrate nervous system takes place in four stages. The first begins when a strip of ectoderm is induced to become neural tissue by the underlying mesoderm, and comes to an end when the newly formed neuroblasts complete their last cell division, an event that marks the 'birth' of the neurons.

This is a truly epochal event because everything that a neuron will ever do in its life is largely determined by the time and the place of its birth. Somehow, these two parameters leave an indelible mark in the young neuron and become a permanent memory for it.

The second phase of neural development is the period in which neurons migrate from their birthplace to their final destination, a target they 'know' because it is somehow 'written' in the memory of their birth.

The third phase begins when neurons reach their definitive residence. From this time onwards, the body of a neuron does not move any more but sends out 'tentacles' that begin a long journey of exploration in the surrounding body. A tentacle (a *neurite*) ends with a roughly triangular lamina (called a *growth cone*), which moves like the hand of a blind man, touching and feeling any object on its path before deciding what to do next. The axons of motor neurons are the longest of such tentacles, and their task is to leave the neural tube for the rest of the body in search of organs that require nerve connections. This is achieved with an exploration strategy that takes place in two stages. In the first part of the journey, the growth cones move along tracks provided by specific molecules, with a preference for those of other axons (which explains why growth cones migrate together and form the thick bundles that we call *nerves*). They do not have a geographic knowledge of their targets, but this is compensated for by an overproduction of cells, which ensures that some of them will actually reach the targets. At this point, the second part of the strategy comes into play. The organs that need to be innervated send off particular molecules, known as *nerve growth factors*, which literally save the neurons from certain death. More precisely, neurons are programmed to commit suicide—that is, to activate the genes of cell death, or *apoptosis*—at the end of a predetermined period, and nerve growth factors are the only molecules that can switch off this self-destruction mechanism. The result is that the neurons that reach the right places survive, and all the others disappear (Levi-Montalcini 1975, 1987; Changeaux 1983).

The fourth phase of brain development begins when the growth cones reach the target areas. At this point, some unknown signal instructs the axon to stop moving and to begin a new transformation. The growth cone loses its flat shape and generates a variety of thin long fingers that are sent off in various directions towards the surrounding cells. When a contact is established, the tips of the finger-like extensions expand themselves and become the round buttons of the *synapses*, the structures that specialize in the transmission of neurochemicals. This turns the neuron into a secretory cell, and from that moment on, the neuron is committed to a life of uninterrupted chemical communication with other cells.

The making and breaking of synaptic connections is the actual wiring of the nervous system and takes place with a mechanism that is based first on molecular recognitions and then on functional reinforcements. Each neuron generates an excess number of synapses, so the system is initially over-connected. The synaptic connections, on the other hand, are continuously broken and reformed, and only those that are repeatedly reconnected become stable structures. Those that are less engaged are progressively eliminated and in the end only the active synapses remain. This mechanism continues to operate long after birth and in some part of the brain it goes on indefinitely, thus providing the means to form new neural connections throughout the life of an individual. According to Donald Hebb (1949), it is this mechanism that lies at the heart of memory, and the results obtained from natural and artificial neural networks have so far confirmed his prophetic idea.

3.3 Codes of Brain Development

Cell adhesion, cell death and cell signalling are major tools of brain development, and in all of them, we can recognize the presence of organic codes. Let us briefly examine a few examples.

1. *Cell Adhesion*
   In the 1940s, Roger Sperry severed the optic nerve of a fish and showed that its fibres grow back precisely to their former targets in the brain. Furthermore, when the eye was rotated 180° in its socket, the fish was snapping downwards at a bait placed above it, thus proving that the connections are extremely specific. This led Sperry (1943, 1963) to formulate the 'chemoaffinity hypothesis', the idea that neurons recognize their synaptic partners by millions of 'recognizing molecules' displayed on their cell membranes. The wiring of the brain is essentially accomplished by molecules that bridge the synaptic cleft and decide which neurons are connected and which are not. They function both as synaptic recognizers and synaptic glue, and recently it has been shown that cadherins and protocadherins are good candidates for these roles. Protocadherins, in particular, have an enormous potential for diversification because their genes contain variable and constant regions like the genes of the immunoglobulins. They could, therefore, provide the building blocks of a neural system that is capable of learning and memorizing and, like the immune system, can cope with virtually everything, even the unexpected (Hilschmann et al. 2001). This suggests that the chemoaffinity hypothesis of Roger Sperry should be reformulated in terms of a code. Rather than listing millions of individual molecular interactions, an organic code can generate an enormous diversity with a limited number of rules, and this is why various authors have proposed that the wiring of the nervous system is based on an *adhesive code* (Readies and Takeichi 1996; Shapiro and Colman 1999).
2. *Cell Death*
   Active cell suicide (apoptosis) is a universal mechanism of embryonic development, one that is used to shape virtually *all* organs of the body. The key point is that suicide genes are present in all cells, and the signalling molecules that switch them on and off are of many different types. This means that the recognition of a signalling molecule and the activation of the suicide genes are two independent processes, so we need to understand what brings them together. Since there are no necessary connections between them, the only realistic solution is that the link is established by the rules of an *apoptosis code*, that is, a code that determines which signalling molecules switch on the apoptosis genes in which tissue.

3. *Cell Signalling*
   Neurons communicate with other cells by releasing chemicals called *neurotransmitters* in the small space (the *synaptic cleft*) that separates their cell membranes. There are four distinct groups of neurotransmitters and dozens of molecules in each of them, but the most surprising feature is that the same molecules are employed in many other parts of the body with completely different functions. Adrenaline, for example, is a neurotransmitter, but it is also a hormone produced by the adrenal glands to spring the body into action by increasing the blood pressure, speeding up the heart and releasing glucose from the liver. Acetylcholine is another common neurotransmitter in the brain, but it also acts on the heart (where it induces relaxation), on skeletal muscles (where the result is contraction) and in the pancreas (which is made to secrete enzymes). Neurotransmitters, in other words, are *multifunctional molecules*, and this suggests that they are used as molecular *labels* that can be given different meanings in different contexts. The most parsimonious explanation is that their function is determined by the rules of an organic code that can be referred to as the *neurochemical code*. The idea that neurotransmitters act like the words of a chemical language is reinforced by the fact that small structural variations can have vastly different meanings. This is very common in language (compare, e.g. the meanings of *dark*, *park* and *bark*), but it is also common in brain signalling. Serotonin, for example, is a normal neurotransmitter, but a slightly modified version of it, such as mescaline, produces violent hallucinations. The same is true for lysergic acid (LSD), which is related to dopamine, and in general for many other chemicals that are structurally similar to neurotransmitters.

In brain development, in conclusion, we see at work mechanisms that have all the defining characteristics of organic codes, and we might as well come to terms with this fact of life.

3.4 The Evolution of Vision

The human retina is made of three layers, one of which contains about 100 million *photoreceptor cells* (rods and cones) that react to light by producing electrical signals. These are sent to the *bipolar cells* of the second layer, which in turn deliver signals to the one million *ganglion cells* of the third layer whose axons form the optic nerve. The 100 million signals of the photoreceptor cells undergo therefore a first processing on the retina, the result of which is one million pulses delivered via the optic nerve to the brain. Here, the signals are sent to the midbrain and, after the *optic chiasm* (where 50% swap direction), are transmitted to the *visual cortex*, at the back of the head, where they are further processed by groups of *cortical cells* arranged in distinct *areas*. It turns out that the operations performed in areas 17, 18 and 19 maintain a certain topological coherence with the visual field of the retina in the sense that adjacent points in the retina are processed by adjacent points in those areas of the visual cortex. In area 17, furthermore, Hubel and Wiesel have found that some cells react only to horizontal movements on the retina, other cells react only to vertical movements and still others to sharp edges (Hubel and Wiesel 1962, 1979). After areas 18 and 19, the visual inputs go on to other cortical areas, but the topological coherence with the retina is rapidly lost, probably because the information on spatial relationships has already been extracted.

The key point, at the higher processing level, is that the brain does not merely *register* the information from the retina but can literally *manipulate* it. When an object is approaching, for example, its image on the retina becomes larger, but the brain still perceives an object of constant size. When the head is moving, the image of an object on the retina is also moving, but the brain decides that the object is standing still. When the light intensity is lowered, the retinal image of a green apple, for example, becomes darker, but the brain compensates for that and concludes that the apple has not changed its colour.

These (and many other) results prove that what we 'perceive' is not necessarily what the sense organs tell us. 'Perceptions', in other words, are distinct from 'sensations'. A sensation is what comes from the senses and has a specific physiological effect (colour, sound, smell, tickle and so on). A perception is what the brain decides to do with the information from the senses, according to its own set of processing rules.

We realize in this way that there are many types of processing going on in the brain, and such a complex hierarchy can only have been the result of a long history, so let us take a brief look at the evolution of vision.

Some of the most primitive eyes are found in flatworms and are little more than clusters of photoreceptor cells that can distinguish day from night. They are also able to detect the direction of the light source, a feat that allows flatworms to swim towards the dark. But flatworm eyes do not have a lens and thus cannot form visual images of the surrounding objects.

The first camera-eye, with a lens that projects an image on the retina, probably appeared in fish. The fish retina already has a three-layered structure (rods and cones, bipolar cells and ganglion cells) and an optic nerve that transmits the visual inputs to the midbrain. In fish, however, all nerve fibres change direction at the optic chiasm, and the midbrain is the final destination of the visual inputs, the place where the signals from all sense organs are converted into orders to the motor organs.

This primitive structure was substantially conserved in amphibians and reptiles, and it was only birds and mammals that started evolving a more advanced design. In their visual system, not all the fibres of the optic nerves crossed direction at the optic chiasm, and the final destination of the visual inputs was moved from the midbrain to the visual cortex and then to other regions of the neocortex. These changes went hand in hand with a gradual transition from an olfactory and tactile mode of life to a life where vision was acquiring an increasingly important role.

The evolution of vision is an outstanding example of the changes that took place in the *cybernetic* brain, more precisely in that part of the cybernetic brain that is in charge of the automatic processing of visual information. The cybernetic brain, however, was only a part of the evolving brain, and we need to consider also the evolution of the bran in its entirety.

3.5 Three Modelling Systems

The results of brain processing are what we normally call feelings, sensations, emotions, perceptions, mental images and so on, but it is useful to have also a more general term that applies to all of them. Here, we follow the convention that all products of brain processing can be referred to as brain *models*. The intermediate brain, in other words, uses the signals from the sense organs to generate distinct *models* of the world. A visual image, for example, is a model of the information delivered by the retina, and a feeling of hunger is a model obtained by processing the signals sent by the sense detectors of the digestive apparatus.

The brain can be described in this way as a *modelling system*, a concept that has been popularized by Thomas Sebeok and that has acquired an increasing importance in semiotics (Sebeok and Danesi 2000). The term was actually coined by Juri Lotman, who described language as the 'primary modelling system' of our species (Lotman 1991), but Sebeok underlined that language evolved from animal systems and should be regarded as a secondary modelling system. The distinction between primary, secondary and tertiary modelling systems has become a matter of some controversy, so it is important to be clear about it. Here, we use those terms to indicate the modelling systems that appeared at three different stages of evolution and gave origin to three different types of brain processing:

1. *The first modelling system*
   This is the system that appeared when the primitive brain managed to produce feelings and sensations. These entities can be divided into two great classes because the sense organs deliver information either about the outside world or about the interior of the body. The first modelling system consists therefore of two types of models, one that represents the environment and one that carries information about the body. Jakob von Uexküll (1909) called these two worlds *Umwelt* and *Innenwelt*, names that express very well the idea that every animal lives in two distinct subjective universes. We can say therefore that *Innenwelt* is the model of the internal body built by the instinctive brain and that *Umwelt* is the model of the external world built by the cybernetic brain of an animal. The brain as we know it—the brain with feelings—came into being when the primitive brain split into instinctive brain and cybernetic brain, and these started producing the feelings and sensations that make up the first modelling system of all triploblastic animals (vertebrates and invertebrates).
2. *The second modelling system*
   Some animals (like snakes) stop chasing a prey when it disappears from sight, whereas others (like mammals) deduce that the prey has temporarily been hidden by an obstacle and continue chasing it. Some can even learn to follow the footsteps of a prey, which reveals a still higher degree of abstraction. This ability to 'interpret' the signals from the environment is based, as we will see, on a new type of neural processing that represents the *second modelling system* of the brain, a system that appeared when a part of the cybernetic brain became an 'interpretive brain'.

3. *The third modelling system*
   The last major novelty in brain evolution was the origin of language, and that too required, as we will see, a new type of neural processing, so it is legitimate to say that language represents a third modelling system.

   There have been, in conclusion, three major transitions in the evolution of the brain, and each of them gave origin to a new type of neural processing that was, to all effects, a new modelling system.

3.6 The Interpretive Brain

The instinctive brain delivers orders to the motor organs and is the directive centre of an animal, responsible for its ability to survive and reproduce. The cybernetic brain is essentially a servomechanism, and it is precisely this function that explains its enormous increase in evolution. The instinctive brain has changed very little in the history of life, and the greatest changes have taken place precisely in the cybernetic tools that animals evolved in order to provide the instinctive brain with increasingly sophisticated servomechanisms.

The neural networks are probably the most powerful of such tools. Their ability to create feedback loops allows them to produce a goal-directed behaviour in a system, but they also have other outstanding properties. In artificial systems, for example, it has been shown that neural networks can provide the basis of *learning* and *memory* (Kohonen 1984), and it is likely that they have similar properties in living systems. It is possible, therefore, that neural networks were the physical tools that evolved learning and memory, but that still leaves us with the problem of understanding the role that learning and memory had in evolution.

Memories allow a system to compare a phenomenon with previous records of similar phenomena, and it is from such a comparison that a system can 'learn' from past experiences. Memories are clearly a prerequisite for learning, but what does learning achieve? What is the point of storing mental representations and comparing them?

So far, the best answer to this problem is probably the idea, proposed by Charles Sanders Peirce, that memories and learning allow animals to *interpret* the world.

An act of interpretation, on the other hand, consists in giving a meaning to something, and this is, by definition, an act of semiosis. Interpretation, therefore, is a form of semiosis, and its elementary components are signs and meanings. According to Peirce (1906), there are three major types of signs in the world, and he called them *icons*, *indexes* and *symbols*:

1. A sign is an *icon* when it is associated with an object because a *similarity* is established between them. All trees, for example, have individual features, and yet they also have something in common, and it is this common pattern that allows us to recognize as a tree any new specimen that we happen to encounter for the first time. Icons, in other words, lead to pattern recognition and are the basic tools of *perception*.

2. A sign is an *index* when it is associated with an object because a *physical link* is established between them. We learn to recognize any new cloud from previous clouds, and any new outbreak of rain from previous outbreaks, but we also learn that there is often a correlation between clouds and rain, and we end up with the conclusion that a black cloud is an index of rain. In the same way, a pheromone is an index of a mating partner, the smell of smoke is an index of fire, footprints are indexes of preceding animals and so on. Indexes, in short, are the basic tools of *learning*, because they allow animals to infer the existence of something from a few physical traces of something else.
3. A sign is a *symbol* when it is associated with an object because a *conventional link* is established between them. There is no similarity and no physical link between a flag and a country, for example, or between a name and an object, and a relationship between them can exist only if it is the result of a convention. Symbols allow us to make arbitrary associations and build mental images of future events (projects), of abstract things (numbers) and even of non-existing things (unicorns).

The part of the intermediate brain that allows an animal to interpret the world can be referred to as the *interpretive brain*, or the *second modelling system* of the brain. It was the result of a specific phase in brain evolution, and we need therefore to understand, at least in principle, how interpretation came into being.

3.7 The Origin of Interpretation

The ability to interpret the world is a form of semiosis, because it is based on signs and meaning, but is it a *new* form of semiosis? More precisely, did interpretation appear only in animals or did it exist also in free-living single cells? We have seen that many organic codes appeared on Earth in the first 3,000 million years of evolution, and this is equivalent to saying that single cells were capable of coding and decoding the signals from the environment. But coding and decoding is *not* the same as interpreting. Interpretation takes place when the meaning of a sign can change according to circumstances, whereas coding takes place when meaning is the fixed result of a coding rule.

The idea that single cells are capable of interpreting the world is still very popular today because single cells have context-dependent behaviour, and it is taken virtually for granted that context dependency can only be the result of interpretation. In reality, it takes only two organic codes to produce a context-dependent response in a cell. A context-dependent behaviour means a context-dependent expression of genes, and this is achieved by linking the expression of genes to signal transduction, that is, by putting together the genetic code with a signal-transduction code (Jacob and Monod 1961). And if it takes only two context-free codes to produce a context-dependent behaviour, one can only wonder at how much more complex the cell behaviour became when other organic codes appeared in the system.

The origins of animals, of embryonic development and of the brain, furthermore, were also associated with new organic codes and were based on coding, not on interpretation. The ability to interpret the world came into being at a later stage, when animals started exploring the potentialities of learning. Neural networks have the ability to form memories, and a set of memories is the basis of learning because it allows a system to decide how to behave in any given situation by comparing the memories of what happened in previous similar situations. A large set of memories, in other words, amounts to a model of the world that is continuously updated and that allows a system to *interpret* what goes on around it.

Such a model, on the other hand, is formed by a limited number of memories, whereas the real world offers an infinite number of possibilities. Clearly, a model based on memories can never be perfect, but it has been shown that neural networks can in part overcome this limit by interpolating between discrete memories (Kohonen 1984). In a way, they are able to 'jump to conclusions', so to speak, from a limited number of experiences, and in most cases, their 'guesses' turn out to be good enough for survival purposes.

This 'extrapolation from limited data' is an operation that is not reducible to the classical Aristotelian categories of 'induction' and 'deduction', and for this reason, Charles Peirce called it 'abduction'. It is a new logical category, and the ability to interpret the world appears to be based precisely on that logic.

We realize in this way that interpretation is truly a new form of semiosis because it is not based on coding but on abduction. What is interpreted, furthermore, is not the world but *representations* of the world, and this means that interpretation can exist only in multicellular systems.

Single cells decode the signals from the environment but do not build internal representations of it and therefore cannot interpret them. They are sensitive to light, but do not 'see'; they react to sounds but do not 'hear'; they detect hormones but do not 'smell' and do not 'taste' them. It takes many cells that have undertaken specific processes of differentiation to allow a system to see, hear, smell and taste, so it is only multicellular creatures that have these experiences.

The evolution from single cells to animals was a true macroevolution because it created absolute novelties such as feelings and instincts (the first modelling system). Later on, another major transition allowed some animals to evolve a second modelling system that gave them the ability to *interpret* the world. That macroevolution gave origin to a new type of semiosis that can be referred to as *interpretive* semiosis, or, with equivalent names, as *abductive* or *Peircean* semiosis.

3.8 The Uniqueness of Language

We and all other animals do not interpret the world but only mental images of the world. The discovery that our perceptions are produced by our brain implies that we live in a world of our own making, and this has led to the idea that there is an unbridgeable gap between mind and reality. Common sense, on the other hand, tells us that we better believe our senses, because it is they that allow us to cope with the world. Our perceptions 'must' reflect reality; otherwise, we would not be able to survive. François Jacob has expressed this concept with admirable clarity: '*If the image that a bird gets of the insects it needs to feed its progeny does not reflect at least some aspects of reality, there are no more progeny. If the representation that a monkey builds of the branch it wants to leap to has nothing to do with reality, then there is no more monkey. And if this did not apply to ourselves, we would not be here to discuss this point*' (Jacob 1982).

Any animal has a modelling system that builds mental images of the world, and we have learned from Darwin that natural selection allows organisms to become increasingly adapted to the environment, that is, increasingly capable of reducing the distance that separates them from reality. Natural selection, in other words, is a process that allows animals to catch increasing amounts of reality. This is because mental images are not about things, but about *relationships* between things, and have been specifically selected so that the relationships between mental images represent at least some of the relationships that exist between objects of the physical world. To that purpose, natural selection can definitely use relationships based on icons and indexes, because these processes reflect properties of the physical world, but it cannot use symbols, because symbols are arbitrary relationships and would increase rather than decrease the distance from reality. Natural selection, in short, is actively working *against* the use of symbols as a means to represent the *physical* world.

Language, on the other hand, is largely based on symbols, and this does give us a problem. The idea that language is based on arbitrary signs, or symbols, is the legacy of Saussure, in our times, whereas the idea that animal communication is also based on signs has been introduced by Sebeok and is the main thesis of zoosemiotics. This extension of semiosis to the animal world, however, has not denied the uniqueness of language. On the contrary, it has allowed us to reformulate it in more precise terms. Such a reformulation was explicitly proposed by Terrence Deacon in *The Symbolic Species* with the idea that animal communication is based on icons and indexes whereas language is based on symbols (Deacon 1997).

Today, this is still the best way to express the uniqueness of language. It is true that some examples of symbolic activity have been reported in animals, but in no way, they can be regarded as primitive languages or intermediate stages towards language. Deacon's criterion may have exceptions, but it does seem to contain a fundamental truth. A massive and systematic use of symbols is indeed what divides human language from animal communication, and we need therefore to account for its origin. How did language come into being?

3.9 The Ape with a Double Brain

In the 1940s, Adolf Portmann calculated that our species should have a gestation period of 21 months in order to complete all processes of foetal development that occur in mammals (Portmann 1941, 1945; Gould 1977). A newborn human baby, in other words, is in fact a premature foetus, and the whole first year of his life is but a continuation of the foetal stage. This peculiarity is due to the fact that the human tendency to extend the foetal period (fetalization) leads to a greater foetus at birth, but the birth canal can cope only with a limited increase of foetal size. During the evolution of our species, therefore, any extension of the foetal period had to be accompanied by an anticipation of the time of birth. The result is that our foetal development became split into two distinct phases—intrauterine and extrauterine—and eventually the extrauterine phase (12 months) became the longer of the two.

It is not clear why this evolutionary result is uniquely human, but it is a historical fact that it took place only in our species. In all other mammals, foetal development is completed *in utero*, and what is born is no longer a foetus but a fully developed infant that can already cope with the environment.

The crucial point is that the last part of foetal development is the phase when most synaptic connections are formed. It is a phase of intense 'brain wiring'. The fetalization of the human body has produced therefore a truly unique situation. In all other mammals, the wiring of the brain takes place almost completely in the dark and protected environment of the uterus, whereas in our species, it takes place predominantly outside the uterus, where the body is exposed to the lights, sounds and smells of a constantly changing environment. In our species, in short, the split between intrauterine and extrauterine foetal development created the conditions for two very different types of brain wiring.

A second outstanding consequence of the fetalization split was an enormous increase in brain size, a phenomenon that was probably caused by embryonic 'regulation'—the ability embryos have to regulate the development of their organs in the critical period of organogenesis. This point is vividly illustrated by a classic experiment. In vertebrate embryonic development, the heart arises from two primordia that appear on the right and left side of the gut, and then migrate to the centre and fuse together in a single organ. If fusion is prevented by inserting an obstacle between them, each half undergoes a spectacular reorganization and forms a complete and fully functional beating heart. The formation of the two hearts, furthermore, is followed by the development of two circulatory systems, and the animal goes through all stages of life in a double-heart condition that is known as *cardia bifida* (DeHaan 1959).

This classic experiment shows that two profoundly different bodies, one with a single heart and the other with two hearts, can be generated *without any genetic change at all*. A modification of the epigenetic conditions of embryonic development is clearly an extremely powerful tool of change and may well be the key to human evolution. The foetal development of our brain has been split into two distinct processes, one within and one without the uterus, and this is a condition that can be referred to as *cerebra bifida* (Barbieri 2010). It is similar to *cardia bifida*, except that in the case of the heart, the two organs arise from a separation in space, whereas in *cerebra bifida*, they are produced by a separation in time.

The *cardia bifida* experiment is illuminating because it shows that the enormous increase in brain size that took place in human evolution could well have been a *cerebra bifida* effect, a duplication of brain tissue caused by the regulation properties of embryonic development.

Extrauterine foetal development and increased brain size, in conclusion, set the stage for a radically new experiment in brain wiring, thus creating the precondition for a uniquely human faculty. Let us not forget, however, that a precondition for language was not yet language. It was only a potential, a starting point.

3.10 The Third Modelling System

The primary modelling system allows an animal to build a representation of the environment, an *Umwelt*, and the second modelling system allows an animal to extract more information from the incoming signals by *interpreting* them. A process of interpretation is an abstraction (more precisely an abduction) that is based on signs, but not all signs are reliable modelling tools. Icons and indexes can indeed favour adaptation to the environment because they reflect properties that do exist in the world, whereas symbols are completely detached from reality. This explains why animals have modelling systems that are massively based on icons and indexes but are virtually incapable of symbolic activity. It does not explain, however, why our species was such an outstanding exception to that rule. How did we manage to communicate by symbols? The solution proposed here is that we did *not* substantially change the first and the second modelling systems that we inherited from our animal ancestors. What we did, instead, was to develop a *third* modelling system.

The human brain is about three times larger than the brain of any other primate, even when body weight is taken into account. This means that the first and second modelling systems that we have inherited from our animal ancestors required, at most, a third of our present brain size. The other two thirds could be explained, in principle, by a further extension of our animal faculties, but this is not what happened. We have not developed sharper eyesight, a more sensitive olfactory system, a more powerful muscular apparatus and so on. As a matter of fact, our physical faculties are in general less advanced than those of our animal relatives, so it was not an improvement of their modelling systems that explains our increased brain volume. It is likely, therefore, that the brain increase that took place in our species was largely due to the development of those new faculties that collectively make up our *third* modelling system, the system that eventually gave origin to language. The brain matter of this system was provided by the extrauterine phase of foetal development, the *cerebra bifida* effect, but that accounts only for the hardware of the third modelling system, not for its software.

The solution proposed here is that our brain used the traditional neural tools that build an 'Umwelt' but used them to build an Umwelt made exclusively of human relationships, a *cultural Umwelt* that exists side by side with the environmental Umwelt. We learned to live simultaneously in two distinct external worlds, one provided by the physical environment and one by the cultural environment. Natural selection, as we have seen, is working against symbols as a means to represent the physical world, but can no longer work against them when they are part of a cultural world that becomes as important as the physical world.

Our third modelling system, in short, evolved in parallel with the first two systems that we have inherited from our animal ancestors, and created a condition whereby we live simultaneously in two environments that not only coexist but somehow manage to merge together into a single reality.

3.11 The Code of Language

Noam Chomsky and Thomas Sebeok are the founding fathers of two research fields that today are known respectively as biolinguistics and biosemiotics and the architects of two major theoretical frameworks for the study of language.

Chomsky's most seminal idea is the concept that our ability to learn a language is *innate*, that children are born with a mechanism that allows them to learn whatever language they happen to grow up with (Chomsky 1957, 1965, 1975, 1995, 2005). That inner mechanism has been given various names—first *universal grammar*, then *language acquisition device (LAD)* and finally *faculty of language*—but its basic features remain its *innateness* and its *robustness*. The mechanism must be innate because it allows children to master an extremely complex set of rules in a limited period of time, and it must be robust because language is acquired in a precise sequence of developmental stages. For this reason, Chomsky concluded that the rules of universal grammar, or the principles and parameters of syntax, must be based on very general principles of economy and simplicity that are similar to the *principle of least action* in physics and to the rules of the *periodic table* in chemistry (Baker 2001; Boeckx 2006).

Thomas Sebeok maintained that language is first and foremost a modelling system, the quintessential example of semiosis, and that 'interpretation' is its most distinctive feature (Sebeok 1963, 1972, 1988, 1991, 2001). He forcefully promoted the Peirce model of semiosis, which is explicitly based on interpretation, and insisted that semiosis is always an interpretive activity. Sebeok underlined that concept in countless occasions and in no uncertain terms: 'There can be no semiosis without interpretability, surely life's cardinal propensity' (Sebeok 2001).

This is the bone of contention between the two frameworks. Is the faculty of language a product of universal principles or the result of interpretive processes? Chomsky insisted that the development of language must be precise, robust and reproducible like the development of any other faculty of the body, and therefore it cannot be left to the vagaries of interpretation. Sebeok insisted that language is semiosis and that semiosis is always an interpretive process, so it cannot be the result of universal principles or physical constraints.

Here, a third solution is proposed. Organic semiosis is a semiosis based on coding not on interpretation, and an embryonic development that follows coding rules is not subject to the vagaries of interpretation. The ontogeny of language, on the other hand, is precise, robust and reproducible even when based on organic codes rather than universal laws. The genetic code, for example, has guaranteed precise, robust and reproducible features in all living system ever since the origin of life.

Language does require rules, but these rules are much more likely to be the result of organic codes rather than the expression of universal principles.

The third solution, in short, is that there was an organic code at the origin of language just as there was a genetic code at the origin of life and a neural code at the origin of mind. It could have been, for example, a code that provided new rules for the brain-wiring processes that take place in the extrauterine phase of foetal development. It is also possible that the codemaker was not the individual brain but a *community* of brains, because language is critically dependent upon *human* interactions in the first few years of life. This is the lesson that we have learned from feral children (Maslon 1972; Shattuck 1981), and the study of 'creole' languages has clearly shown that the major role in the making of new linguistic rules is played by children (Bickerton 1981).

It must be underlined that today we have no evidence in favour of a foundational code of language. This is pure speculation, at the moment, but it does have a logic. All great events of macroevolution were associated with the appearance of new organic codes, and language *was* a macroevolution, so it makes sense to assume that in that case too nature resorted to the same old trick, to creation by coding.

4 Conclusion

Organic codes appeared throughout the history of life, and their origins were closely associated with the great events of macroevolution. Organic semiosis—the semiosis based on organic codes—has been the sole form of semiosis on Earth for the first 3,000 million years of evolution, and it was that form that provided the codes for the origin of the brain. Once in existence, however, the brain became the centre of a new macroevolution that brought feelings and instincts into being, thus giving origin to mind. In the course of time, furthermore, it gave origin to interpretive semiosis, in vertebrates, and then to cultural semiosis, in our species. The brain, in short, created the mind, and our problem is to understand *how* that happened. Today, the scientific models that have been proposed on this issue can be divided into three major groups:

1. The *computational theory* is the idea that lower-level brain processes, such a neuron firings and synaptic connections, are transformed into feelings by neural processes that are equivalent to *computations*. Brain and mind are compared to the hardware and software of a computer, and mental activity is regarded as a sort of data processing that is implemented by the brain but is in principle distinct from it, rather like a software is distinct from its hardware (Fodor 1975, 1983; Johnson-Laird 1983).
2. The *connectionist theory* states that lower-level brain processes are transformed into higher-level brain events by neural networks, that is, by webs of synaptic connections that are not the result of computations but of explorative processes. The reference model, here, is the computer-generated neural networks that simulate the growth of the synaptic web in a developing brain (Hopfield 1982;

Rumelhart and McClelland 1986; Edelman 1989; Holland 1992; Churchland and Sejnowski 1993; Crick 1994).

3. The *emergence theory* states that higher-level brain properties emerge from lower-level neurological phenomena, and mind is distinct from brain, because any emergence is accompanied by the appearance of new properties (Morgan 1923; Searle 1980, 1992, 2002).

The main thesis of this paper is that the brain produces the mind by assembling neural components together with the rules of a neural code, very much like the cell produces proteins with the rules of the genetic code (Barbieri 2006). This implies that feelings are no longer *brain objects* but *brain artefacts*. It implies that feelings are not side effects of neural networks (as in connectionism), that they do not come into existence spontaneously by emergence and that they are not the result of computations, but of real manufacturing processes. According to the code model, in short, feelings and instincts are *manufactured artefacts*, whereas according to the other theories, they are *spontaneous products* of brain processes.

This does make a difference, because if the mind were made of spontaneous products, it could not have *rules of its own*. Artefacts, instead, do have some autonomy because the rules of a code are not dictated by physical necessity. Artefacts, furthermore, can have *epigenetic* properties that add unexpected features to the coding rules. The autonomy of the mind, in short, is something that spontaneous brain products cannot achieve whereas brain artefacts can.

References

Alberts, B., Bray, D., Lewis, J., Raff, M., Roberts, K., & Watson, J. D. (1994). *Molecular biology of the cell*. New York: Garland.

Baker, M. (2001). *The atoms of language. The mind's hidden rules of grammar*. New York: Basic Books.

Barash, Y., Calarco, J. A., Gao, W., Pan, Q., Wang, X., Shai, O., Blencowe, B. J., & Frey, B. J. (2010). Deciphering the splicing code. *Nature, 465*, 53–59.

Barbieri, M. (1981). The ribotype theory on the origin of life. *Journal of Theoretical Biology, 91*, 545–601.

Barbieri, M. (1985). *The semantic theory of evolution*. London/New York: Harwood Academic Publishers.

Barbieri, M. (1998). The organic codes. The basic mechanism of macroevolution. *Rivista di Biologia-Biology Forum, 91*, 481–514.

Barbieri, M. (2003). *The organic codes. An introduction to semantic biology*. Cambridge: Cambridge University Press.

Barbieri, M. (2006). Semantic biology and the mind-body problem-the theory of the conventional mind. *Biological Theory, 1*(4), 352–356.

Barbieri, M. (2008). Biosemiotics: A new understanding of life. *Naturwissenschaften, 95*, 577–599.

Barbieri, M. (2010). On the origin of language. *Biosemiotics, 3*, 201–223.

Bickerton, D. (1981). *The roots of language*. Karoma: Ann Arbour.

Boeckx, C. (2006). *Linguistic minimalism*. New York: Oxford University Press.

Boutanaev, A. M., Mikhaylova, L. M., & Nurminsky, D. I. (2005). The pattern of chromosome folding in interphase is outlined by the linear gene density profile. *Molecular and Cell Biology, 18*, 8379–8386.

Changeaux, J.-P. (1983). *L'Homme Neuronal*. Paris: Librairie Arthème Fayard.
Chomsky, N. (1957). *Syntactic structures*. The Hague: Mouton.
Chomsky, N. (1965). *Aspects of the theory of syntax*. Cambridge, MA: MIT Press.
Chomsky, N. (1975). *The logical structure of linguistic theory*. Chicago: University of Chicago Press.
Chomsky, N. (1995). *The minimalist program*. Cambridge, MA: MIT Press.
Chomsky, N. (2005). Three factors in language design. *Linguistic Inquiry, 36*, 1–22.
Churchland, P. S., & Sejnowski, T. J. (1993). *The computational brain*. Cambridge, MA: MIT Press.
Crick, F. (1994). *The astonishing hypothesis: The scientific search for the soul*. New York: Scribner.
Deacon, T. W. (1997). *The symbolic species: The co-evolution of language and the brain*. New York: Norton.
DeHaan, R. L. (1959). *Cardia bifida* and the development of pacemaker function in the early chicken heart. *Developmental Biology, 1*, 586–602.
Dhir, A., Emanuele Buratti, E., van Santen, M. A., Lührmann, R., & Baralle, F. E. (2010). The intronic splicing code: Multiple factors involved in ATM pseudoexon definition. *The EMBO Journal, 29*, 749–760.
Edelman, G. M. (1989). *Neural darwinism. The theory of neuronal group selection*. New York: Oxford University Press.
Flames, N., Pla, R., Gelman, D. M., Rubenstein, J. L. R., Puelles, L., & Marin, O. (2007). Delineation of multiple subpallial progenitor domains by the combinatorial expression of transcriptional codes. *The Journal of Neuroscience, 27*(36), 9682–9695.
Fodor, J. (1975). *The language of thought*. New York: Thomas Crowell Co.
Fodor, J. (1983). *The modularity of mind. An essay on faculty psychology*. Cambridge, MA: MIT Press.
Gabius, H.-J. (2000). Biological information transfer beyond the genetic code: The sugar code. *Naturwissenschaften, 87*, 108–121.
Gabius, H.-J., André, S., Kaltner, H., & Siebert, H.-C. (2002). The sugar code: Functional lectinomics. *Biochimica et Biophysica Acta, 1572*, 165–177.
Gamble, M. J., & Freedman, L. P. (2002). A coactivator code for transcription. *TRENDS in Biochemical Sciences, 27*(4), 165–167.
Gilbert, S. F. (2006). *Developmental biology* (8th ed.). Sunderland: Sinauer.
Gould, S. J. (1977). *Ontogeny and phylogeny*. Cambridge, MA: The Belknap Press of Harvard University Press.
Hebb, D. O. (1949). *The organization of behaviour*. New York: John Wiley.
Hilschmann, N., Barnikol, H. U., Barnikol-Watanabe, S., Götz, H., Kratzin, H., & Thinness, F. P. (2001). The immunoglobulin-like genetic predetermination of the brain: The protocadherins, blueprint of the neuronal network. *Naturwissenschaften, 88*, 2–12.
Holland, J. A. (1992). *Adaptation in natural and artificial systems*. Cambridge, MA: MIT Press.
Hopfield, J. J. (1982). Neural networks and physical systems with emergent collective computational abilities. *Proceedings of the National Academy of Sciences USA, 79*, 2554–2558.
Hubel, D. H., & Wiesel, T. N. (1962). Receptive fields, binocular interaction and functional architecture in the cat's visual cortex. *Journal of Physiology, 160*, 106–154.
Hubel, D. H., & Wiesel, T. N. (1979). Brain mechanisms of vision. *Scientific American, 241*(3), 150–182.
Jacob, F. (1982). *The possible and the actual*. New York: Pantheon Books.
Jacob, F., & Monod, J. (1961). Genetic regulatory mechanisms in the synthesis of proteins. *Journal of Molecular Biology, 3*, 318–356.
Jessell, T. M. (2000). Neuronal specification in the spinal cord: Inductive signals and transcriptional codes. *Nature Genetics, 1*, 20–29.
Johnson-Laird, P. N. (1983). *Mental models*. Cambridge, MA: Harvard University Press.
Knights, C. D., Catania, J., Di Giovanni, S., Muratoglu, S., et al. (2006). Distinct p53 acetylation cassettes differentially influence gene-expression patterns and cell fate. *Journal of Cell Biology, 173*, 533–544.
Kohonen, T. (1984). *Self-organization and associative memory*. New York: Springer.

Levi-Montalcini, R. (1975). NGF: An uncharted route. In F. G. Worden (Ed.), *The neurosciences – Paths of discoveries*. Cambridge, MA: MIT Press.
Levi-Montalcini, R. (1987). The nerve growth factor 35 years later. *Science, 237*, 1154–1162.
Lotman, J. (1991). *Universe of the mind: A semiotic theory of culture*. Bloomington: Indiana University Press.
Marquardt, T., & Pfaff, S. L. (2001). Cracking the transcriptional code for cell specification in the neural tube. *Cell, 106*, 651–654.
Maslon, L. (1972). *Wolf children and the problem of human nature*. New York: Monthly Review Press.
Morgan, L. C. (1923). *Emergent evolution*. London: Williams and Norgate.
Nicolelis, M., & Ribeiro, S. (2006). Seeking the neural code. *Scientific American, 295*, 70–77.
Peirce, C. S. (1906). The basis of pragmaticism. In C. Hartshorne & P. Weiss (Eds.), *The collected papers of Charles Sanders Peirce* (Vols. I–VI). Cambridge, MA: Harvard University Press. 1931–1935.
Perissi, V., & Rosenfeld, M. G. (2005). Controlling nuclear receptors: The circular logic of cofactor cycles. *Nature Molecular Cell Biology, 6*, 542–554.
Pertea, M., Mount, S. M., & Salzberg, S. L. (2007). A computational survey of candidate exonic splicing enhancer motifs in the model plant *Arabidopsis thaliana. BMC Bioinformatics, 8*, 159.
Portmann, A. (1941). Die Tragzeiten der Primaten und die Dauer der Schwangerschaft beim Menschen: ein Problem der vergleichen Biologie. *Revue Suisse de Zoologie, 48*, 511–518.
Portmann, A. (1945). Die Ontogenese des Menschen als Problem der Evolutionsforschung. *Verhhandlungen der Schweizerischen Naturforschenden Gesellschaft, 125*, 44–53.
Readies, C., & Takeichi, M. (1996). Cadherine in the developing central nervous system: An adhesive code for segmental and functional subdivisions. *Developmental Biology, 180*, 413–423.
Rumelhart, D. E., & McClelland, J. L. (1986). *Parallel distributed processing: Explorations in the microstructure of cognition*. Cambridge, MA: MIT Press.
Searle, J. R. (1980). Minds, brains and programs. *Behavioural Brain Science, 3*, 417–457.
Searle, J. R. (1992). *The rediscovery of the mind*. Cambridge, MA: MIT Press.
Searle, J. R. (2002). *Consciousness and language*. Cambridge: Cambridge University Press.
Sebeok, T. A. (1963). Communication among social bees; porpoises and sonar; man and dolphin. *Language, 39*, 448–466.
Sebeok, T. A. (1972). *Perspectives in zoosemiotics*. The Hague: Mouton.
Sebeok, T. A. (1988). *I think I am a verb: More contributions to the doctrine of signs*. New York: Plenum Press.
Sebeok, T. A. (1991). *A sign is just a sign*. Bloomington: Indiana University Press.
Sebeok, T. A. (2001). Biosemiotics: Its roots, proliferation, and prospects. In: K. Kull (Ed.), Jakob von Uexküll: A paradigm for biology and semiotics. *Semiotica, 134*(1/4), 61–78.
Sebeok, T. A., & Danesi, M. (2000). *The forms of meaning: Modeling systems theory and semiotic analysis*. Berlin: Mouton de Gruyter.
Segal, E., Fondufe-Mittendorf, Y., Chen, L., Thastrom, A., Fiels, Y., Moore, I. K., Wang, J. P., & Widom, J. (2006). A genomic code for nucleosome positioning. *Nature, 442*, 772–778.
Shapiro, L., & Colman, D. R. (1999). The diversity of cadherins and implications for a synaptic adhesive code in the CNS. *Neuron, 23*, 427–430.
Shattuck, R. (1981). *The forbidden experiment: The story of the wild boy of Aveyron*. New York: Washington Square Press.
Spemann, H. (1901). Entwicklungphysiologische Studien am Tritonei I. *Wilhelm Roux'Archiv für Entwicklungsmechanik, 12*, 224–264.
Sperry, R. W. (1943). Visuomotor coordination in the newt (*Triturus viridescens*) after regeneration of the optic nerve. *Journal of Comparative Neurology, 79*, 33–55.
Sperry, R. W. (1963). Chemoaffinity in the orderly growth of nerve fibers patterns and connections. *Proceedings of the National Academy of Science USA, 50*, 703–710.
Strahl, B. D., & Allis, D. (2000). The language of covalent histone modifications. *Nature, 403*, 41–45.
Tomkins, M. G. (1975). The metabolic code. *Science, 189*, 760–763.

Trifonov, E. N. (1987). Translation framing code and frame-monitoring mechanism as suggested by the analysis of mRNA and 16s rRNA nucleotide sequence. *Journal of Molecular Biology, 194*, 643–652.

Trifonov, E. N. (1989). The multiple codes of nucleotide sequences. *Bulletin of Mathematical Biology, 51*, 417–432.

Trifonov, E. N. (1996). Interfering contexts of regulatory sequence elements. *Cabios, 12*, 423–429.

Trifonov, E. N. (1999). Elucidating sequence codes: Three codes for evolution. *Annals of the New York Academy of Sciences, 870*, 330–338.

Tudge, C. (2000). *The variety of life. A survey and a celebration of all the creatures that have ever lived.* Oxford/New York: Oxford University Press.

Turner, B. M. (2000). Histone acetylation and an epigenetic code. *BioEssay, 22*, 836–845.

Turner, B. M. (2002). Cellular memory and the histone code. *Cell, 111*, 285–291.

Verhey, K. J., & Gaertig, J. (2007). The tubulin code. *Cell Cycle, 6*(17), 2152–2160.

von Uexküll, J. (1909). *Umwelt und Innenwelt der Tiere*. Berlin: Julius Springer.

Woese, C. R. (1987). Bacterial evolution. *Microbiological Reviews, 51*, 221–271.

Woese, C. R. (2000). Interpreting the universal phylogenetic tree. *Proceedings of the National Academy of Science USA, 97*, 8392–8396.

Woese, C. R. (2002). On the evolution of cells. *Proceedings of the National Academy of Science USA, 99*, 8742–8747.

The Descent of Humanity: The Biological Roots of Human Consciousness, Culture and History

Angelo N.M. Recchia-Luciani

Abstract The notion of species-specific modelling allows us to construct taxonomies of mental models, based on the concept of *qualia*, such as posing 'invariant requests to neural processes', supporting networks of which are subject to selective pressures. The selection is based on their respective capacity to differently adapt to behaviour patterns, which neural networks control. For extremely premature births, thanks to foetalization, in *Homo sapiens sapiens*, specific neural groups are offered for selection in early critical periods of development and in a social environment. As a consequence, far beyond any other primate, new cognitive devices are developed, which lead to a high level of abstract thinking. Therefore, the reproposition of the cultural-historical psychology is important. Foetalization and education are the two pillars that give rise to the human being's ability to accumulate a perceivable and collective knowledge, which is precluded to other animal cultures. These are the roots both of consciousness and of the specific mechanisms that give rise to transmissibility and variability and adaptability of the human cultures. The key to this evolutionary quantum leap is the advent of a new class of replicators: memes, defined as informational patterns of a signic nature with a metaphorical, relational organization; memes are the basic framework in the structure of personality both in individuals and in social groups.

Keywords Qualia • Consciousness • Awareness • Mental representations • Abstract thinking • Metaphor • Biosemiotic epistemology • Conscious and unconscious metaphors • Memetics

---

A.N.M. Recchia-Luciani (✉)
Università degli Studi della Basilicata,
Polo di Matera – Via San Rocco, 75100 Matera, Italy
e-mail: arklcn@tin.it

L. Swan (ed.), *Origins of Mind*, Biosemiotics 8, DOI 10.1007/978-94-007-5419-5_3,

1 Three Important Ideas from Neuroscience

According to the first biosemiotic thesis (Kull et al. 2009), life occurs when, in addition to exchanges of energy and matter, there are also exchanges of information. Comparative psychologists have analysed the ability, found in both young and adult monkeys, chimpanzees and human beings, of making a conceptual construction of classes. The rhesus macaques (*Macaca mulatta*), thanks to simple associations, are capable of classifying objects on the basis of some invariant characteristics. Unfortunately, they cannot go any further: Thompson (Thompson and Oden 2000) describes them as 'paleo-logical', meaning that they are incapable of perceiving relations-between-relations. The macaques can recognize similarities and differences among the objects but only inside the elements of one single class (first-order classifying).

Chimpanzees (*Pan troglodytes*) are, on the contrary, capable of recognizing similarities and differences in more than one class; therefore, they can distinguish between pairs of items. For example, they can collect a set of keys and differentiate 'real' metal keys from coloured plastic toy keys. Thus, chimpanzees can make an elementary second-class classification and, accordingly, conceive analogies.

Thanks to work in primatology, we have found out that the analogical conceptual capacity, which is exclusive to chimpanzees and human beings, is not spontaneous. In both cases it appears only after its *teaching*. The pupil has to be educated in using a symbolic system (a language or a token system) and only after the training he is able to learn representations (*encoding*) and operate on them (*manipulation*).

In this way, a 5-year-old chimpanzee can select objects, gathering them in two different sets, the inner contents of which are similar or identical. These results have been obtained by running tests of similarity or correspondence to a sample (*matching-to-sample task*, MTS). Other tests, the *preference-for-novelty tasks* (which record the behavioural reaction times to a new visual or auditory stimulus), show that the similarity in the relationships is perceived, both in human children and baby chimpanzees, before the training. Human beings and chimpanzees seem therefore 'predisposed' to perceive relationships of similarity and understand analogies—an ability that the rhesus monkeys, although capable of recognizing the physical identity, never acquire, not even in the adult stage.

In his *Darwin's Dangerous Idea*, Daniel Dennett (1995) proposes the metaphor of the 'Tower of generate-and-test': an imaginary structure where each floor is inhabited by creatures in a different developmental stage. The first stage is comprised of the 'Darwinian creatures', who evolve by natural selection and whose behaviour is defined by their genes. Then come 'Skinnerian creatures', susceptible to operant conditioning.[1] The next are the 'Popperian creatures', who are the first to show an 'inner' capacity to emulate reality. Dennett quotes Popper: 'Emulation permits our hypothesis to die in

[1] Operant conditioning is one of the key concepts in behavioural psychology. It is a form of learning, in which a behaviour is modified, thanks to the reinforcement by the consequences of the behaviour itself.

our stead'. The 'Popperian creatures' show a kind of self-representation of the outer world.

Different stages of emulation build models of increasing complexity of the external world. With distinct potentialities and several adaptive capacities, Maturana and Varela (1985) have accustomed us, for years, to the concept of co-evolution: species do not 'adapt' themselves passively to an 'immovable' environment; rather, they change it and even generate it through their existence. Dawkins has broadened the idea of phenotype (the totality of an organism's observable characteristics or traits), with his concept of an extended phenotype (Dawkins 1982), which describes all the effects that a gene exerts on the external world (effects that, in turn, may influence the genes and determine their chances of being replicated). Dawkins pointed out that each organism may influence the behaviour of another organism.

Furthermore, we could hypothesize that the phenotype 'recognizes' itself not only via the body and the behaviour of its species in respect to others but also in the environmental modifications produced by its species. So, when a new type of 'emulator' is available, the species does not simply 'emulate' new worlds but, literally, it generates new ones.

On this basis, it is easy to understand why it is so interesting to know which new models of the world have caused the appearance of our species: a turning point in natural history with enormous consequences for the life of our planet. The appearance, and above all the selection (in the newborn's brain), of cerebral structures aimed at generating and manipulating *metaphors* constitutes *a* (may be *the*) fundamental passage in the natural history of humans—as I have discussed in my previous works (Recchia-Luciani 2005, 2006, 2007, 2009, 2012).

Marcello Barbieri sums up in a few phrases the story of the so-called foetalization theory: 'In 1926, Luis Bolk, professor of anatomy at the Amsterdam University, proposed in the "fetalization theory" the idea that the origin of man was due to the extension of foetal or juvenile features to the adult stages of life (Bolk 1926). The idea was not new [...] but it was Luis Bolk who turned that idea into a compelling doctrine by the sheer number of data with which he supported it' (Barbieri 2010). In 1940, Adolf Portmann calculated that our species should have a gestation period of 21 months in order to complete the processes of foetal development that occur in all other mammals (Portmann 1941, 1945): a newborn human baby, in other words, is in fact a premature foetus, and the whole first year of his life is but a continuation of the foetal stage. This allows the scientist to propose a model that could explain how this single variation can be generative of such wide consequences. A conceptual model named *cerebra bifida*, where an event in the scale of time, rather than in that of space, produces a complete subversion/expansion of the epigenetic potentialities.

In embryonic vertebrate development, the heart arises from two primordial sections, and simply cutting the point where the fusion normally takes place will cause each half to form a complete and fully functional beating heart, which is then followed by the development of two circulatory systems. As an experimental outcome, this double-heart condition, compatible with survival, is known as *cardia bifida* (DeHaan 1959). The essential element emphasized by Barbieri is the absence of any genetic change. Here, in fact, the remarkable modification takes place exclusively through an alteration of the epigenetic conditions of the embryonic development.

The gradual dilatation of our foetal development period, together with the constraint of the birth canal, has split the foetal development of our brain into two distinct processes, the first within and the second outside the uterus, whereas in all other mammals it has remained a single internal process. This splitting of the foetal brain development is called by Barbieri *cerebra bifida.* As happened in the embryonic heart experiment of DeHaan, we are dealing with a separation in time rather than a separation in space. In both cases, the same series of genes could have produced very different results only by operating under two different environmental conditions: a conclusion widely supported by innumerable examples made in the field of embryonic development.

Whether related to other important acquisitions of modern neurobiology, this idea plays a prominent role in the comprehension of the origin of our species and in particular of our specific typologies of 'reality emulation'.

The first concept we will try to explain is *connectionism.* In its original version, *modularism* (Fodor 1983) presupposed a cognitive architecture arranged in terms of *modules*, that is, structures capable of converting inputs into representations. The modules (or 'devices', like Chomsky's famous linguistic device, or mental organs) had great meaning in the historical development of the cognitive sciences, but they were inadequate to build satisfactory models of the working mind/brain. Actually, the old modular models were based on the hypothesis of mind/brain serial functioning. So, instead of eliminating the concept of modules (the specialized areas of the brain are an undisputed model in clinical practice), so-called connectionism made a comeback.

The great speed with which the classical computer—with sequentially connected computational units—handles electrical signals, manipulating one line of code at a time, is very different from the parallel processing of the central nervous system which, while not intrinsically fast, is strongly interconnected, allowing for greater information processing power overall (Hebb 1949).

Thus, if the brain is divided into specialized areas (modules), it is possible that these areas provide 'superior' functions only when and if they are connected and functioning in strict coordination.

At birth we dispose of the majority of our neurons. During the developmental process, the neurons, as we will analyse in detail below, are selected, and this selection causes the loss of a great deal of units. 'Losing' so many neurons is not a pathological fact; rather, only what is necessary is chosen and everything else is thrown away.

Yet, many nervous cells are not sufficient alone: the postnatal development of the nervous system depends on two processes that are essential in making short and very distant brain connections. The first process is the realization of the synapses: specialized microscopic gaps that permit a neuron to pass an electrochemical signal to another cell (neural or otherwise). The formation of synaptic networks is followed by the phenomena of selection, reorganization and redefinition. The networks in the beginning are redundant and then are 'pruned'.

At the microscopic level the synapses allow for communication among close neurons in order to create the 'local' neural networks. At the macroscopic level, neural networks are possible only when myelination enables the anatomic and functional connectivity between different brain regions.[2] In humans, little myelin exists in the brain at the time of birth. Myelination continues throughout the adolescent stages of life until adult age.

We have to consider, however, that when foetal growth is completed inside the protected and humid darkness of the uterus, it is rather different than spending the last year of development—with all its critical stages—totally dependent upon a milieu. Hence, it is easy to explain the enormous differences between the species, though being genetically very close: there is a famous computation that states that human beings and chimpanzees share 94% of their genes in common (Demuth et al. 2006).

2 Natural Selection into the Brain: Developmental Critical Stages and Neuronal Group Selection

The concept of *developmental critical stages* needs to be explained: it is the second concept coming from neuroscience that we need in our model.

David Hubel and Torsten Wiesel won the Nobel Prize in 1981 (shared with Roger Sperry, who studied the hemispheres' asymmetry) for their fundamental studies on the leading role of experience in defining the cerebral architecture of the visual system (Hubel and Wiesel 1963; Wiesel and Hubel 1963a, b). The genes, more than establishing stiff and immutable schemes, seem to 'predispose' some architectures to be 'open' to several possibilities, which are afterwards defined and selected by the environment and experience. This selection, made on the basis of what is really necessary and useful, has given new strength to the co-evolutionary concept of *fitness* (the environment is the cause and effect of the evolution of the organisms and vice versa).

In the precocious stages of postnatal development, there are some 'critical periods' wherein experience produces deep and decisive effects on the brain's organization. The same sensorial deprivation in successive ages does not yield the same results.

These considerations have a peculiar meaning in *Neural Darwinism*, where Gerald M. Edelman puts forth a theory called 'neuronal group selection' (TNGS)

[2] Myelin is an electrically insulating material that forms a layer, the myelin sheath, around the nerve fibres allowing for their autonomy and specificity, functioning as structures for the transmission of the nerve impulse. The larger neural networks (e.g. among two distant cerebral lobes) are possible only when the myelination enables the anatomic and functional connectivity between different brain regions.

(Edelman and Mountcastle 1978; Edelman 1987). This is the third neuroscientific concept we need for our model.

In 1972, Edelman shared the Nobel Prize with Porter for having explained the *immunological specificity* by applying the theory of *natural selection* (*selectional* theory vs. *instructive* theory). Simply put, it was impossible to explain the specificity of the antibodies by referring to genetic determinism: there were not enough genes that could explain the vast number and variety of the antibodies present in organisms. Therefore, as already evident at a level of individuals and species, the 'right' antibodies were the result of a process of selection.

Edelman extended the same core issue to the mechanisms that define the neuronal architectures: neurons are organized in networks that are capable of specific functions, thanks to selection phenomena that act on an array of strongly redundant elements. The redundancy is easily demonstrable on many levels of organization: genes, neurons and synapses are redundant, and the selection is entirely a *post*-event phenomenon and hence, by definition, *epigenetical*.

The groups of neurons, and not the single neuron, constitute the selection unit: (a) in the embryonic and postnatal *developmental* stage, (b) by behaviourally determined *experiential* selection and (c) thanks to *re-entry* phenomena: 'Reentry is the continual signalling from one brain region (or map) to another and back again across massively parallel fibres (axons) that are known to be omnipresent in higher brains' (Edelman 2006, p. 28).

The neural reduction is a well-known phenomenon that characterizes both child and adult cerebral development. Without entering into details, there are data showing that the increase of neural network specificity correlates with a reduction in the number of neurons and sometimes of synapses. This does not depend on an injury but on a learning process that occurred previously. Edelman himself, together with Gally (Edelman and Gally 2001), demonstrated the central role of the degeneration processes in the facilitation of the biological evolution of systems.

3 The Others

As already mentioned above, the newborn depends on adults (foetalization); many theoretic and experimental data assert the centrality of *groups* within which we complete our development, which we called 'foetal', although postnatal. Among the many, Kanzi's story deserves to be told (Savage-Rumbaugh and Lewin 1996; Segerdahl et al. 2005).

Born on the 28th of October 1980 at Yerkes Field Station (at Emory University), Kanzi (which in Swahili means treasure) is a male pygmy chimpanzee (or *bonobo,* officially classified as *Pan paniscus*). The baby ape was moved to the Language Research Center at Georgia State University to be left to the care and study of the primatologist Sue Savage-Rumbaugh. Taken away from his mother and adopted by a more dominant female, Matata, Kanzi accompanied her to useless sessions where she was taught *Yerkish*, an artificial, non-verbal language, appositely developed for this aim. *Yerkish* is not spoken: it employs a keyboard whose keys contain *lexigrams*, abstract symbols corresponding to words. Matata was not a good pupil, and the lessons were not meant for Kanzi: one day, while Matata was away, Kanzi began spontaneously using the keyboard lexigrams, becoming the first known ape to have learned aspects of language in a natural way rather than through direct training.

Kanzi's story teaches us a lot: the first aspect pertains to the critical stages of development. Kanzi learned a language because he had evidently completed the development of the neural structures that enabled this incredible achievement.

Secondly, this neural infrastructure had 'met' a milieu where its selection had been possible, thanks to a series of positive 'supports', before the infrastructure itself could be 'eliminated' (selectively). This also happens to *feral children* (human children who, throughout the entire critical stage of development, do not receive any 'stimulus' from the milieu). Clearly, the same thing had occurred to Matata. She had been submitted to a more specific training but 'after the time limit'. Matata had secured the complete attention of the 'others' (the primatologists) but too late. The role of the 'others', and of the milieu, represents the second great lesson from the exemplar story of the famous pygmy chimpanzee.

What we call our *self* is not an object but a function that depends on many elements. The *self* structure is *dialogical*. And the identity is a *gift from the other*.

The group or set of 'significant others'—all the people who are or have been important to us (or even who could be, e.g. the desire we feel for a stranger we find attractive or for a future child)—is the matter that makes 'I' and 'me', with every possible variation in theme. We can creatively copy traits, invert, sum and overlap them to completely alien elements, elements with an idiosyncratic structure that depends on the relationships which mark us during our lifetime.

What we become is, in equal measure, up to us and to the others: by continuously building the story of what we are, what we have been and—mainly—what we are going to be.

The brain has some necessary mechanisms to 'introject' itself inside the social group membership, but that is not all. The self is constituted by dynamic 'voices' that configure 'meaning nuclei'—internal or external to the same self.

From this comes that, one self and an individual mind are, dynamically and continuously, regenerated. Both are emerging functions of a system of which the hierarchical level is higher than the single individual that forms it.

Me represents the possibility of looking to ourselves from the other's point of view.

The voice heard—with its possibility of a control based on feedback—is an inner dialogue model, not with two, but with many voices: the voices of all the *significant* others resonate inside us.[3] Hubert Hermans, building on Bakhtin's lesson, speaks about the multivoicedness and dialogicality of the self, challenging 'the notion of unity of the self and the distinction between self and others' (2001).

Luigi Pirandello used to speak of masks when he was referring to multifaceted personalities, chosen by us or imposed by society. He employed a theatre metaphor

[3] The word is both used as 'important' and 'able to give a meaning'.

to describe a psychological phenomenon: the assumption of 'roles' in a social contest. In this model of the psyche and self, these roles are called '*positioning*' (Hermans et al. 1992; Hermans and Kempen 1993). The part of ourselves that acts as an observer plays the role of a supervisor and puts itself in a meta-position. That part of us considers the facts of our lives as if they concerned the life of someone else. As we all know, this natural and 'philosophical' observer is not the part that represents us most of the time.

In the 'normal' psychic functioning, the self-stability does not stop from being variable both in time and context (i.e. positioning). On the contrary, the self-'richness' (of one specific personality structure) can be proportionate to the size of the cohorts of the 'others' with whom our meta-position (the observer) can dialogue, because the meaning is social and presupposes relationships and power relations.

The observer and his meta-position have a central relevance as he keeps the organization and adaptability of the self, which we attribute to so-called mental health.

The cohort of voices (or *multivoicedness*) has a hierarchical organization, thus very dynamic: in fact, we are not exactly the same person when we work or enjoy our leisure time. Its rigidity, caused, for example, by an inner 'authoritarian' voice which dominates permanently, generates pathologies, as in multiple personality disorder (Mininni 2003).

The parts of the self are committed in a nonstop and polyphonic conversation: this is why we speak about *dialogic self.* The celebrated Stanislavsky's theatre system consists of learning how to modulate speech, thanks to its omnipresent *subtext* ('Every sentence that we say in real life has some kind of subtext, a thought hidden behind it', Vygotskij 1934, Ed. It. 2004 p.389, my translation).

Our being a *symbolic species* (Deacon 1997) is due to a specific neural organization, although its nature is essentially social. There is, in fact, a notable literature about feral children (Recchia-Luciani 2006). Without a social group, in the critical stages of development, language cannot develop and neither can a 'human' mind nor self. Biology—genetics—does not change here: but this 'natural experiment' indicates a development outside cultural evolution that does not produce human beings with modern consciousness.

Hence, being human depends on the interaction between the world around/environment (*Umwelt*) and our own world (*Eigenwelt*), together with the important contribution of the social/cultural world, the world *with the others* (*Mitwelt*).

Identity emerges from the interaction with others and finds its origin in the possibility, exclusive of the human thought, of elaborating symbols.

Giambattista Vico (1744) was the first to call the metaphor a *cognitive device* rather than a rhetorical artifice. In a *metaphorical theatre*, the actors are *signs*, *symbols* actually. It is the scenery of a *consciousness* that still today, with Julian Jaynes (1976), is considered 'postlinguistic' with an evolutionary significance (Jaynes 1976).

For this fundamental author, the metaphor, in its more general meaning, is the language cornerstone: 'the use of a term for one thing to describe another because of some kind of similarity between them or between their relations to other things'. Just 4 years later, Lakoff and Johnson (1980) affirmed that the metaphor is principally a way of conceiving of one thing in terms of another. The metaphor serves as a vehicle for *understanding*. Without understanding, no experience can exist, at least as far as the linguistic domain is concerned. We mean a language as the human possibility of using whatever kind of code, rather than simple verbal ability.

We always (and only) understand 'something' as 'something else': starting from the direct and personal experience of the physical body, on this planet, in our three-dimensional environment with its gravity and in our collective environment, primarily made of our own species.

Metaphor is a cognitive instrument. From its activity depends the existence of the homonymous rhetorical form, not the contrary. We are talking about metaphor or '*trope*' as a *figure of speech in which words are used in a sense different from their literal meaning*, in linguistics and semiotics, a *sign*.

Self-constitution is therefore corporeal, mental and emotional, always referring to the *significant others*. Communication is a contextual, situated negotiation outcome and depends always on what has been put 'at stake'. In a linguistic game we build *metaphorical landscapes* where the actors, symbols in competition, move around: messages are different interpretations of the reality, fighting for the primacy (Lawley and Tompkins 2000).[4] Interpretation is not necessarily rational or adaptive but could depend on economical or power factors (the military missions are *wars* or *peacekeeping missions*?).

4 Models of Worlds

Human thought has the exclusive possibility of elaborating symbols. However, symbols are only one of the potential categories of signs. We have already examined different concepts provided to the signification relation or rather to the sign (Recchia-Luciani 2012, refer to for a more articulate analysis). Beyond de Saussure and Peirce and their schools of thought affirmed in the twentieth century (semiotics and semiology), we want to highlight the conceptual effort of Thomas Sebeok and Marcel Danesi to overcome the conflicts (Sebeok and Danesi 2000; Danesi 2008).

Sebeok and Danesi gave a new name to the fundamental components of the sign, which is defined as the relation [A stands for B] (it can be written [A=B]). Part [A] is the *form* and part [B] the *referent*. The link between the two components, their own relation [A=B], produces *a model*. Peirce classified the relationships between signs and objects, as *icons,* the signs' ways of having a semblance to their objects; *indexes,* factual connections (in spatial and temporal terms like in the co-occurrence, or by cause) to their object; and *symbols,* that need a habit or rule for their interpretation (Peirce 1931–1958).

In Sebeok's and Danesi's modelling systems theory—or MST—using symbols is specifically human within a more general theory of semiosis. A *referent* is whatever is attributed with a *form*. For the aims of this work, we need to report some basic implications of this theory; the first approach is to consider signs, symbols and human

[4] The theory of language games has been developed by Ludwig Wittgenstein in his *Philosophical Investigations* (1953).

consciousness as steps in the evolutionary natural history of life: 'Species-specific understanding of the world is indistinguishable from the forms used to model it (the modeling principle), as some of its essential implications. The modeling principle implies simply that for something to be known and remembered, it must be assigned some form. The variability principle implies that modeling varies according to the referent and to the function of the modeling system' (Danesi 2008, p. 291). Symbols allow the definition of language proper: as stated by Deacon (1997), not 'a whatever system of communication', even when organized by a specific syntax, but rather a 'system of communication based on symbolic reference (just as words refer to things) which contemplates combinatory rules, including a system of synthetic logic relations across the same symbols'. Other animal species have not developed proper languages, unable as they are to 'get how word combinations make reference to things'.

As introduced by, with our classification of the cognitive capabilities—progressively more complex—of monkeys, anthropomorphic primates and man, what is extremely important, in order to understand how the world models of the different species of animals vary, is the structurally hierarchical nature of signs. Let us come back with Deacon to the primatologists' crucial work. In analysing the experiments of Sue Savage-Rumbaugh and Duane Rumbaugh with the chimpanzees Sherman and Austin, Deacon describes a crucial achievement obtained during the test: 'The animals have discovered that the relation between a lexigram and an object is a function of the relation with the other lexigrams, and not only of the correlated occurrence of lexigrams and objects. This is the essence of a symbolic relation'.

The advent of symbolic relations is a complex, total change in the modelling strategy of a species, in terms of both understanding and memorization. The full co-occurrence of relations, which are typical of both the iconic and indexical models, becomes completely unnecessary: the co-occurrence of relations (and not of things themselves) gives the possibility of making categorial speculations among the few possible alternatives. As Favareau put it, in summarizing a position that he shares with Sebeok and Deacon, '*we [humans] "manipulate representations" (and not the things themselves)*' (Favareau 2008).

5 The Mystery of Qualia

As already cited in the section titled 'three important ideas from neuroscience', we could say that the brain, primarily the immature brain of the newborn, seems prone to model itself on the characteristics underlying both the physical world and the environment (*Umwelt*), thanks to personal experience (*Eigenwelt*) and through the social/cultural world mediation, the *world with the others* (*Mitwelt*).

As of the age of 4–5 months, children become very curious. They are not able to speak, but we can easily prove their loss of interest: they lose their interest in the object. The *preference-for-novelty tasks* are based on that observation: with a timed video recording of the child's activities, it is possible to measure precisely the average time of the observation. Imagine showing a child a red ball: at the beginning, driven by curiosity, the child will watch it with interest. At this point, the ball is hidden behind a screen; then it reappears. Soon afterwards the child gets bored with this game, and the comeback of the red ball will receive less and less attention and quicker glances. Yet, if we replace the red ball with a yellow one, the attention will come back immediately with longer-lasting stares. The child is evidently surprised and starts seeking the missing red ball. He does not seem to 'believe' that the red ball has turned into a yellow one. The child seeks an object that exists because it is permanent. The permanence of the object is linked to the permanence in time of one (or more than one) of its properties. For our senses—and for classical physics too—the concept of reality is unequivocally linked to the concept of properties, for example, in a sensory domain.

Although surprising, classical experiments from ethology pioneers such as Niko Tinbergen can support this fact. In ethology, a *superstimulus* is an artificial stimulus to which there is an existing response tendency or any stimulus that elicits a response more strongly than the natural stimulus for which it evolved. Konrad Lorenz noticed how birds can brood eggs similar to theirs but only if the eggs are larger. Niko Tinbergen (Tinbergen 1951, 1953) and his group made some systematic observations in the species producing dappled eggs. Their tests showed how plaster eggs, with more defined markings, larger markings or more saturated colour, were preferred by most of the species. Hence, they studied the specific characteristics of the stimuli that caused the food claim in the herring gull chicks and built a famous collection of artificial stimuli.

The chicks of the herring gull peck on their parent's beak (which is yellow with a red dot) begging for them to regurgitate food that constitutes their meal. In Tinbergen's tests, the herring gull's yellow beak with a red dot was replaced by a realistic three-dimensional model with a larger red dot, afterwards by a yellow beak with a larger dot bereft of the head and finally by a sharp red stick with three yellow stripes only: clearly, the more distant model from their veridical parent. Still, the chicks kept asking for food (pecking more frequently) with increasing insistence, moving from the first to the last model. In brief, given an accurate three-dimensional reproduction, the accentuation of one or more invariant properties of the stimulus (the dot dimension, the number of the dots or the colour saturation on the egg or beak surface) produces a larger response that is possible to measure.

Also in respect to human psychology, a lot of work has been carried out on the superstimuli: what we want to stress here is how specific neural networks, capable of managing the instinctive behaviour already present at birth in many animal species, can respond to stimuli that are not perceived in a generic or global way but with specific attention towards physical properties immutable in time, properties towards which the species use suitable sensory channels.

In general, we can consider the behavioural responses as 'innate' or 'instinctive' wherever their realization does not require any learning through experience. These responses are already present at the moment of birth. Normally, the indispensable responses to pure surviving are of this kind.

The chick of the herring gull does not 'learn' how to peck on its parent's beak asking for food: where a pathology or a mutation prevents this behaviour, death will be the unavoidable outcome. We can imagine something similar for the innate 'rooting reflex' in the newborn babies. But also many acquired behaviours, not present at birth and requiring some experience, are structured to respond to stimuli prone to react to invariant properties, properties towards which the species has adequate sensory channels (which produce sensations[5]) or more complex nervous structures capable of complex perceptions.[6] Timbergen's work on herring gulls becomes fundamental, because it is proof for the existence of an innate neural network, tuned to specifically perceive the redness of red of the presence or absence of a dot on their parent's beak. The redness of red is an archetypal *sensory quale*.

In 1929, Clarence I. Lewis (Lewis 1929) introduced a term bound to cause a lasting havoc in the scientific and philosophical community. In his book 'Mind and the world order', he employed for the first time the Latin term *quale* (in the plural *qualia*), as 'recognizable qualitative characters of the given (Lewis 1929, p. 121) that usually refers to mental states with characteristics of highly distinct subjectivity, or to phenomenal aspects of mental life accessible solely through introspection'. The choice of the Latin term refers, as does the term, to *qualities* or *sensations* considered in a form that is *isolated* from their effect on the behaviour. Since the beginning, *qualia* have been considered irreducible qualities of states of mind: perceptive experiences or corporeal sensations according to somebody but also states of mind depending upon emotions, feelings and moods, for others. The most classic example is the 'redness' which means the quality of being red independently from the objects of reality. The 'red' *quale* and its 'redness' (just being red) associates red roses with that peculiar state of the traffic lights that force us to stop. They are considered specific properties of the sensory experiences. Of course, we have at our disposal some instruments allowing us to share experience. These instruments are constituted by a *theory of mind*: the comprehension of the other is due to the recognition—in oneself—of states of minds analogous to ours: to those of the known subject. In this first stage, we refer to sensory qualities only: therefore, the qualia we are talking about are domains given to us by the evolution, as instruments particularly suitable to make our species adapt in (and together with) specific ecological environments of life on this planet. These are the *sensory qualia*.

In our hypothesis, both in the case of the awareness and innate-instinctive behaviours and in the case of awareness and acquired behaviours, when a response is induced by a specific sensory quality or by a specific constellation of different sensory qualities, its 'recognition' occurs in the neural networks where the senses are posing *'consistent invariant demands on neural processes'*.[7]

Apt neural networks can be so indispensable to surviving that they must be genetically determined (so they will govern innate-instinctive behaviours) or

[5] Labelled as modifications of the nervous system induced by the relations with the environment mediated by the sensory organs.

[6] Labelled as a complex synthesis of simple sensory elements in forms containing meaning.

[7] Deacon, ibidem p. 329.

selected by learning from experience (so they will govern learned behaviours). As Hubel and Wiesel demonstrated through tests with kittens made blind, the same process of 'seeing' is partially 'learned'.

We have already referred to the metaphor as a typically human cognitive device: in this, as in other recent works (Recchia-Luciani 2006, 2007, 2009, 2012), we pose again a classification of these psychic phenomena, which we consider attainable only through autoanalysis.

Actually, there is a distinction between *sensory qualia* and *metaphorical qualia*, both phenomena of our mental life and introspectively accessible and both able to become 'objective' only within the context of recognition and mutual comparison, that is, the intersubjective validation. In the facts, as we all know, coming to an agreement on a perception (the meaning of a perceived object) is easier for concrete objects than for abstract ones. The sensory organs have to assure the greatest achievable adherence to the 'material' and 'tangible' features of the objects: to some of their invariant properties. If we cannot see very well, we consult an eye specialist and do not try to 'interpret' the reality through a 'new type of eyesight'.

Ludwid Eduard Boltzmann (1905, quot. in Antiseri 1986), physicist and mathematician (as well as philosopher), well known for his laws on the kinetic theory of gases and the second law of thermodynamics, described the brain as the organ that builds the images of the world, in 1905. The sensory perception is focused on invariant features of data provided by the sensory organs, the sensory qualia: they coincide with the constant properties of the objects that can be detected since early childhood.

Sensory organs evolve, thanks to the selection of particular transducers and of the peripheral and central nerve networks able to deal with the information coming from them.[8] But why developing, evolving 'new senses'? Because new 'sensory models' can turn out as adaptive, this means the capacity to improve the species fitness.[9] This is the reason why different visual apparatuses have evolved in different species: natural history has seen them appear many times and independently.[10] Animals with eyesight have probably shown better capacities of surviving and reproducing than the blind animals (or those with sight defects). This is the 'selective pressure': the appearance of a new feature can give a competitive advantage in the struggle for existence, and who is able to survive can breed more and better. The snake world is made (mainly) of smells, the bat world is made of echoes, and our world... above all of other human beings.

[8] A transducer conveys energy from one point to another one altering some features, so it 'converts' input energy of one form into output energy of another; it 'transduces' a signal into another kind of signal. For instance, a microphone transforms variations of air pressure into an electric signal. Each transducer is characterized by a peculiar mathematic function, named a transfer function.

[9] In biology, fitness is the estimate of reproductive success (number of offspring) of an individual or a genotype. It is the estimate of the birth rate, as the adaptation shows itself through an increase in birth rate.

[10] As can be demonstrated by the independency of the genes' families that control the various types of 'eye'.

A *speciation* is a biological event through which a new species 'originates' in relatively short periods of time (short if compared to the geological eras!) and exhibits features different from those of the species from which it has originated.

When an animal species produces an 'orientation system' based on a signal (for instance, electromagnetic waves for eyesight, or volatile molecules for the olfaction), which become important information for surviving, the accomplished evolutionary step can also mark a speciation, if we considered the great adaptive advantages it is able to generate.

Generally, organisms do not have a direct relationship with the world but with their own perceptions of it. Not only sensory perceptions: in human beings capable of elaborating symbols, a perception can concern, for instance, the interpretation given to a certain situation. The brain is a *reality emulator* (Llinas 2001), inside which our orientation is linked to our map of the world.

For this reason we will propose here the achievement of a new 'mental organ', the *metaphorical device*, as the essential threshold for the speciation of the *Homo sapiens sapiens*: a new species with a new type of reality emulator able to generate language and thought, allowing us to develop, besides awareness, a consciousness in the strict sense of the word. Functions which are based on the perception of new kinds of invariant properties: functions which are based on new types of qualia.

At the basis of the combinatory power of proper languages, there is the choice of the elements likely or unlikely to be combined, replaced and manipulated, which produce new levels of correspondence defined by linguistics as '*semantic traits*', for example, the absence or presence of a certain property. Semantic traits share with other kinds of qualia a feature: they pose *'consistent invariant demands on neural processes'*.[11]

6 Defining 'Consciousness', Defining 'Unconscious'

In the present chapter, we propose that the species *Homo sapiens sapiens* owes its existence to a new system of reality emulation, to a new *image-of-the-world device*. This image-of-the-world device includes the functional properties of *awareness* and *consciousness*. For this reason, we will now look in detail at the operational definition of the term *consciousness*. The scientific, technical use of these terms poses a problem.

A comprehensive and detailed examination of the polysemic spectrum of meanings of these complex concepts in different branches goes beyond the scope of this work. Here, as in other works, we choose to adopt a modified, enriched version of the famous Julian Jaynes' model of consciousness (Jaynes 1976, 1986).

Julian Jaynes' consciousness has several distinctive features: *spatialization, excerption, the analogue I, the metaphor me, narratization and conciliation.*

Spatialization is not 'simple' perception of space (basics of apparatus—image of world existing in the nervous systems of less complex animals), because here, the term is referred to as 'mind space'.

---

[11] Deacon, ibidem p. 329.

Excerption is being conscious of particulars, which come to represent the concepts they are examples of.

Analogue I and *metaphor me* introduce a multifaceted point of view and identify, besides a speaking *I*, a 'listener' and a 'critical' *me*.

In this dialogue, a *narratization*, the *analogue 'I'* (me envisioned from within) and the *metaphor me* (me envisioned from within observed from outside), is the point of view of the others about ourselves.

Among the most important functions, there is the *metaphorization* to generate the *I* and the *me*, who are constantly committed in a dialogue which creates a story, a narration: a *narratization*.

Conciliation is the function that produces congruent worlds, thanks to which consciousness becomes sometimes blind, 'denying' the objects that are irreconcilable with our world vision: its functionality is guaranteed by *suppression* and *concentration*.

In other, more recent works, we have other good operational (i.e. functional) definitions of the term *consciousness*. Gerald Edelman, immunologist and neuroscientist, made a distinction between *primary consciousness* and *higher-order consciousness*. *Primary consciousness* is the awareness of the present without 'having concepts of the past and of the future' (Edelman 1987). *Higher-order consciousness* allows for self-recognition of our actions and feelings, construction of a personal identity and awareness of the past and future: it consists of an immediate awareness of mind episodes *without* the involvement of any sensory organs or receptors. This form of consciousness is self-reflexive: it gives to human beings the *consciousness of being conscient* and an *explicit perception of time*.

António Damásio proposes a *core consciousness*: the self in the here and now, and an *extended consciousness*, provided with a sense of identity and a sense of self in 'historical' time, with an 'awareness of both a past and a future with respect to oneself and the world' (Damásio 1994). For Damásio, the self is the protagonist of consciousness. Damásio also declares that the *self* is structured on many levels: in fact, we have a *proto-self*, substantially corporeal and biological, that constitutes the base of a *core self* and at the highest level an *autobiographical self* (narrative, historicized and abstract).

We have mentioned Edelman and Damásio because both consider (besides a more developed form of consciousness on the evolutionary level) the presence of less complex and powerful world-modelling functions. Edelman speaks of the primary consciousness, Damásio of the core consciousness. There is nothing similar in Julian Jaynes.

This is the reason why we have added a definition of awareness, exclusively limited to the presence of an object inside a sensory-perceptual-motor domain (and vegetative, hormonal and immune) of an individual.[12] Awareness and consciousness

[12] The choice of the term *awareness*, related to this 'consciousness *without* an analogue I which narratizes in a mind space', is due to its explicit use made by the mystical traditions and the religious or laical meditation techniques. These practices tend to a sort of 'extinction' of the continuous self dialogue, often pursued by different means that 'saturate' the sensory channels with contents towards which the highest concentration is directed. 'Against' the pervasiveness of the abstract thought, here, the sensory content is proposed.

share the functions of *spatialization, excerption and conciliation*; spontaneous is its tendency to *suppression and concentration even extreme.* It is not like that for the *analogue I, the metaphor me* and *narratization*, because we consider these functions as metaphorical expressions whose introduction coincides with consciousness in the strict sense of word. This *awareness* (close to Edelman's *primary consciousness* and Damásio's *core consciousness*) is the protagonist of Jaynes' mind world in the long-lasting eras of the bicameral civilizations: where it was the Gods' voice suggesting choices and taking decisions (Jaynes 1976).

But why give a greater authoritativeness to Jaynes' model of consciousness? There are many reasons to support this choice. The first one is the articulation, the level of detail of its functional model. Jaynes forms himself as a psychologist, and the precision and specificity of his functional descriptions are not comparable to Edelman and Damásio.

The second is the importance given to the role of the metaphor as a cognitive 'device'. This setting recalls, although unwittingly, Vico (the Neapolitan philosopher, who was the first to call the metaphor a cognitive device, is never quoted, but this is quite obvious given his not so great popularity in America during the 1970s).

The third reason is the recognition of the central role of social groups in the genesis and structure of the conscious function. Edelman and Damásio seem to be looking for consciousness mostly 'in' the brain.

The fourth and last motivation reminds us of Kanzi's story: we have already said that the analogical conceptual capacity, which is exclusive to chimpanzees and human beings, is not spontaneous, but it appears only by learning through training. The pupil has to be educated to use a symbolic system to operate encodings and manipulations. A predisposition is not sufficient: to realize it education is needed. And here Jaynes is explicit: according to this author, consciousness 'come after language' (Jaynes 1976, p. 66), a mind space generated by speech. Learning a new symbolic system allows the mind to use new forms of abstract thought.

An operational, functional definition of consciousness calls for the study of its antonym: the *unconscious*. First, we need to define unconscious.

Here, we see unconscious as meaning a mental state that exists outside of conscious focus.

Alternatively, the unconscious is a mental state where learning has happened implicitly.

In other cases, being unconscious is a complex mental and physical state controlled by the autonomic nervous system (ANS, sometimes defined using the old term of 'vegetative') and/or peptides with complex functions (neuroimmunoendocrine).

In the final decades of the twentieth century, the great Chilean psychoanalyst Ignacio Matte Blanco (1975, 1988) focused his research on the monumental task of describing the cognitive processing of Freud's unconscious, which only partly coincides with the first of the possible definitions provided above.

Matte Blanco used the tools of formal logic, derived from set theory, starting from Dedekind's definition of the infinite set. 'A set is infinite when and only when it can be put in bi-univocal correspondence with a proper part of it' (Matte Blanco 1975, p. 33).

Matte Blanco did not draw up long lists of 'properties of the unconscious' but systematically demonstrated that the *infinite sets of the unconscious* were governed by two fundamental principles: the *principle of symmetry* and the *principle of generalization.*

In relation to the principle of symmetry, the unconscious performs a 'symmetrization of asymmetrical relations'. For example, the mother-child relationship is asymmetrical because it is opposite, 'the child is the mother of his mother', is not a real statement, but it becomes real in the unconscious, where the asymmetrical relationship and its opposite are both 'true'.

Symmetrical logic and asymmetrical logic coexist, just as consciousness and the unconscious coexist. What changes, depending on the cognitive tasks, is the degree to which these two elements of bi-logic combine.

In a relationship that has been symmetrized, the absence of asymmetry (right and left, above and below, back and forwards) makes it impossible to conceptualize space.

The absence of a concept of space makes it impossible to conceptualize time (which is always an act of spatialization).[13] Learning that is without space and time is given infinite space and time. It is everywhere, forever. It has no history. Learning without history and context is not open to changes. It is an informational pattern that is structured to be repeated and is characterized by stability and protected against change.

Such an informational pattern—either unconscious since its origins or become as much—is able to induce behaviour in relation to which it exists at a high level of logic.[14]

The principle of symmetry is combined with the principle of generalization, where Unconscious logic does not take account individuals as such, it deals with them only as members of classes, and of classes of classes.

A single element in a set—the individual member—and the class to which this element belongs coincide in essence. This is akin to metonymy, where the part is identified with the whole.

Even a superficial analysis clearly seems to show that the principle of generalization is crucial for the mind to categorize things. The cognitive processing done by the unconscious 'automatically' generates classes and categories.

Symmetric cognitive processing coincides in its purest form, in Matte Blanco's approach, with 'being', while asymmetric cognitive processing coincides in its purest form with 'becoming' or, more precisely, with the agency of 'events'. 'Feeling', with the possibility that it offers us to become one with the object of our knowledge by overcoming the barriers of separation, corresponds to placing the symmetrical way of being within rational thought (predominantly 'asymmetric').

---

[13] But not of the perception of space and the orientation within it!

[14] 'There are patterns to human moves. 1.1 Moves are events: they happen, they take time, they begin and end. 1.11 Events are types, not tokens. That is, they can *re*occur: the same event can happen repeatedly' (Bencivenga 1997, pp. 5–6).

7 The Other Qualia

In our evolutionary hypothesis, the senses must have come before any form of awareness and, even more so, before any type of consciousness. Sensory qualia have a value in that they are positive or negative, good or bad. Yet, what is this in relation to? As is often the case, it is in relation to survival and reproductive success. It is in giving a qualitative value to what the senses perceive that Damásio's somatic markers become so critical (Damásio 1994).

An *emotion*[15] always has some sense—good, ugly and so on—although not always a linguistic meaning. We are all more than aware that a powerful emotion leaves us 'speechless'. Of course, it is something that is found in many animal species that do not even speak! Indeed, emotion has always played an essential role in the survival of mammals.[16] It has an immediate and absolute selective value since, if we are able to learn, after birth, that things that are good and bad for us exist, then our independence from the surrounding environment will be far superior to that which might be predicted or expected merely from our genes and from the 'rigid' phenotype created by our genes.

The emotion from sensory qualia in mammals is somatic, carnal, bodily and necessary for survival. It is a rich and constant source of essential information in an environment that is constantly changing.

The 'pure' sensory quale is selected for its ability to construct worlds; it is the form of biological cognition. In humans, with their consciousness and awareness, sensory qualia are necessary for assigning meaning and value to the *metaphorical quale* that is built on them.

Metaphorical qualia are given a far more complex connotative meaning, namely, the *semantic differential.*

The semantic differential has a dual origin since the value assigned is either 'good' or 'bad' (just like what the emotions do for the senses) as well as being semantic.[17] Thus, it can provide a meaning that is genuinely spoken or, expressed differently, defined by language. Such meaning is tied to or located in the context or learned from a 'source of tradition' (related to the family, culture, one's peers and so on) that has been passed on.

---

[15] Damásio defined emotions as publically observable responses, while feelings were private mental experiences (Damásio 1999).

[16] In MacLean's *triune brain* model, mammals 'invent' emotion and the limbic system.

[17] Osgood et al. (1957) created the semantic differential technique to identify the different qualitative attributes that are specific to different cultures and that give meaning to abstract concepts. These abstract concepts are assigned an arbitrary score, using a scale from 1 to 7 where the opposite ends of the scale indicate opposing adjectives. For example, on a scale where 1 is good and 7 is bad, where do you place 'honourability'? By analysing the information gained in questionnaire format using this technique, it was found that the positive or negative values assigned have 'prevalent' or 'dominant' social tendencies that changed both from culture to culture and even within the same culture.

In biological and evolutionary terms, assigning new meanings—the process of re-signification—perhaps as a result of contextual changes or a renegotiation, produces an *exaptation.*[18]

Adding meaning to experience or bodily sensations is what makes the value attributed to metaphorical qualia so richly variable, in that they have a meaning, *connotation* and *extension.*[19]

Since our *metaphorical quale* has, to some degree, shared meaning, it enables understanding and communication.[20] However, this means that the four famous properties that Dennett found to be commonly assigned to qualia in philosophical debates about the mind disappear (Dennett 1988, 1991, 1994). Indeed, the way qualia are described here, they are no longer *ineffable,*[21] *intrinsic*[22] or able to refer to a wholly *private* experience.

To share a sensory experience, we can use analogy (red-like traffic lights) or a linguistic description (electromagnetic radiation with a wave length of 700 nm).

Yet, in both cases we are totally unable to 'render' the direct perception of red, for example, to someone who has been blind since birth. In the first case we fail because we cannot produce a sensation. In the second, because without the sensation, the blind person cannot understand what the linguistic description is referring to. In cases where such limits do not exist, many devices—from mirror neurons to formal discussions, all of which are based on the perception of various types of invariant qualities—display a descriptive ability and relational properties and they make shared experiences possible.

Our definition of consciousness also removes Dennett's fourth property since qualia are directly or immediately apprehensible in consciousness. In our case, sensory qualia do not necessarily meet this criterion.

In the same vein as Edelman, our definition of qualia is totally biological, functional and rigorously based on evolution. There is nothing similar to special qualia without functional properties. Qualia are not accidental occurrences. They are defined because of their capacity to select neural networks that improve fitness. They do not form part of a representation, unless they are metaphorical.

Moreover, qualia are not a fundamental element but a late acquisition both when they come from *analysing* sensations and, even more so, when they are defined linguistically.

[18] Also known as preadaptation, this is when some trait evolved due to selective pressure to perform one function, but then it unpredictably came to serve a new function. The classic example is a bird's feather. These originally evolved in dinosaurs as a means of keeping warm, but then their function changed to allow birds to fly.

[19] These technical terms are problematic because in the current use of semiotics in scientific practice *reference = denotation = intension*, while *sense = connotation = extension*. This is different from how they are used in Peirce and in the logical tradition where *denotation = extension*, while *intension = connotation.*

[20] Vygotskij assigned language two functions, namely, *building a model of the world* and then *communicating it* (Vygotskij 1934, 1978).

[21] That is, it can be described using words.

[22] In other words, without relational properties.

Infantile syncretic-synthetic perceptions precede all forms of analysis. A baby starts by perceiving an object—a ball—as a whole, including even the name used to identify this object among its properties. In perceiving this ball, the baby does not perceive an 'abstract class of ball objects', and he does not perceive the abstract 'roundness as a constant property' nor the 'redness as a property of red objects'. This is all fairly clear if we consider that these are 'abstract' properties, that is, they are derived from abstraction processes.

Learning the ability to analyse goes beyond 'syncretic' perception, allowing the constant nature of properties to be perceived. On the one side, there are the objects, while on the other there is the capacity to identify a property. This ability is far from simple and certainly does not occur in the same way for everyone. When Vygotskij and Luria (Luria 1976) showed geometric shapes to their illiterate Uzbek countrymen, they did not identify them as rectangles or trapezoids but as window frames or specific amulets!

It is these very analytical abilities that, through introspection, allow us to focus our attention on the objective of a pure emotion, without an object.[23] If in a given experience, the invariant property is the emotional one, then we once again find 'qualitative traits that are recognizable from the given'.

Once again, this is *a mental state with highly distinct characteristics of subjectivity*. These are once again major *aspects of our internal mental life that can only be reached through introspection*. What we have here is the *emotional quale*.

Damásio defined *emotions* as observable, public responses and *feelings* as private, mental experiences (Damásio 1994, 1999). In our system, Damásio's feelings are a type of metaphorical qualia since they require the hippocampus as a neural structure and consciousness as a psychic function. They are qualia because they are subjective mental states that we cannot see in others but that we can only perceive in ourselves.

If an emotion becomes a feeling when expressed consciously—when we are able to understand it—the senses alone are no longer sufficient to describe it. Consciousness uses *metaphorical qualia* as raw material for creating its stories or narratives. Sometimes, these relate to our feelings, when we *describe* what *we feel*.

The *emotional quale* is something quite different. It is an *emotional quality accessible through introspection that is distinct from the object that caused it*.

Having done this, we can focus our attention on ourselves. As we have noted at various points above, consciousness is self-reflexive, being a dialogue between two parties: 'I' and 'me'.

We can perceive and understand *the actual subject of perception*. By placing our *analogue I* in front of the *metaphorical me*, we achieve genuine *self-reflexive awareness*.

Our mind puts in place a perceptive illusion, clearly advantageous for fitness, which literally represents the perception of a *continuum* and of a *oneness* attributed

[23] Psychiatry distinguishes between *anxiety* where this is no object and the *fear* of something specific.

to our self. These are two properties that each individual attributes to him- or herself, yet—we might say—without any physical and chemical or biological foundation. From the moment we have consciousness, we have a *qualic self.*

The qualic self is the agent in charge of those complex causal connections we call our actions. 'Arbitrarily', but advantageously (in evolutionary terms), the self is perceived as unique and constant. By constant, we mean invariant or, in other words, the property that makes identification possible, the 'essence'. This is the key part of a typically human *image-of-the-world apparatus* that is able to reconstruct a causal chain and thus understand an event using the human capacity for understanding through metaphorical device.

So, what are our *qualia*? What are they made of or, expressing this question technically, what is their *ontological status*? This is an important question, because we live in the world of perceptions that we get from our different qualia. Colours, smells and sounds have proven to be especially suited to improving the fitness of species that developed transducers, sensory systems and brain maps able to perceive them.

Their existence is intrinsically subjective since *they only exist in the interaction between the world and the perceiver*. However, their importance for fitness does not make them unique elements; on the contrary, these are shared elements that generate a universe of senses and perceptions and the ability to move in all individuals of the same species.

We maintain that *Homo*—and more specifically *Homo sapiens sapiens*—developed a new category of 'sensory' devices and that this system is characterized by the capacity to represent something in terms of something else. This step formed part of the natural history of the species and was a wholly biological and functional step, equivalent to what happened with the normal 'senses'. This specific ability to represent—as happens for the qualitative elements of sensory experience—has a primarily subjective existence and is accessible through introspection. The validation and objectification of this ability comes from recognizing and comparing against the other (validation is intersubjective), using the functions identified as 'theory of mind'.

In the same way as our senses construct our world, the metaphor sense literally constructs that part of the physical world of humans that we build using our symbols. Just like for images, smells and sounds, this word is born from the interaction between the world and the perceiver.

8 Signs, Metaphors and Cultures

Merlin Donald, in his book *Origins of the Modern Mind* (2001), hypothesizes different stages of development, with 'a structural change in cognitive organization, as well as a profound cultural change. A complex of new cognitive modules accompanied each adaptation.' He posits entire 'levels' of emerging properties and more recent cognitive modules that are 'physically bounded somewhere, often in external memory'. For Donald, *external symbolic storage* 'must be regarded as a hardware change in human cognitive structure, albeit a nonbiological hardware change' (P. 18).

The *mental organs* that allowed memory to be organized in a new way are the ones that made it possible to understand things through the use of metaphors. These are the same organs that enabled man to invent and systematically use tools. Tools are always prostheses, that is, artificial devices apt not only to replace missing or diseased parts of the body (e.g. a medical prosthesis) but also clearly to boost function. Alternatively, sticking to the broad sense, we can say they allow humans to do things that biology has not 'foreseen'. Continuing this line, an aircraft is really just a 'winged prosthesis'.

The tools—the prostheses—are non-verbal metaphors. They are objects that 'represent something else'. The first hominid that took this step was *Homo habilis.*[24] This species lived 2.4 to 1.5 million years, and tools found with the remains of members of this species suggest it had the capacity to imagine a potential use for them. For instance, it is plausible that the first tools imitated, to some extent, the biological abilities of other animals since they were not weapons but utensils to tear meat from prey that had been killed. We could say the first tools were 'superteeth'.

What mental organs are needed to 'imagine' tools (even before building them, with the right materials)? The answer is mental devices able to process 'tropes', which is a technical term to refer to metaphor. The existence of these mental devices requires not only brains but groups of organisms with social organization.

Symbolic interactionism,[25] *cultural-historical psychology*[26] and many other schools of thought have shown us that, in order to achieve truly refined forms of abstract thought and consciousness as we defined it, a very special prosthesis is needed, namely, a written language based on a phonetic alphabet.

The capacity of the mind to understand something in terms of something else is at the origin not only of language but also of all sign systems. Metaphor is the essential function for understanding; it is what enabled abstract thought and language. According to Lorenz, 'in all these cases [of animal tradition] the transmission of knowledge is dependent on the presence of the object. Only with the evolution of abstract thought and human language does tradition, through the creation of free symbols, become independent of the object. This independence is the prerequisite of the accumulation of supra-individual knowledge and its transmission over long periods, an achievement of which only man is capable' (Lorenz 1973, Eng. Ed. (1977) p. 165).

Ethology has documented that culture is not an invention of mankind. Each complex of notions and of practices that forms the heritage of a social group can be defined as a culture.

---

[24] This idea is a moot point as it seems *Australopithecus garhi* was using tools some 100–200 thousand years earlier.

[25] This was the dominant theory in the work of G. H. Mead, who is considered to be one of the founders of social psychology (Mead 1934).

[26] Vygotskij and his followers.

Thus far, *Homo sapiens sapiens* is the only species we know of where 'cultural traditions' are not constantly linked to the object, in other words: regularly break down when the objects they refer to are not there for a whole generation (adapted from Lorenz, ibid.).

The idea that evolution plays a part not only in biology but also in human culture is an idea that is almost as old as Darwinism itself. As early as 1880, Huxley saw 'theories' as 'species of thought' subject to natural selection. Daniel Schacter (2001) has brought us the work of the German biologist Richard Semon (already known for the concept of engram), *Die mnemischen Empfindungen in ihren Beziehungen zu den Originalempfindungen*, translated into English in 1921 with the title *The Mneme*. Starting in the 1970s, many people have tried to understand the constant changes in human behaviour as part of cultural evolution, which was in itself based on selection (Cavalli-Sforza and Feldman 1973; Cloak 1975; Boyd and Richerson 1985, Calvin 1996).

Dawkins (1976, 1982) made the concept of a nongenetic *replicator* famous. The gene is the biological unit of inheritance. Is there such a thing as inheritance in cultures? Dennet (1995) argued that Darwin's dangerous idea is such a powerful concept that biological evolution looks like a 'special case'. As he himself asserts (Dennett 1999), the idea that cultures evolve is so obvious that it must be considered a truism.

Heraclitus, back in his time, stated that everything flows (*pánta rhêi*). If we perceive 'things' (i.e. fixed objects), in addition to 'processes', it is just because we compare everything—especially our sensations and perceptions—with the relative duration of our existence. Expressed differently, nothing is truly unchangeable. Even 'objects' have 'histories', although the events in these histories are so slow that it is hard to perceive them.

By contrast, cultures, particularly over the last few hundred years, have changed quite quickly, and we see them more as events than as stable entities. Cultures evolve through *memes*.

Let us define the *memes: information structures* (informational patterns), of a *cognitive* or *behavioural type,*[27] that are held in an individual's memories[28] and are able to be *copied* to the memories of other individuals, because they are units for replications or *replicators*. This property—the possibility to be copied—is common to both genes and memes, making them both replicators. The one from whom the pattern is copied and, equally, the one that will obtain the copy is the *carrier*. This is the basis for *memetics*.

According to Dawkins' theory on replication, melodies, songs, rhymes, urban legends, 'catchphrases', 'famous phrases' from books and the media, epic poems, stories, jokes, sayings and aphorisms are classic examples of memes. Using our definition, other more fitting examples include regulations and laws.

---

[27] We refer to '*cognitive* informational patterns' as explicit declarative memories and '*behavioural* informational patterns' as implicit-procedural memories. This is a well-known, important technical distinction in psychology.

[28] That is, in brain structures.

Just like for genes, all of the criteria of *copying fidelity*, *fecundity* and *longevity apply* (Heylighen 1993–2001). Copying fidelity, fecundity and longevity all refer to the *semantic content* of the meme rather than to its 'container' (technically speaking, its formal and syntactic characteristics). This work uses the hypothesis that memes, as information structures, are assumptions for interpreting reality. Assumptions emerge, as new ideas, in the brains of individuals or small groups and undergo a selection process. As usual, the stakes are survival and reproductive success. The key point here is the information content, rather than its external form.

9 What Type of Signs Are Memes?

Semiotics rightly criticized the meme for being little more than a primitivized concept of sign that is ignorant of de Saussure and Peirce (Deacon 1997; Kull 2000, Benitez-Bribiesca 2001) *(An underdeveloped special version*, Kilpinen 2008; *A degenerate sign,* Kull 2000). Henson (1987) argued that memetics neglects evolutionary psychology, ignoring the psychological and behavioural consequences that replicated informational patterns have on their carriers.

In terms of memes, it is still necessary to overcome the primary hurdle, namely, defining the *ontological status of memes* and some of their essential properties. *What are memes?* Are memes the 'thing' that transfers from one brain to another? This interpretation of the meme can be attributed precisely to Dawkins and, more generally, to the line of thinking that Eldredge defines as 'ultra-Darwinism'. When one talks about the 'selfish gene', one is merely assigning an 'object' or, better still, a concept an *intentional stance*, as if it were a person! The definition of replicator refers to a structure with the sole purpose and interest of producing copies of a given informational pattern.

Genes or memes in and of themselves are not information but structured signals that become information only as part of the system that uses them. Culture is very prone to historical analyses. That is why we need to define which entities (parts of the system) endow cultures with the characteristics of a hereditary system. There is no history without memory of the past and the possibility of change in the future. Cultures require a form of memory that is able to produce both continuity between different generations (what ensured survival and reproductive success should not be easily changed) and allow some variability since this is fundamental to adaptation when conditions change.

This is why *copying fidelity*, *fecundity* and *longevity* are so central to the history of evolution. Of course, not having defined replicators (or, more technically, not having given them an *ontological status*), it is impossible to know the *nature of the processes* being studied.

Memes have been compared to viruses. Viruses are not independent forms of life. In some cases they are little more than 'DNA containers' that are 'injected' into fully functional cells, which then experience a disruption in their functioning.[29] After infection, the whole system that the cell needs to produce its own proteins only produces copies of viral DNA.

By analogy, memes would be able to 'infect' the brain. However, viruses and memes, defined as such, are replicas not replicators (Deacon 1999). They are replicas whose nature and size are unknown, and they are characterized by such limited copying fidelity that they even prevent evolutionary selection processes from occurring (Dawkins, introduction to Blackmore 1999). In this model, we are not even able to understand whether they can be compared to the genotype or the phenotype in a genetic analogy.

Semiotics accuses memetics of focusing exclusively on signs or, in some of the more radical arguments, on vehicles of signs (Peirce's *representamen*). This occurs because a sign is not characterized by its physical characteristics. Some photons in the visible light spectrum might not have any meaning 'by themselves'. However, they might have a very precise meaning for you, if they are produced by the rear brake lights of a fast-moving car that is only metres in front of your car hood.

The basis of the difficulties with the concept of meme, as it has been defined by his supporters thus far, lies in an anti-semiotic view of the information content. Stephen Gould (1997) defined *Darwinian fundamentalists* as the champions of the selfish gene or meme. Although 'personalizing' genes and memes could make them more popular, describing the role of a replicator without considering the complex system it is part of, whether it is biology or culture, means curtailing the chance of understanding how it works.

Placing the utmost importance on the copy (re-presentation, representation) mechanism—as shown by the choice of the name 'replicator', incidentally without any reference either to the security mechanisms that protect the content or to those that can guarantee variability—demonstrates an inability to understand its most profound function.

The function of replicators is more interpretative than replicative. It involves guiding the development processes in the fulfilment of the living, not only in making a phenotype (more or less extensive!) concrete, but in its *ontogenic structural drift*. 'Every ontogeny as an individual history of structural change is a structural drift that occurs with conservation of organization and adaptation' (Maturana and Varela 1985, Eng. Ed. (1987) pp. 102–103).

This is so because 'signs evolve, and they have pragmatic consequences, by virtue of which they are selectively favored to remain in circulation or become eliminated over time. It is by virtue of the memetic analogy to genetic evolution that we may discover the dynamical logic still required for a complete theory of semiosis, and not just a semiotic taxonomy' (Deacon 1999). In the same work, Deacon clarified that at the heart of semiotics is not so much the study of signs but

---

[29] 'RNA containers', in other cases.

the study of the very process of *semiosis*. Semiosis tries to identify and describe the repetitive patterns that enable the recognition of what generates sense in hierarchical systems.

This organization is known as *stratified order*, and what is being studied are levels of organization of reality, starting from biology and reaching psychology—individuals and groups—right up to the behaviours that regulate the society, economy and global ecosystems that have an impact on the planet as a whole.

What differentiates *hard science*, which is concerned with things, and the soft or historic sciences, which are concerned with processes, is precisely this: the identification of invariant patterns of what we have for a long time called the unchangeable laws of nature. Where do signs come from? How can we overcome the sphere of first-order relations (those between object and sign) so that we can initially perceive and then manipulate the relations-of-relations?

Peirce taught us that nothing is intrinsically meaningful, without an interpretation, and the interpretation is up to the system that 'receives' the information, not to the system that generates it.

All information needs to be placed in context, not only in human language. Jablonka et al. (1998; quoted in Kull 2000) refer to *four hereditary systems*: epigenetic (*epigenetic inheritance system*, EIS), genetic (*genetic* I S), behavioural (*behavioural* I S) and linguistic (*linguistic* I S). In these systems, information processing entails, respectively, the regeneration of cellular structures and metabolic networks (EIS), DNA replication (GIS) and social learning (BIS, LIS), with the latter two *being through the use of symbols*. Life itself 'starts' with the creation of the first metabolic networks, which are autonomous and able to self-replicate. These are complex phenomena, typically emerging, systemic and hierarchically structured.

Genetic and *epigenetic* phenomena determine structures that co-evolve with the environment in a constant regeneration of *meaning*. In Maynard Smith and Szathmáry's identification (1997, 1999) of the major transitions in evolution, *systems of 'unlimited' heredity* play a fundamental role. Their main feature is modularity, which is defined in terms of fundamental components, called *replicators*, in reduced numbers, but that can be assembled in different sequences to produce an indefinitely large number of different replicable structures. When the replicable structures are different sentences, we can have an indefinitely large number of different *meanings* (Maynard Smith and Szathmáry 1997).

Maynard Smith and Szathmáry argue that we only know of two examples that fall completely within the definition of *unlimited inheritance systems*: *genetic code* and *language*. We argue—along with authors like Bynumin, Havelock, Jaynes, Lord, Luria, Milman, Ong, Parry and Vygotsky (in alphabetical order!)—that it should be genetic code and language and thought backed by a written phonetic alphabet.

This clarification is needed because memes do not infect one brain after the other but are *structured informational patterns* that are essential for the '*Information Contextualizing System*' (Mininni 2008) that we call consciousness.

10 Asymmetrical Metaphors of Consciousness, Symmetrized Metaphors of the Unconscious

In the *symmetrization and generalization* section, we stated the following: the *absence of spatialization makes it impossible to conceptualize time* (which is always an act of spatialization). A learning process without any space and time has infinite space and time and is no longer *a process* but a *'static' cognitive entity that controls and governs behaviour.*

These *semantic informational patterns—memes—*emerged because of variability mechanisms followed by selection processes, protected by 'accidental' mutations, thanks to their stability. Through their behaviour they can induce the 'physical' transfer of matter and energy.

The conquest—as a species and as individuals—of complex systems of signs entails a major, dual evolutionary advantage: first, emulating reality with so much more capacity and power than what came before and, secondly, overcoming of limits connected to the possession of memory systems that can only process sensory input.

Their initial genesis (but not further development) requires biological and genetic variation. Their subsequent evolution (e.g. the step from an icon and analogy-based system to a fully symbolic system) could have occurred entirely within the sphere of cultural evolutions, where the informational patterns are memetic rather than genetic.

Genes are important replicators not only for the material support they use but especially because they can be used for *encoding, depositing and* retrieving *biological informational patterns.* This is just like *memes* in memory.

The three characteristics listed below apply to both the replicators of biological evolution and the replicators of cultural evolution. Indeed, such aspects are typical of genes and memes:

(a) Need a mechanism that allows transformations with controlled *variability* and hence generation (which originally is individual or the expression of a small number of subjects)
(b) When they are generated, must undergo a selection process, based on their selective value, as demonstrated in terms of fitness[30]
(c) After being selected, need stability and thus a mechanism to protect against accidental variability

Here, we suggest that the mechanism that protects information from 'accidental' changes, and makes memetic informational patterns relatively stable, is *their becoming unconscious.*

---

[30] *Intersubjective validation* based on the local laws of reason can use an original, subjective idea to create an idea that is widely accepted by a community, which adopts it and uses it 'objectively'.

Perfectly adaptive behaviour—'excellent' and worthy of imitation—can be unconscious just like maladaptive behaviour that might lead to great suffering.

In both cases, the behaviour is based on an implicit-procedural learning process, which is outside consciousness and awareness and, above all, is highly repetitive behaviour.

The informational pattern that constitutes the higher level of logic, that mental state that controls behaviours, is 'protected' against changes, is stable and does not evolve further.

It is a learning process without time or space; one could say 'infinite'. It is also truly repetitive. Without such repetition, we would not be dealing with an 'unconscious pattern', whether of 'excellence' or 'suffering'. It is the unconscious pattern that 'contains' learning.

What prevents implicit-procedural learning from changing? Implicit-procedural learning is outside of the realm of consciousness and awareness; hence it is subject to unconscious cognitive processing. The biological/evolutionary value of this functional mechanism lies in *removing spatiality and temporality traits from the informational patterns* that represent *forms of adaptation*, fit for the conservation of the organization of living beings. As such, these forms are protected by preventing further 'accidental' evolution.

This is the *ontological status of memes: informational patterns of a signic nature with a metaphorical relational organization, with individual generation and social selection. Their stability is ensured by them becoming unconscious, namely, 'ahistorical', in individuals, groups, organizations and institutions.*

When we learn a procedure, what we have learned is 'in our body' and we do not think about it any longer. It is like riding a bike. The learning is in the pattern: the pattern has a history. The history of genes is a *genesis*[31]: the history of memes is a *memesis*.

When genes are organized in the tight meshes that we call chromosomes, they are not 'functioning', but in this form they are 'protected' and less prone to environmental influences.[32] The same can be said about the information of our semantic memory (Lawley and Tompkins 1996).[33]

In terms of the informational patterns for biological conservation and adaptation, genes and chromosomes determine the specific modality that guarantees the generation, mutability under controlled conditions and conservation of what has been selected. By the same token, the signs in metaphorical relationships—subject to the principles of symmetry and generalization—represent informational patterns for the generation, mutability under controlled conditions and conservation of what has been selected in human cultures.

[31] From Ancient Greek γένεσις (genesis, 'creation, beginning, origin'). Retrieved from http://en.wiktionary.org/wiki/Genesis

[32] They do not produce RNA—and indirectly proteins—in this physical state.

[33] This is a powerful metaphor from David Grove.

For this reason, human cultures, in contrast to animal cultures, no longer depend on the constraints of the sensory field, nor do they depend on the physical presence of objects, whether in a single individual or social organizations.

These fundamental sign-based informational patterns, of a metaphorical nature, using the 'metaphor cluster' mechanism (as explained by cognitive linguistics), generate the structure of the character and the personality both for individual forms and for social organizations.

In cultures, there are both strong traditions and transition stages, like Kuhn's scientific revolutions or any other transformation periods in history.

We can hypothesize a general and unified theory of knowledge. We must give signs a natural history, understand how they can really be compared to forms of life that are capable of generation, development and death and how they can guide the evolutionary structure of cultures.

This theory becomes, thanks to cognitive linguistics, an epistemological proposal, as it contains the possibility, offered by the metaphor as a cognitive entity responsible for comprehension, to measure the *degree of truth* (Lakoff and Johnson 1980). This is a powerful criterion, able to assess the level of reliability achieved by each possible statement, regardless of the specific field of knowledge it refers to, whether it is scientific, humanist, technical or artistic, a criterion from cognitive linguistics, adequate to generate a *biosemiotic epistemology*.

References

Antiseri, D. (1986). Epistemologia evoluzionistica: da Mach a Popper. *Nuova civiltà delle macchine online, 1*(13), 111.

Barbieri, M. (2010). On the origin of language. A bridge between biolinguistics and biosemiotics. *Biosemiotics, 3*(2), 201–223.

Bencivenga, E. (1997). *A theory of language and mind.* Berkeley: University California Press.

Benitez Bribiesca, L. (2001). Memetics: A dangerous idea. Interciencia: *Revista de Ciencia y Technologia de América* (Venezuela: Asociación Interciencia), *26*(1), 29–31.

Bolk, L. (1926). *Das Problem der Menschwerdung*. Jena: Gustav Fischer.

Boltzmann, L. (1905). Über die Frage nach der objektiven Existenz der Vorgänge in der unbelebten Natur. In *Populäre Schriften*. Leipzig: Barth.

Boyd, R., & Richerson, P. J. (1985). *Culture and the evolutionary process.* Chicago: University of Chicago Press.

Cavalli-Sforza, L., & Feldman, M. (1973). Cultural versus biological inheritance: Phenotypic transmission from parents to children. *Human Genetics, 25*, 618–637.

Calvin, W. (1996). *The Cerebral code: Thinking a thought in the mosaics of the mind.* Cambridge, MA: MIT Press.

Cloak, F. T. (1975). Is a cultural ethology possible? *Human Ecology, 3*, 161–182.

Damásio, A. R. (1994). *Descartes' error emotion, reason, and the human brain.* New York: Avon Books.

Damásio, A. R. (1999). *The feeling of what happens, body, emotion and the making of consciousness.* London: Heinemann.

Danesi, M. (2008). Towards a standard terminology for (bio)semiotics. In M. Barbieri (Ed.), *Introduction to biosemiotics*. Dordrecht: Springer.

Dawkins, R. (1976). *The selfish gene.* Oxford: Oxford University Press.

Dawkins, R. (1982). *The extended phenotype*. Oxford: Oxford University Press.
Deacon, T. W. (1997). *The symbolic species. The coevolution of language and the brain*. New York: W.W. Norton & Company.
Deacon, T. W. (1999). Editorial: Memes as signs. The trouble with memes (and what to do about it). *The Semiotic Review of Books, 10*(3), 1–3.
DeHaan, R. L. (1959). *Cardia bifida* and the development of pacemaker function in the early chicken heart. *Developmental Biology, 1*, 586–602.
Demuth, J. P., Bie, T. D., Stajich, J. E., Cristianini, N., & Hahn, M. W. (2006). The evolution of mammalian gene families. *PLoS ONE, 1*(1), e85.
Dennett, D. C. (1988). Quining qualia. In A. Marcel & E. Bisiach (Eds.), *Consciousness in contemporary science*. Oxford University Press: Oxford.
Dennett, D. (1991), *Consciousness explained*. Boston: Little, Brown & Co.
Dennett, D. C. (1994). Instead of qualia. In A. Revonsuo & M. Kamppinen (Eds.), *Consciousness in philosophy and cognitive neuroscience*. Hillsdale: Lawrence Erlbaum.
Dennett, D. C. (1995). *Darwin's dangerous idea: Evolution and the meanings of life*. New York: Simon &Schuster.
Dennett, D. C. (1999, March 28). The evolution of culture. The Charles Simonyi lecture, Oxford University, Feb 17, 1999. *Edge*, 52.
Donald, M. W. (2001). *A mind so rare: The evolution of human consciousness*. New York: W. W. Norton & Company.
Edelman, G. M. (1987). *Neural Darwinism. The theory of neuronal group selection*. New York: Basic Books.
Edelman, G. M. (2006). *Second nature: Brain science and human knowledge*. New Haven/London: Yale University Press.
Edelman, G. M., & Gally, J. A. (2001). Degeneracy and complexity in biological systems. *Proceedings of the National Academy of Sciences of the United States of America, 98*(24), 13763–13768.
Edelman, G. M., & Mountcastle, V. M. (1978). *Mindful brain: Cortical organization and the group-selective theory of higher brain*. Cambridge, MA: MIT Press.
Favareau, D. (2008). The evolutionary history of biosemiotics. In M. Barbieri (Ed.), *Introduction to biosemiotics*. Dordrecht: Springer.
Fodor, J. A. (1983). *Modularity of mind: An essay on faculty psychology*. Cambridge, MA: MIT Press.
Gould, S. J. (1997). Darwinian fundamentalism. *New York Review of Books, 44*(10), 34–37.
Hebb, D. O. (1949). *The organization of behavior: A neuropsychological theory*. New York: Wiley.
Henson, K. (1987, August). Memetics and the modular-mind. *Analog*.
Hermans, H. J. M. (2001). The dialogical self: Toward a theory of personal and cultural positioning. *Culture & Psychology, 7*, 243–281.
Hermans, H. J. M., & Kempen, H. J. G. (1993). *The dialogical self: Meaning as movement*. San Diego: Academic.
Hermans, H. J. M., Kempen, H. J. G., & van Loon, R. J. P. (1992). The dialogical self: Beyond individualism and rationalism. *American Psychologist, 47*, 23–33.
Heylighen, F. (1993–2001). Memetics. In *Principia cybernetica web*. Retrieved from http://pespmc1.vub.ac.be/MEMES.html
Hubel, D. H., & Wiesel T. N. (1963). Receptive fields of cells in striate cortex of very young, visually inexperienced kittens. *Journal of Neurophysiology, 26*, 994–1002. Retrieved from http://jn.physiology.org/cgi/reprint/26/6/994
Jablonka, E., Lamb, M., & Eytan, A. (1998). 'Lamarckian' mechanisms in Darwinian evolution. *Trends in Ecology and Evolution, 13*(5), 206–210.
Jaynes, J. (1976). *The origin of consciousness in the breakdown of the bicameral mind*. Boston: Houghton Mifflin Company.
Jaynes, J. (1986). Consciousness and the voices of the mind. *Canadian Psychology, 27*(2), 128–148.
Kilpinen, E. (2008). Memes versus signs. On the use of meaning concepts about nature and culture. *Semiotica, 171*(1/4), 215–237.

Kull, K. (2000). Copy versus translate, meme versus sign: Development of biological textuality. *European Journal for Semiotic Studies, 12*(1), 101–120.

Kull, K., Deacon, T., Emmeche, C., Hoffmeyer, J., & Stjernfelt, F. (2009). Theses on biosemiotics: Prolegomena to a theoretical biology. *Biological Theory: Integrating Development, Evolution, and Cognition, 4*(2), 167–173.

Lakoff, G., & Johnson, M. (1980). *Metaphors we live by*. Chicago: University of Chicago Press.

Lawley, J., & Tompkins, P. (1996, August). And, what kind of a man is David Grove? *Rapport,* Issue 33. Retrieved from http://www.cleanlanguage.co.uk/articles/articles/37/1/And-what-kind-of-a-man-is-David-Grove/Page1.html

Lawley, J., & Tompkins, P. (2000). *Metaphors in mind: Transformation through symbolic modelling*. London: The Developing Company Press.

Lewis, C. I. (1929). *Mind and the world order*. New York: C. Scribner's Sons.

Llinas, R. (2001). *I of the vortex: From neurons to self.* Cambridge, MA: MIT Press.

Lorenz, K. (1977). *Behind the mirror, a search for a natural history of human knowledge*. New York: Harcourt Brace Jovanovich. (Original Ed. Lorenz, K. (1973). *Die Rückseite des Spiegels. Versuch einer Naturgeschichte des menschlichen Erkennens*. München/Zürich: Pieper).

Luria, A. R. (1976). *Cognitive development its cultural and social foundations*. Cambridge, MA: Harvard University Press. (Original Ed. Lurija, A. R. (1974). *Istori eskoe razvitie poznavatel'nyh processov*. Moskva: M.G.U).

Matte Blanco, I. (1975). *The unconscious as infinite sets: An essay in bi-logic*. London: Duckworth.

Matte Blanco, I. (1988). *Thinking, feeling, and being: Clinical reflections on the fundamental antinomy of human beings and world.* London/New York: Routledge.

Maturana, H., & Varela, F. (1985). *El Árbol del Conocimiento: Las bases biológicas del entendimiento humano.* Santiago: Editorial Universitaria. (Engl. Ed. Maturana, H., & Varela, F., (1987). *The tree of knowledge: The biological roots of human understanding*. Boston: Shambhala).

Maynard Smith, J., & Szathmáry, E. (1997). *The major transitions in evolution.* New York: Oxford University Press.

Maynard Smith, J., & Szathmáry, E. (1999). *The origins of life: From the birth of life to the origin of language*. Oxford: Oxford University Press.

Mead, G. H. (1934). *Mind, self and society: From the standpoint of a social behaviorist.* Chicago: University of Chicago Press.

Mininni, G. (2003). *Il discorso come forma di vita.* Napoli: Guida.

Mininni, G. (2008). La mente come orizzonte di senso. In M. Maldonato (Ed.), *L'Universo della Mente*. Roma: Meltemi.

Osgood, C. E., Suci, G. J., & Tannenbaum, P. H. (1957). *The measurement of meaning*. Urbana: University of Illinois Press.

Peirce, C. S. (1931–58). *The collected papers of C. S. Peirce*, vols. 1–6, ed. Charles (C. Hartshorne & P. Weiss Eds.); vols. 7–8, (A. W. Burks Ed.). Cambridge, MA: Harvard University Press.

Portmann, A. (1941). Die Tragzeiten der Primaten und die Dauer der Schwangerschaft beim Menschen: ein Problem der vergleichen Biologie. *Revue suisse de zoologie, 48*, 511–518

Portmann, A. (1945). Die Ontogenese des Menschen als Problem der Evolutionsforschung. *Verhandlungen der Schweizer Naturforschenden Gesellschaft, 125*, 44–53.

Recchia-Luciani, A. N. M. (2005). Menti che generano metafore e metafore che generano coscienze, In Per una genealogia dell'autocoscienza Soggettività, esperienza, cognizione (2ª parte, M. Cappuccio Ed.), *Élites*, 4/2005, pp. 21–34. Soveria Mannelli: Rubbettino.

Recchia-Luciani, A. N. M. (2006). Biologia del dispositivo metaforico. In S. Ghiazza (Ed.), *La metafora tra letteratura e scienza*. Bari: Servizio Editoriale Universitario.

Recchia-Luciani, A. N. M. (2007). Biologia della Coscienza. In M. Maldonato (Ed.), *La Coscienza – come la biologia inventa la cultura*. Napoli: Alfredo Guida Editore.

Recchia-Luciani, A. N. M. (2009) Memorie oltre le generazioni. Memi, segni e neuroscienze cognitive per un'ipotesi evolutiva della cultura. *Chora, 16*(7), 89–95, Milano: Alboversorio.

Recchia-Luciani, A. N. M. (2012). Manipulating representations. *Biosemiotics, 5*(1), 95–120.

Savage-Rumbaugh, E. S., & Lewin, R. (1996). *Kanzi: The Ape at the brink of the human mind.* New York: Wiley.

Schacter, D. (2001). *Forgotten ideas, neglected pioneers: Richard Semon and the story of memory.* Philadelphia: Psychology Press.

Sebeok, T. A., & Danesi, M. (2000). *The forms of meaning: Modeling systems theory and semiotics.* Berlin: Mouton de Gruyter.

Segerdahl, P., Fields, W. M., & Savage-Rumbaugh, E. S. (2005). *Kanzi's primal language: The cultural initiation of apes into language.* London: Palgrave/Macmillan.

Thompson, R. K. R., & Oden, D. L. (2000). Categorical perception and conceptual judgments by nonhuman primates: The paleological monkey and the analogical ape. *Cognitive Science, 24*(3), 363–396.

Tinbergen, N. (1951). *The study of instinct* ("Based on a series of lectures given in New York, 1947, under the auspices of the American Museum of Natural History and Columbia University"). Oxford: Clarendon Press.

Tinbergen, N. (1953). *The herring gull's world.* London: Collins.

Vico, G. (1744). Principj di una scienza nuova. In *Opere* (A. Battistini Ed., It. Trans.). Milano: Mondadori.

Vygotskij, L. S. (1934). *Myšlenie i reč'. Psihologičeskie issledovanija.* Moskvà-Leningrad: Gosudarstvennoe social'no-èkonomiceskoe izdatel'stvo. (It. Ed. Vygotskij, L.S. (1990) *Pensiero e linguaggio. Ricerche psicologiche.* Bari: Laterza).

Vygotskij, L. S. (1978). *Mind in society.* Cambridge, MA: Harvard University Press. (It. Ed. Vygotskij, L. S. (1978) *Il Processo Cognitivo.* Torino: Boringhieri).

Wiesel, T. N., & Hubel, D. H. (1963). Effects of visual deprivation on morphology and physiology of cells in the cat's lateral geniculate body. *Journal of Neurophysiology, 26*, 978–993. Retrieved from http://jn.physiology.org/cgi/reprint/26/6/978

Wiesel, T. N., & Hubel, D. H. (1963). Single-cell responses in striate cortex of kittens deprived of vision in one eye. *Journal of Neurophysiology, 26*, 1003–1017. Retrieved from http://jn.physiology.org/cgi/reprint/26/6/1003

Wittgenstein, L. (1953). *Philosophical investigations.* Malden: Blackwell.

From Non-minds to Minds: Biosemantics and the *Tertium Quid*

Crystal L'Hôte

Abstract I present and evaluate the prospects of the biosemantic program, understood as a philosophical attempt to explain the mind's origins by appealing to something that non-minded organisms and minded organisms have in common: representational capacity. I develop an analogy with ancient attempts to account for the origins of change, clarify the biosemantic program's aims and methods, and then distinguish two importantly different forms of objection, *a priori* and *a posteriori*. I defend the biosemantic program from *a priori* objections on the grounds that the standard of explanation presupposed by them is inappropriate and leads to absurdities if consistently applied. Once the way is cleared of *a priori* objections, the success of biosemantics turns on the strength of *a posteriori* objections, that is, on the program's empirical adequacy. Here, its prospects are less clear, but I offer reasons, by analogy with chemical combination and other everyday phenomena, to think that minded beings and their representational capacities might well have their origin and explanation in non-minded beings. An evolutionary origin and explanation of mind is plausible, at least as far as naturalistic accounts and explanations go.

But only Nature's aspect and her law,
Which, teaching us, hath this exordium:
Nothing from nothing ever yet was born.
– Lucretius

1 *Ex Nihilo Nihil Fit*

Philosophers have long struggled to make sense of change and generation in the natural world. How and why – by what fundamental principles – do plants and animals grow and decay? How and why do such beings come to exist in the first

C. L'Hôte (✉)
Associate Professor of Philosophy,St. Michael's College, Colchester, VT, USA
e-mail: clhote@smcvt.edu

L. Swan (ed.), *Origins of Mind*, Biosemiotics 8, DOI 10.1007/978-94-007-5419-5_4,

place? Two early responses to these perennial questions are especially influential. Instead of providing a positive account of natural change and generation, the sixth-century BC philosopher Parmenides argued that change and generation are ultimately illusions, that all that exists is ultimately One, and that only "Being Is" (Freeman 1984b).[1] From the Parmenidean perspective, it is ultimately unnecessary to explain change and generation, then, however much an explanation of the illusion is in order. In the same century, Heraclitus had reached a conclusion that, on the face of it, is opposed to Parmenides' conclusion: that only change is real and that all constancy is illusion. For Heraclitus, "The sun is new each day" (Freeman 1984a).[2]

These two responses to the problem of natural change are opposed only on the face of things. If change were fundamental in the manner that Heraclitus suggests, then change would be at most a self-explanatory explainer and would itself be beyond explanation. Heraclitus and Parmenides both agree, then, that change is inexplicable, even if their reasons for thinking so differ. And they both reach their respective conclusions by hyperextending a reasonable-enough principle of metaphysics and explanation: *ex nihilo nihil fit*.[3] From nothing, nothing is produced. Just as Parmenides failed to see how there could be change unless it had already been there in the fundamental nature of things as a first principle (arche), Heraclitus failed to see how there could be constancy in the world if change is the first principle. The two philosophers were agreed, then, that change comes from only change and only change comes from change.

On the face of it, such ancient attempts to come to grips with change and generation in the natural world bear little on the present topics, that is, the nature and origin of mindedness in the natural world and the adequacy of a biosemantic account. But an inspection of the contemporary debate about mindedness reveals deep similarities with the ancient debate about change. In particular, the *ex nihilo* principle is ever as much at work. In the contemporary context, however, a hyperextension of the same principle yields two superficially opposed accounts of the nature and origins of mindedness: dualistic supernaturalism and panpsychic naturalism.

Dualistic supernaturalism follows broadly from two claims, the claim that (a) mindedness cannot come from non-mindedness and (b) the natural world is originally and fundamentally non-minded. It follows that mindedness can have only a supernatural explanation, if any at all. And a panpsychic naturalism follows if the first claim, that (a) mindedness cannot come from non-mindedness, is instead coupled with the claim that (b) mindedness has a natural explanation. From these, it follows that mindedness must be a fundamental and even pervasive feature of the natural world, that mindedness must always already have been there in the world, in the

[1] "Being has no coming-into-being and no destruction… And it never Was, nor Will Be, because it Is now, a Whole all together, One, continuous" (Freeman 1984b). The paradoxes of Zeno (of Elea) bolster, a Parmenidean metaphysics.

[2] More provocatively, "In the same river, we both step and do not step, we are and we are not" (Freeman 1984a).

[3] This principle is commonly attributed to Parmenides.

earliest prokaryotes, the primordial muck, and the stardust.[4] So, just as the claim that change cannot come from non-change figures into the accounts of both Parmenides and Heraclitus, the claim that mindedness cannot from non-mindedness figures into both dualistic supernaturalism and panpsychic naturalism. Opposites cannot come from opposites.

Fifth-century (BC) Platonic dualism and Aristotelian hylomorphism can be understood as avoiding the extremes of Parmenides and Heraclitus by putting a proper limit on the *ex nihilo* principle. According to Aristotle, there is change in the natural world, but there is not only change. As an acorn grows into an oak tree, for example, it undergoes a material change, but its form remains the same. In a similar way, the contemporary biosemanticist avoids the extremes of dualistic supernaturalism and panpsychic naturalism by limiting the scope of the *ex nihilo* principle. According to the biosemanticist, mindedness cannot come from non-mindedness, but minded beings can nonetheless come from non-minded beings (organisms).[5]

The biosemanticist means to establish not only that the evolution of minded from non-minded beings is logically possible or conceivable but that an evolutionary account of minded beings is plausible. She means not only to show that a bridge across this gap can be built but to build it. To this end, she identifies a *tertium quid* (third thing), in this instance a property that is common to both non-minded and minded organisms and constitutes a relevantly explanatory link.[6] According to Peter Godfrey-Smith, Fred Dretske, Ruth Millikan, Karen Neander, David Papineau, and others, that *tertium quid* is a common capacity for normative representation.[7] Accordingly, even non-minded, simple cellular organisms and their tiniest parts represent features of the external environments in a manner that is robust enough to allow for the possibility of error and, consequently, in a manner that is plausibly continuous with our own higher-level capacity for representing the environment via perception and complex, truth-evaluable thought.[8]

[4] To be sure, panpsychism has many forms. For instance, some panpsychists will maintain that there is mentality everywhere and to the same degree, while others will maintain that the degree varies from place to place; some maintain that mentality exists at the subatomic and cosmic levels and everything in between, while others will maintain that it can be found only at the level of ordinary medium-size objects. It becomes the burden of panpsychist who attributes mentality to the subatomic, medium-sized, and cosmic to explain the relation between the low-level and high-level mindedness of single entities.

[5] Likewise, the claim that green *apples* can become red *apples* is distinct from the claim that that green can become red.

[6] Of course, not any common feature will suffice; although both non-minded and minded organisms occupy space-time, this common feature sheds no light on the manner by which minded organisms might have evolved from non-minded organisms.

[7] For instance, see Millikan (1984, 1989, 1993), Neander (1991a, b, 1995), and Papineau (1987, 1993, 1997).

[8] By contrast, nonnormative representations cannot be wrong. For instance, although we may misinterpret the meaning of the rings of a tree – thinking that it is older than it really is – this is not because the rings have misled us or somehow lied. The rings provide a nonnormative representation of the age of the tree. Likewise, we may interpret lightning to mean that a storm is on its way, but the lightning has not erred or made a mistake if no storm occurs.

2 Basic Biological Representation and the Biosemantic Program

Consider the phenomenon of basic, biological representation as it occurs in anaerobic marine bacteria, as cited by an early Dretske:

> Some marine bacteria have internal magnets (called magnetosomes) that function like compass needles, aligning themselves (and, as a result, the bacteria) parallel to the earth's magnetic field. Since these magnetic lines incline downwards (towards geomagnetic north) in the northern hemisphere (upwards in the southern hemisphere), bacteria in the northern hemisphere, oriented by their internal magnetosomes, propel themselves toward geomagnetic north.[9]

As it happens, these bacteria survive and thereby secure an opportunity to reproduce and pass on heritable traits because they generally head toward geomagnetic north. If the bacteria were to instead head south in their normal environment (the northern hemisphere), they would also head toward oxygenated surface water, which would kill them. As the biosemanticist sees it, there is every reason to think that the (biological) function of the tiny magnetosomes is to prevent this by representing to the bacteria relevant features of the environment. Biosemanticists disagree about exactly which features of the environment are represented by the magnetosome, for example, proximal magnetism or distal oxygenation, but they agree that a magnetosome that steers a normally situated bacterium away from geomagnetic north and toward surface water has malfunctioned.

According to the biosemanticist, representations of this basic biological sort occur throughout the world of living organisms and play vital evolutionary roles. Importantly, however, the biosemanticist is no panpsychist. The biosemanticist does not advance the thesis that marine bacteria or their magnetosomes possess minds but rather that their basic representational capacities are similar to and plausibly continuous with our higher-level representational and intentional mental capacities. To be sure, our mental capacities are more plausibly continuous with the capacities of monkeys than bacteria. But if the biosemanticist is to explain how mindedness originated, if she is to put her finger on the Big Bang of mindedness, then she must show that minded organisms are plausibly continuous with organisms that are patently non-minded.

The biosemanticist also eschews pansemanticism. Only living organisms and some of their parts and subsystems are attributed the relevant sort of representational capacity. Stones are not. So, although the biosemanticist acknowledges that a pile of stones may represent a hiking trail and that each blade of grass might be made to mean something to someone, she notes that minded creatures like us have assigned these meanings to these entities. By contrast, the meaning that magnetosomal position has for the marine bacteria – whether "north this way!" or "oxygen-free water over here!" – has not been assigned. The meaning of magnetosomal position is

[9] Dretske (1994, p.164).

"original" and consequently of the same (relevant) sort exhibited by our own mental states.[10] Finally, from the fact that only living organisms have the relevant sort of representational capacities, it does not follow that all living organisms do – for example, plants. That is, the biosemanticist does not endorse what Godfrey-Smith (1996) calls the "strong continuity thesis," according to which all living beings and/or their subsystems thereby display a degree of mindedness.[11]

A successful biosemantic account must show how basic biological representations and, ultimately, minded organisms might fit into and come-to-be in a world without them, and it must do so without appealing to any besides patently naturalistic phenomena. These accounts have the following form, where only patently naturalistic notions can be used to complete the biconditional[12]:

R represents O iff __________.

Of course, an account aimed at naturalizing the most fundamental representational phenomena cannot presuppose representational phenomena – whether basic or higher-level – on pain of circularity. This is a challenge. Whether high-level human thoughts or low-level bacterial indicators, representations are essentially about something. The bacterium's magnetosomal position represents something, whether oxygen or magnetic north, and our thoughts point beyond themselves to whatever it is that we are thinking about. As Brentano (1874) argues, it is precisely because our thoughts and other intentional states essentially point beyond themselves that they are resistant to a naturalistic explanation, and the same is true of more basic representational phenomena. Again, a naturalistic account of representation can appeal to only patently naturalistic phenomena, and patently naturalistic phenomena – stars, stones, and molecules – do not point beyond themselves.[13] They are semantically inert.

It is here that the biosemantic appeal to natural selection and to the etiological notion of a proper biological function does its most impressive work. Roughly speaking, the proper biological function of a trait is just whatever that trait did or

[10] For example, it is not by assignment or convention that states of the amygdala represent danger. Neither we nor tiny homunculi assign meanings to our mental states, arguably, on pain of regress.

[11] Importantly, the strong continuity thesis to a plant does not have the same degree of mindedness as is possessed by a human, nor does it attribute nonliving entities mindedness, a la varieties of panpsychism. See Stillwaggon Swan and Goldberg (2010a) for an illuminating argument in favor of the strong continuity thesis, one that appeals to the work of the biochemist Gordon Tomkins. Unfortunately, space does not allow for a comparison of biosemiotic and biosemantic analyses of Tomkins' view on metabolic coding systems.

[12] R = representation, O = object of representation.

[13] Moreover, the relation that exists between a mental representation and its object is unlike any ordinary physical relation. Ordinary physical relations such as *on top of* and *next to* are sensitive to the time and location of would-be relata. But mental representations readily stand in the aboutness relation even to the distant past and to the causally inefficacious future, as well as to things that will never exist: dream vacations and world peace. No ordinary physical relation has what does not exist as a relatum.

brought about in the past that enabled the relevant species to survive and reproduce, that is, whatever that trait contributed to the species fitness. For instance, it is the proper function of our hearts to pump blood because pumping blood conferred a selective advantage on our ancestors. The proper biological function of a trait – whether that trait is structural or behavioral – is not what the trait actually does at present but what it ought to do, in light of its selection history: hearts that skip beats are not functioning properly. And it is not the proper function of hearts to beat loudly, even if beating loudly now proves useful on occasion, since beating loudly does not explain the persistence and existence of hearts and the organisms that have them.[14]

The biosemantic account of representation makes use of the notion of a proper biological function. On Ruth Millikan's analysis, for instance, to say that the magnetosome represents oxygen-free water is just to say that it is the proper function of the magnetosome to coordinate the bacteria with oxygen-free water. And to say this is to say no more (and no less) than that being coordinated with oxygen-free water conferred a selective advantage on the ancestors of present-day marine bacteria. Representation is just a manner of biological function.

Note that, on this approach, the selection history of a species imposes constraints on what representational traits subsequently mean. For instance, the position of a present-day marine bacterium's magnetosome could not mean oxygen-free water (say) unless ancestral bacteria were selected because they had magnetosomes that coordinated them with oxygen-free water. Other marine bacteria must have failed to survive and reproduce because they lacked such magnetosomes. Also note that magnetosomes could not have been selected for coordinating the bacteria with oxygen-free water unless oxygen-free water existed in the ancestral environment. In short, it is the proper biological function of present-day magnetosomes to represent oxygen-free water only if (a) ancestral bacteria with oxygen-free-water-coordinating magnetosomes were selected over bacteria without them and, what this presupposes, that (b) oxygen-free water existed in the environment of ancestral bacteria.

In this general fashion, biosemantics provides a naturalistic account of the representational capacities of basic biological structures. Its hope is that the naturalization of basic representational capacities will demystify the phenomenon of representation more generally, that an increased attention to basic representations will make more plausible a naturalistic approach to the higher-level representational phenomena. That evolutionary principles and concepts can be used to explain aboutness or proto-aboutness at a basic biological level provides some reason to think that they might also be used to account for the aboutness of higher-level intentional states. That some human neural phenomenon represents some feature of the world might be

[14] For instance, even if a loud heartbeat confers survival benefits by enabling a physician to detect health problems with a stethoscope, it is not plausibly the biological function of the heart to beat loudly enough that its beating can be detected by a stethoscope. Stethoscopes could not have had any causal influence on our Pleistocene ancestors. Similarly, it is not the biological function of our fingers to type on a keyboard, even if the ability to type confers some selective advantage today.

explained by its being its proper evolutionary function to coordinate the relevant organism with that feature of the world.[15] For instance, most biosemanticists maintain that our neural states or processes are about edges or food or danger just in virtue of the fact that directing our ancestors toward or away from these items conferred some selective advantage; others will maintain that even complex and higher-level beliefs and desires are amenable to this manner of naturalistic analysis.

One common objection to the biosemantic approach to mental representations is that its reliance on history seems to rule out the possibility of thoughts about telephones or computers, not to mention thoughts about future or nonexistent phenomena (such as world peace). Plainly, we do think about telephones, even if our Pleistocene ancestors were not coordinated with telephones in a way that conferred selective advantage. Replies to this objection are available. For example, Millikan argues that it is the proper function of our brains (say) to coordinate us with and represent telephones and other modern technologies for the same reason it is the proper function of a chameleon's skin to coordinate it with the color it sits on, even if no ancestral chameleons never encountered that specific color.

A second common objection is that biosemantic program is built upon a conceptually unstable foundation: the notion of a proper function. In many cases, the proper function of a trait is neither clear nor determinate. Although Millikan's take on the magnetosome case is unequivocal – the magnetosome represents oxygen-free water since being coordinated with oxygen-free water contributed to species fitness – its function might reasonably be specified in other ways. If an unsuspecting bacterium were nabbed from the northern hemisphere and transported to the southern hemisphere, it would likely steer the bacterium to oxygenated water and destruction. If Millikan's specification of the magnetosomes' proper function is correct, then it has thereby malfunctioned. But surely it is reasonable to think that pointing the bacterium in the direction of oxygen-free water in such strange and dizzying circumstances is too much to expect, that the magnetosome has not malfunctioned if it fails to do this. On such grounds, some biosemanticists contend that the proper function of the magnetosome is simply to point the bacterium toward magnetic north. Ultimately, however, these in-house disagreements do not seem to diminish the program's overall promise for bridging the gap between non-minds and minds.

3 Two Forms of Objection: *A Priori* and *A Posteriori*

In this section, I distinguish two major forms of objection to biosemantics, *a priori* and *a posteriori*, both of which deny that the program succeeds in bridging the explanatory gap between non-minds and minds. I defend the biosemantic program against the *a priori* form of the objection, arguing that the standard of explanation

[15] Stillwaggon and Goldberg analyze this insight in (2010b).

presupposed by it is inappropriate. Whether or not the biosemantic program meets its aims entirely depends on the strength of *a posteriori* objections.

According to objections of the *a priori* type, the biosemantic program makes no progress whatsoever toward bridging the explanatory gap between non-minds and minds. Even providing a naturalistic account of the representational capacities of basic biological organisms like marine bacteria does nothing to strengthen the case. The import of first form of objection is *not* that non-minded and minded beings are not alike enough for minded beings to plausibly come from non-minds; in principle, it would be possible to meet that sort of objection by showing that non-minds and minds are more similar than it initially appears. The force of the *a priori* objection is not that the similarity between non-minded beings (bacteria) and minded beings (humans) is not strong or relevant enough to make continuity plausible but that the biosemantic program makes no progress at all in showing how mindedness might evolve out of a non-minded world. The force of the a priori objection is that the gap between non-minds and minds remains as wide as it ever was. *Tertium non datur.*[16]

In this way, the *a priori* form of objection evokes Lewis Carroll's beloved "Tortoise and Achilles," in which the recalcitrant tortoise refuses to accept the validity of modus ponens without a deductive proof of it. Since every proof that Achilles offers the tortoise presumes the validity of modus ponens, their conversation both goes nowhere and is potentially endless. Still, the dialogue comes to a close after some parting jokes about Zeno. Among the morals of the dialogue is a Humean one: that, even in a valid argument, the inferential link between premises and conclusion rests ultimately upon what is nonlogical. Even logical justification must come to an end somewhere.[17]

This is all the more true in inductive and empirical contexts like the present one. To be sure, doubting that the sun will rise tomorrow is not unreasonable, especially in philosophical contexts. However, it is unreasonable to doubt this while endorsing other claims like it, for example, that the ground beneath will remain solid. *A priori* forms of objection fail in this way. They simply presume that nothing besides mindedness (or something supernatural) could give rise to or explain mindedness and so run up against most theories of explanation.[18] And if the theory of explanation that is thereby presumed is applied consistently, it follows from it that most of the things we recognize as successful metaphysical accounts and explanations actually fail. For instance, it is widely accepted that water can have its origin and explanation in non-water (hydrogen and oxygen) and also that a dessert is made by combining flour, sugar, butter, and so on. To be sure, from the fact that flour, sugar, and butter are combined, it does not follow with any logical necessity that a dessert will result. But to suppose that this must be the case in order for the account to succeed is to hold

[16] That is, no third thing obtains. This rule of (a priori) reasoning is commonly called the "law of excluded middle."

[17] See Haack (1976).

[18] See Achinstein (1983).

a reasonable account about the workings of the empirical world to an inappropriate standard. As Hume argues, the cause-effect relationship is not a logical one.

Similarly, to resist the claim that non-minded beings can evolve into minded beings on the grounds that there is a necessary gap between them – that there cannot be a *tertium quid* as a matter of logical necessity – is to apply an inappropriate standard, with an absurd result. If consistently applied, this same standard would have us deny that desserts are accounted for by their ingredients (plus labor and so on), on the grounds that the dessert was not there before it was baked. As in the work of Parmenides and Heraclitus, the reasonable-enough principle *ex nihilo* is here hyperextended.

With the way cleared of the *a priori* objection, which exercises quiet but considerable influence, the second form of objection to the biosemantic program emerges as the most serious. The upshot of *a posteriori* objections is the same as that of the *a priori* objection: the metaphysical and explanatory gap between non-minds and minds has not been bridged by the biosemantic program. The link between non-minds and minds had not been forged by the biosemanticist's *tertium quid* – representational capacity. However, *a posteriori* forms of objection acknowledge that the biosemantic program makes some progress, that it narrows the gap even if it does not close the gap. Accordingly, a naturalistic, evolutionary account of the origin of mindedness is possible in principle. It is just that more steps are needed to show that non-minds and minds are sufficiently alike and plausibly continuous.

Such matters are not settled by demonstrative arguments so much as they are settled by comparing cases. For instance, green is quite unlike the blue and yellow that combine to make it, cakes looks nothing like their ingredients, and so forth. Yet, there is no significant gap. Why is it any less plausible that our representational capacities are continuous with the representational capacities of lower-level organisms? A better analogy is provided by accounts of the origins of life. Here, it is the going view that a naturalistic account has been provided. At the least, life has been produced in laboratories, organic from inorganic materials, and the possibility that life could have come from inorganic matter through natural processes was established as early as 1953, via the Miller-Urey experiment. It would be unreasonable for the observer of the experiment to insist that what she witnessed was not the production of a form of life, even if she could reasonably deny that life on Earth actually originated this way. And it would be similarly unreasonable for the observer of a labor and delivery to deny that what she saw was the birth of a baby, however miraculous it seemed.

So, a naturalistic explanation of the coming-to-be of life has been supplied. Certain "how" and "why" questions have been answered, school children can replicate the experiments, and, however unlike the lives of amino acids our own lives may seem, it is demonstrably possible for organic materials to come from inorganic. Although there is not time to sit by and watch for minded organisms to evolve from non-minded organisms, no demonstration is needed. The archeological record and the research of our best thinkers do well enough to show that minded organisms did in fact evolve from non-minded organisms, if anything did ever evolve from anything. Our experiment is superior to that which was conducted by Miller and

Urey, because it is not an experiment. And if evolution accounts for our biological origins, why deny that it also accounts for the origins of all of our skills and capacities, including mindedness? It would be as strange to resist the conclusion that the evolution explains the origins of our capacity for sight. As Millikan puts it, "To suspect that the brain has not been preserved for thinking with or that the eye has not been preserved for seeing with – to suspect this, moreover, in the absence of any alternative hypotheses about causes of the stability of these structures – would be totally irresponsible."[19]

Yet even Millikan, whose program is the most ambitious of all, denies that "bacteria and paramecia, or even birds and bees, have inner representations in the same sense that we do."[20] Although the bacterium's magnetosome enables it to represent its environment, it does not thereby perceive or think. Indeed, all are agreed that there are significant differences between low-level biological representations and human perception and thought. Still, Millikan urges that our (mental) representations are explicable in terms of lower-level representations plus other equally naturalistically explicable features – for example, non-self-representing elements and storage – and that features that distinguish thoughts and other mental representations from low-level biological representations are also amenable to an evolutionary explanation. Still, it would be sufficient for the purposes of an overall naturalism if these supplementary features were amenable to any naturalistic explanation, whether evolutionary, chemical, physical, or other, and there is every reason to think that they are.

4 The Deeper Disagreement

On inspection, a naturalistic account of the origins of representation and mind is more plausible and complete than a naturalistic account of the origins of life. We have more reason to believe that minded beings naturally evolved from non-minded beings than we do to think that life emerged from nonlife. To deny *a priori* the possibility that minds could evolve from non-minds, or that life could emerge from nonlife, is to miss the meaning and point of naturalistic explanations, which are assessed by empirical adequacy.

On the other hand, to suppose that a naturalistic explanation is the only legitimate form of explanation, that all of our "how" and "why" questions are thereby answered, is also to miss the point of a naturalistic explanation. Indeed, the more substantial disagreement between those who maintain that the link between non-minds and minds is bridgeable by such means as the biosemanticist offers, and those who deny this, seems to concern the nature of explanation. I have suggested that the evolutionary account of the origin and explanation of mindedness, supported by the biosemantic

---

[19] Millikan (1989, p. 285).

[20] Millikan (1989, p. 288).

program, is a plausible account, as far as naturalistic accounts and explanations go. But I submit that not all of our (legitimate) "why" and "how" questions have been answered thereby. If we accept, as I think we should, that accounts and modes of explanation – mechanical, intentional, synchronic, and so forth – have different purposes and excellences, that they are more or less useful for our various ends, then the possibility that biosemantics advances our understanding of the relationship between non-minds and minds does not preclude the possibility that entirely different modes of explanation are necessary to answer our most burning questions.

References

Achinstein, P. (1983). *The nature of explanation*. New York: Oxford University Press.

Brentano, F. (1995/1874). *Psychology from an empirical standpoint*. 2nd English edition. London: Routledge.

Dretske, F. (1994). Misrepresentation. In S. Stich & T. Warfield (Eds.), *Mental representation: A reader* (p. 157–174). Cambridge: Blackwell.

Freeman, K. (1984a). Heraclitus. In *Ancilla to the pre-Socratic philosophers*. Cambridge: Harvard University Press.

Freeman, K. (1984b). Parmenides of Elea. In *Ancilla to the pre-Socratic philosophers* (pp. 41–46). Cambridge: Harvard University Press.

Godfrey-Smith, P. (1996). *Complexity and the function of mind in nature*. Cambridge: Cambridge University Press.

Haack, S. (1976). The justification of deduction. *Mind, 85*(337), 112–119.

Millikan, R. (1984). *Language, thought, and other biological categories*. Cambridge: MIT Press.

Millikan, R. (1989). Biosemantics. *The Journal of Philosophy, 86*, 281–297.

Millikan, R. (1993). *White Queen psychology and other essays for Alice*. Cambridge: Bradford Books, MIT Press.

Neander, K. (1991a). Functions as selected effects. *Philosophy of Science, 58*, 168–184.

Neander, K. (1991b). The teleological notion of function. *Australasian Journal of Philosophy, 69*, 454–468.

Neander, K. (1995). Misrepresentation and malfunction. *Philosophical Studies, 79*(2), 109–141.

Papineau, D. (1987). *Reality and representation*. Oxford: Blackwell.

Papineau, D. (1993). *Philosophical naturalism*. Oxford: Blackwell.

Papineau, D. (1997). Teleosemantics and indeterminacy. *Australasian Journal of Philosophy, 76*, 1–14.

Stillwaggon Swan, L. S., & Goldberg, L. J. (2010a). Biosymbols: Symbols in life and mind. *Biosemiotics, 3*(1), 17–31.

Stillwaggon Swan, L. S., & Goldberg, L. J. (2010b). How is meaning grounded in the organism? *Biosemiotics, 3*(2), 131–146.

Cybersemiotics: A New Foundation for a Transdisciplinary Theory of Consciousness, Cognition, Meaning and Communication

Soren Brier

Abstract The modern evolutionary paradigm combined with phenomenology forces us to view human consciousness as a product of evolution as well as accept humans as observers from the 'inside of the universe'. The knowledge produced by science has first-person embodied consciousness combined with second-person meaningful communication in language as a prerequisite for third-person fallibilist scientific knowledge. Therefore, the study of consciousness forces us theoretically to encompass the natural and social sciences as well as the humanities in one framework of unrestricted or absolute naturalism, viewing the conscious lifeworld with its intentionality as well as the intersubjectivity of culture as a part of nature. But the sciences are without concepts of qualia; will and meaning and the European phenomenological-hermeneutic 'sciences of meaning' do not have an evolutionary foundation. It is therefore interesting that C.S. semiotics—in its modern form of a biosemiotics—was based on an evolutionary thinking and ecology of sign webs. But Cybersemiotics shows that it is also necessary to draw on our knowledge, from science and the technologically founded information sciences, systems theory and cybernetics to obtain a true transdisciplinary theory.

1 Introduction to the Scientific Problem of Awareness and Experience

When you open the skull and investigate the brain neurophysiologically and include the nerves from the sense organs and those going to the muscles, the sciences have not managed to find any qualia, experience, emotions or awareness, but only

S. Brier (✉)
Department of International Business Communication,
Copenhagen Business School
e-mail: sb.ibc@cbs.dk

L. Swan (ed.), *Origins of Mind*, Biosemiotics 8, DOI 10.1007/978-94-007-5419-5_5,

electrochemical impulses, transmitter molecules, hormones and functional structures of neurons, glia and muscle cells. New brain-scanning techniques make it possible to see which parts of the brain are used in what kinds of perceptions, actions and moods by following the increased blood flow to the active parts, as the brain uses a lot of oxygen. We can also induce certain feelings, moods and sensation qualities, or the memory of them, which people report orally, when we stimulate the brain electrically or do and say certain things to people. We can, through electrical stimulation of nerves, make limbs move and organs do their function. We can also from the outside register and describe the interaction between sense stimuli and behaviour in meticulous experiments with humans and other living beings as has been done since the heyday of Skinner's radical behaviourism and the European ethology of Lorenz and Tinbergen. But no matter how refined our empirical scientific approaches become, we cannot find any experiences in the brain. It does not matter if it is our own brain or that of other animals. The felt awareness seems to be found on another level of abstraction (Hinde 1970). Something central about the brain's function as an organ escapes us (McGinn 2000: 66–68, Hofstadter 2007; Penrose 1997; Searle 2007). So far, our only access to the first-person experiences is through meaningful verbal or written communication from the experiencing person (Heil 2004: 3). This is our main problem.

Among other things it means that language and culture are 'in the way'. We cannot experience other people's experiences directly. What people experience when performing certain behaviours, we only know about from their own reports, though we can see what part of the brain they use or how they behave externally as well as internally, physiologically. The paradox of modern attempts to work towards a 'science of consciousness' is that we have no direct scientific empirical access to the experiential qualities of will, intentions and meaning on which to build such a science (Edelmann 2000: xi). As a philosopher of science, it seems to me that this is why we have the qualitative phenomenological, hermeneutical and discourse theoretical methods of the humanities and the social sciences. But they are not really considered to be scientific by the natural sciences (Bennet and Hacker 2007); only the brain sciences are.

But as responsible and experientially aware social citizens, we are not identical to our brains (Edelmann 2000: 1), although we do need them in order to stay conscious. But we seem to be a more complex integrative product of physical, chemical, biological, social, mental, semiotic and communicative systems producing and produced by culture and language, of which the brain and the body surely are important components, but so is the ability of living systems to produce experience, and think about and communicate them through language. This is the problem, which some formulate as an *explanatory gap* (Thompson 2003: vii, Levine 1983).

There is no agreement on how to formulate this explanatory gap problem (Rorty 1980: chap. 1), so I will suggest a working hypothesis here: The attempt to explain consciousness from the scientific physico-chemical as well as informational and computational paradigms runs into the claims of phenomenological paradigms that our knowledge or process of knowing is based on an experiential world (what Husserl called a 'lifeworld'), prior to any culturally developed scientific explanations. His method was to attempt to put these influences in parenthesis or bracketing (Epochè) to try to get to the pure phenomena or the 'thing in itself' (Husserl 1997, 1999) through a systematic peeling away of their symbolic layers of meanings until only the thing itself as 'originally' meant and experienced remains.

Husserl's problem was that our consciousness and intentionality always are infected with intersubjective linguistic and culturally mental conceptions and ontological assumptions of the situation at hand, so in order to get to the pure phenomenon, we must seek beyond those obstacles. We thus conclude that even phenomenology has trouble getting to experience itself. This basic phenomenological position is shared by Edmund Husserl, Maurice Merleau-Ponty and Charles Sanders Peirce.[1] His development of a triadic[2] phaneroscopy is the point of departure for his semiotics.

Our gap problem is that these scientific and the phenomenological paradigms are in Kuhn's (1970) terminology 'incommensurable'. They do not have the same epistemological and ontological conceptions. They have two different maps of reality: This is my *philosophy of science working hypothesis of what is at the root of the explanatory gap*. This dovetails with argumentation by, Penrose (1997: 101) whom from his physicalistic but non-computational paradigm writes his final viewpoint, as 'Awareness cannot be explained by physical, computational or any scientific terms'.

My suggestion of a cure is to contribute to the crafting of a transdisciplinary framework—inspired by Luhmann and Peirce—wide and deep enough to contain both paradigms and thus enlarge our ontological conception of reality beyond Penrose's. I have called the framework Cybersemiotics, as it attempts to combine the two major attempts to unify theories of cognition and communication with the intersubjective, systematic and consistent systems of knowledge: (1) the informational-cybernetic and (2) the semiotics-phenomenological-hermeneutical meta-paradigms.

2 Is Consciousness a Part of Reality?

A basic problem in our culture's systematic knowledge production is that the natural and social sciences as well as the humanities do not agree on a common definition of reality. We talk about the physical, mental and social realities, but do not really know how to fit them together into a larger conception. Instead they each compete to take ownership of defining reality.

---

[1] I find these three authors most relevant for the problem I here want to discuss, and there are multiple references to these writers in the reference list, whom I have selected as the most interesting defenders of the phenomenological transdisciplinary view.

[2] When analysing Peirce's work, it is clear that his three categories are foundational to his whole semiotic and pragmaticist paradigm that was developed over many years. Peirce attempted to prove mathematically that triadic relations cannot be broken down to duals, but it has never been widely accepted. But I find the phenomenological argumentation very convincing and currently supported by many other developments in science. But the fundamentality of the triadic thinking has been the stumbling block for many scholars failing to accept Peirce's paradigm. But one should not underestimate how deep reflections of logic—including the logic of relations, time, reality, continuity, moment, perception and meaning—are connected to this groundbreaking invention of Peirce. Joseph J. Esposito (1980) *Evolutionary Metaphysics: The development of Peirce's Theory of Categories* describes this quest in a most profound way.

This power struggle has been a problem ever since Otto Neurath (Neurath 1983) introduced the logical positivistic idea of a unified science based on physicalism. The physical world is here considered to be the given. Critiques from the social sciences and the humanities have never stopped since. Its most alternative reaction has been to produce radical forms of social constructivism, disclaiming any kind of positivistic truth claims (Colling 2003). Most radical social constructivists consider political ideological as well as cultural conceptions of reality to be the primary reality, of which science and the phenomenological lifeworld is only one product out of many. But phenomenology from the Husserlian and Peircean traditions insists on a third view, namely, that the experiential phenomenal world is the given reality and the truth is to be found in analysing its structure, be it as intentionality schemata (i.e. the Husserlian tradition) or basic categories of cognition in the form of sign types, which are then developed into a semiotics (i.e. the Peircean tradition).

The eternal foundation that Husserl (1997, 1999) was seeking in the pure intentional structures or forms of conscious awareness became for Peirce semiotic dynamical ways of knowing that emerged through Peirce's concept of continuity (synechism) from firstness as 'may-bes' and developed into 'would-bes' in thirdness through the evolution of reasonableness:

> Once you have embraced the principle of continuity, no kind of explanation of things will satisfy you except that they grew. The infallibilist[3] naturally thinks that everything always was substantially as it is now. Laws at any rate being absolute could not grow. They either always were or sprang instantaneously into being by a sudden fiat like the drill of a company of soldiers. This makes the laws of nature absolutely blind and inexplicable. Their why and wherefore can't be asked. This absolutely blocks the road of inquiry. The fallibilist won't do this. He asks, may these forces of nature not be somehow amenable to reason? May they not have naturally grown up? After all, there is no reason to think they are absolute. If all things are continuous, the universe must be undergoing a continuous growth from non-existence to existence. There is no difficulty in conceiving existence as a matter of degree. The reality of things consists in their persistent forcing themselves upon our recognition. If a thing has no such persistence, it is a mere dream. Reality, then, is persistence, is regularity.[4] In the original chaos, where there was no regularity, there was no existence. It was all a confused dream. This, we may suppose, was in the infinitely distant past. But as things are getting more regular, more persistent, they are getting less dreamy and more real (Peirce CP 1.175).[5]

To Peirce, firstness is an unbroken continuity of pure mind or feeling, quality and tendencies to become existent in what Peirce called secondness. Thus, Peircean semiotics in its development as biosemiotics presents a third way between the natural and the social sciences.

The social sciences and humanities have felt dominated by biologistic-scientistic-reductionist explanations of experience and behaviour of human beings like Dawkins'

---

[3] Already before Popper, Peirce had a fallibilist theory of science. There is no absolute proof of truth in science.

[4] Which is what Peirce calls 'habits' and an expression of his category of thirdness.

[5] As convention goes, this refers to Peirce, C.S. (1994), which is the collected paper (CP).

(1989) selfish genes, memetics (Blackmore 1999) and E.O. Wilson's sociobiology and his later attempt to make a unified view from it (Wilson 1999). What this reductionist meta-scientific paradigm is supposed to mean is most clearly spelled out in Edward O. Wilson's *Consilience: The Unity of Knowledge* (1999). Taking up the torch from logical positivism, Wilson predicts that most of the humanities will be replaced by hard scientific knowledge, just like neuroscience will eventually tell us what conscious experience is. Consilience, literally a 'jumping together' of knowledge, has its roots in the ancient Greek concept of logos, which is the vision of an intrinsic orderliness governing the Cosmos. The problematic view, much science and analytic philosophy has inherent, is that logos is comprehensible by formal logical processes only. A reason to believe that Peirce's semiotics can move us out of this predicament is that he combines his view of semiotics and logic in an evolutionary pragmaticist framework. He writes:

> Logic will here be defined as formal semiotic. A definition of a sign will be given which no more refers to human thought than does the definition of a line as the place which a particle occupies, part by part, during a lapse of time. Namely, a sign is something, A, which brings something, B, its interpretant sign determined or created by it, into the same sort of correspondence with something, C, its object, as that in which itself stands to C. It is from this definition, together with a definition of 'formal', that I deduce mathematically the principles of logic.[6]
>
> (C.S. Peirce 1980: 20–21 & 54.)

For Peirce, pure mathematics is more fundamental than logic, and in combination with phenomenology is the foundation of his metaphysics, as we have already shown. This view clashes with the received view of science, which does not include phenomenology. As a function of the 'logos and unity of science' view, the received mathematical and deterministic version of science (Penrose 1997: 2) denies the validity of all claims and practices other than its own. In this way, it turns science into a kind of war machine, destroying all other discourses and points of view, a tendency which the physicist and philosopher Paul Feyerabend (1975) was aware of. The same critique applies to the information and computer science-based cognitivistic explanations of human social coordination and communication (Brier 2008a). But natural science was confronted by the social sciences in what is called the 'linguistic turn' in philosophy of science and various forms of constructivism, from solipsistic radical ones to social constructivisms (Brier 2009a), all undermining the objective authority of science's explanations of how the world works. This ignited what has so often been called the 'science wars', of which not much good emerged aside from a realization among some researchers of the necessity to construct a new integrative transdisciplinary framework, in which all can work together in a fruitful way.

---

[6] Peirce considered pure mathematics to be a more fundamental discipline than logic. According to Peirce, logic comes from mathematics and not the other way around as some researchers and philosophers believe. His thinking seems to be close to that of Penrose (1997) here, but the semiotics Peirce creates is beyond anything imagined in Penrose's paradigm.

Nicolescu (2002) is one of the rare examples of a quantum physicist engaged in a non-reductionist transdisciplinary philosophy of Wissenschaft.[7] One fact that has been emerging from the science wars with the social sciences and the humanities is the realization that the natural sciences were dependent on the language they were formulated in and that language, world view and mentality are deeply interconnected. Thus, we are back to Neurath's basic ideas, since we have given up on the idea of a special objective scientific language combining logic and mathematics to unite all Wissenschaft. Thus, theories of language, cognition and conditions for signification had to be integrated into the interpretation of scientific data. This is another reason for introducing Peirce's semiotics (Peirce 1931–1935), which was a research project mainly conducted from 1865 to 1910 in order to provide an understanding of the logic of scientific method. The result was his semiotic, phenomenological and pragmaticistic view of knowledge aimed at providing insight into the methodological commonalities found in all attempts to produce scientific knowledge, or what one could formulate as the semiotic processes of science. The project ended as a semiotic paradigm with a new transdisciplinary ontology and epistemology. As Emmeche writes:

> A logical implication of the ontological-phenomenological basis of Peirce's semiotics … points to an interesting continuity between matter, life and mind, or, to phrase it more precise, between sign vehicles as material possibilities for life, sign action as actual information processing, and the experiential nature of any interpretant of a sign, i.e., the effects of the sign upon a wider mind-like system.
>
> (Emmeche 2004: 118)

The problem of explaining the awareness of sensory information and its qualia, how we come to interpret sense experience and how it is connected to subjectivity is also a problem at the basis of philosophy of science, as well as questions of truth and meaning and how science is placed between them or may contribute to integrating them.

3 Philosophy of Science's Problem of a Science of Consciousness

Thus, the hard problem of why we have qualitative phenomenal experiences is not a superficial question; rather, it is one that demands that we dig deep down into the prerequisite for our way of producing knowledge, world views and explanations. Bennett and Hacker (2007: 4) underline that

[7] For lack of a better word, a *transdisciplinary paradigm* is what I will call what we aim for. The concept *transdisciplinary science* is supposed to cover the sciences, as wells as humanities and social sciences, much like the German word '*Wissenschaft*' or the Danish word '*videnskab*'. Basarab Nicolescu has written the *Manifesto of Transdisciplinarity* (2002), where he explores or rather develops the consequences of a transdisciplinary view of the world and the sciences.

> Conceptual questions antecede matters of truth and falsehood. They are questions concerning our forms of presentation, not question concerning the truth or falsehood of our empirical statements… when empirical questions are addressed without the adequate conceptual clarity, misconceived questions are bound to be raised, and misdirected research is likely to ensue… any incoherence in the grasp of the relevant conceptual structures is likely to be manifested in incoherence in the interpretation of the results of experiments.

Thus, in this chapter, I will suggest a way to deal with these problems through a philosopher of science's reflection on the limitation of coherence and consistency in our generally accepted but specialized epistemological and ontological frameworks in the natural, life, information and social sciences as well as the humanities.

The first move towards constructing a transdisciplinary framework (or meta-paradigm) including the natural sciences, phenomenology and a paradigm of semiotic-linguistic constructionism is to accept that natural, life and social scientific knowledge as well as knowledge in the humanities is created in intersubjectively meaningful communicative action by embodied living systems and that we are unable to give any final proof of its truth. This is in accordance with Popper's (1972) and Peirce's (1931–1935) idea of fallible objective knowledge. This view is also based on the fact that meaningful intersubjective communication is still—like first-person consciousness—not yet scientifically explainable or technologically realizable in meaningful linguistically communicating robots. Furthermore, we need to be aware that the life sciences have their own perspective, which we also need to integrate, since all the conscious beings we know today are embodied in living, autopoietic systems. No computers, AIs or robots can produce conscious awareness presently. AI is still not AC (artificial consciousness).

The intersubjective and the autopoietic embodied subjective awareness of differences that make a difference combined with semiotically based communication is a prerequisite for all intersubjective productions of knowledge. All scientific knowledge demands embodied minds meaningfully sharing interpretation of sense experiences through signs. Robots do not make science on their own, only as tools for humans, because they do not have experiential bodies.

Meaning is thus in a way created before and outside the realm of natural science, as we know it today. Meaning is primarily dealt with in ordinary social language and its paralinguistic bodily influenced signals. The subjective and intersubjective cultural meaning is explicitly removed from the foundational framework of the classical positivistic influenced concept of science for its strive towards knowledge of universal character mostly in the form of deterministic or statistical laws. In order to obtain objectivity in the empirical sciences, it is usually taken for granted that one must remove any influence of the subjective and cultural ideas of reality. This fact presents one aspect of the problem of a scientific explanation of consciousness, as subjective awareness and meaningful communication are not really deeply reflected in the concept of scientific objective knowledge. Heelan (1983 and 1987) has spent a lifetime investigating and arguing for the relevance of hermeneutics and phenomenology for the understanding of scientific observation and the interpretation of data, which is also the main point of Gadamer's (1989) main work.

4 Integrating the Four Views on Consciousness in the Cybersemiotic Star

Cybersemiotics suggests then that we have four different approaches to the understanding of cognition, communication, meaning and consciousness. First are the exact natural sciences. Second are the life sciences. Third are the phenomenological-hermeneutic interpretational qualitative 'sciences'. And forth is the sociological discursive-linguistic cultural view. We are here inspired by Wittgenstein's (1953) pragmatic linguistic view, but not only that. The point in the Cybersemiotic paradigm is that it views the production of knowledge from the middle, where we, as embodied, are aware of semiotic and communicating living systems and create knowledge in a cultural and ecological surrounding. This means that we cannot attribute more importance to one of the four systems of knowledge than any of the others without committing a reductionism or an unfounded one-sided simplification of reality. Thus, the four approaches are all equally important. This philosophy is parallel to Bruno Latour's break with modernity in his book *We Have Never Been Modern (*Latour 1993) and also inspired by Merleau-Ponty (1962). I work with four main paradigms, where Latour works primarily with the dichotomy between nature and culture.

In Latour's actor-network theory (ANT) and philosophy of science (Latour 1993, 2004), explaining consciousness only through the brain as a natural entity is nearly an impossible idea because what are considered 'natural entities' by science, for Latour, are 'hybrids' and they achieve their existence for us through a semiotic network of actants. But Latour does not deny that they have a 'Ding an sich' existence as independent reality. We should not forget that Bruno Latour's (1993 and 2004) theory of hybrids and actor-network theory are based on a semiotics, inspired by Greimas' actant model that is a semiotic combination of material existence and social role as created by a narrative. Latour views science as one narrative of the working of nature among many possible narratives based on the data we have so far. But not all stories about nature have been shown to be viable. Latour's view is thus of a semiotic processual kind. Its semiotics is not really a Peircean version (Brier 2008b), but a special brand of Saussurian semiology developed by Greimas and further formed by its inclusion in Latour's realistic vision of a communicative/semiotic network of humans, things (including technology and cultural artefacts), living and dead natural entities we relate to and which are organizations in the way that they act back on the social and change it (the HIV virus is an example) (Latour 2007: 10–11). Despite the fact that many call Latour a social constructivist and a postmodernist, he insists on being a realist and that the normative view of ANT is that it should contribute to a better social order, not to breaking things down (Latour 2007). This places him closer to Peircean semiotics than Saussurian semiology.

Science is a cultural product. It is a technology that we use in order to see, understand and manipulate the natural world on which our existence is dependent. The tool of scientific discourse based on empirical investigations makes us able to describe the part of reality we need to handle and in that process ascribe meaning to it and its processes. That certainly does not mean we are able to describe all of

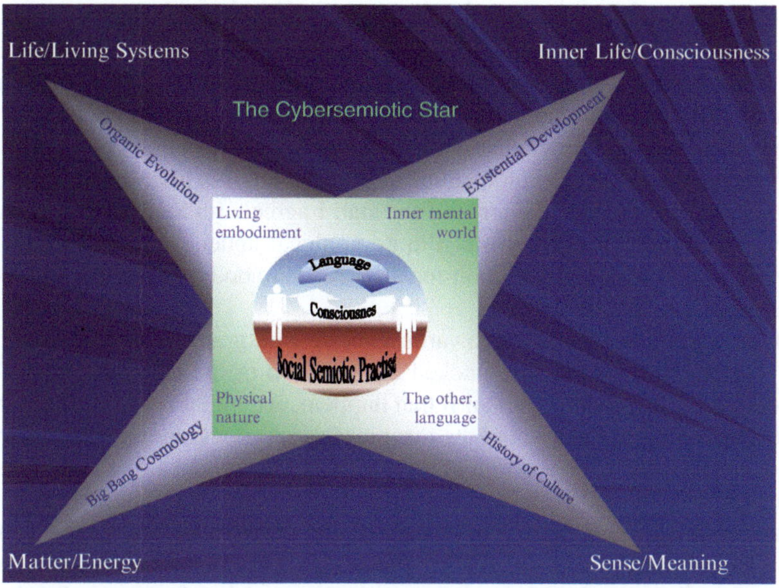

Fig. 1 The Cybersemiotic star: A diagram of how the communicative social system of embodied minds' four main areas of knowledge arises. Physical nature is usually explained as originating in energy and matter, sometimes also information, living systems as emerging from the development of life processes (such as the first cell). Social culture is explained as founded on the development of meaning and power in language and practical habits, and, finally, our lifeworld is explained as deriving from the development of our individual lifeworld and consciousness. In spiritual and religious frameworks it often ultimately conceptualized as originating from an objective transcendental spirit or as a soul coming from a personal creator or God

nature or give consistent meaning to all we have described so far, such as the relation between brain, culture and consciousness.

The idea of Fig. 1, called the Cybersemiotic star, and the epistemological turn it is illustrating is to escape the great explanatory burden of reductionistic mainstream science, which aims to explain both life and consciousness from its basic assumption of energy and mathematical mechanistic laws. The Cybersemiotic philosophy of natural, life and social sciences as well as humanities sees their different types of explanations moving from our present state of sociolinguistically common-sense-based conscious semiosis towards self-organized and highly specialized autopoietic knowledge systems. Each of them develops towards a better understanding of the prerequisites of language, culture and our self-conscious subject, and their production of systematic knowledge in a time perspective.

There are four forms of historical explanations invoked here: (1) the cosmological (physico-chemical), (2) the biological (biosemiotics and biosciences),[8] (3) the historical (sociocultural) and (4) the subjective perception of a lifetime, or experienced time.

[8] Cartwright (1997: 165) and Shimony in footnote in Penrose (1997) also argue for the independence of biological knowledge.

The *Cybersemiotic star* illustrates the equal importance of the four basic approaches, and from the model a few other points can be made. To be a realist about the possibility of science giving us usable knowledge about reality is to accept the reality of language, autopoietic embodied minds, culture and noncultural environments as well as the idea that our knowledge springs from processes of interaction between them. But that is something quite different from believing in reductionist explanations from one of the arms of the star. I agree with Steffensen and Cowley that we must move towards a much more nonlocal understanding of mind. What they call '…a transdisciplinary non-local approach to bodily, cognitive and interactional processes' (2010: 348).

The natural sciences work towards making one grand cosmogenic explanation.[9] But so far we have not cracked the problem of the emergence of life and consciousness in evolution, so until that happens, we might have to accept that an all-encompassing explanation of the meaningful conscious communication process cannot be provided from any one of the corners of the model alone. I argue further for this in the rest of the chapter. As we cannot reduce our scientific explanations to one grand story and claim it to be the one and only reality, my theory is that we have to juggle and work with all four types of knowing at the same time. This puts us in a new situation and changes the research questions about consciousness, as I will argue for further in the rest of this chapter.

The reason science works on the assumption that the physical world has no sense experience or meaning at all, but only natural laws,[10] is that scientists are brought up to think that to indulge in the opposite ontological assumption would make our search for knowledge religious or political, as these are the two major meaning-producing systems we know. Science fought its way out of the powerful grip of religion in the Enlightenment and later out of totalitarian political ideologies like Nazism and Communism.

Steering clear of religion and political world views, what are we then to call the meaning interpreting disciplines in the social sciences and humanities? This problem is well-known, and answers have been developed within phenomenology, phaneroscopy (Peircean triadic semiotic phenomenology) and hermeneutics, the ultimate philosophical version of which was developed by Gadamer (1989). Gadamer's book is clearly developing a philosophy for the humanities and the qualitative social sciences. Are we then going to accept meaningful interpretation as part of our view of consciousness and legitimate objective knowledge? I cannot see how we can ignore this fundamental human process of cognition, since meaningful human communication is a prerequisite for the possibility of science. If we want to give scientific answers about the nature of consciousness, we must integrate some version of hermeneutics into a transdisciplinary theory of knowing.

---

[9] But George F.E. Ellis (2004: 622) also accepts that there are four different worlds, though his fourth is mathematical abstract reality and not linguistic intersubjectivity.

[10] A conundrum described in 1944 by Schrödinger (1967/2006: 163) in his *What is Life?* which was first printed in 1944.

In this case, we need to move from talking about a science of consciousness to calling what we deal with a *Wissenschaft* of consciousness, as this German concept includes natural as well as social sciences and humanities in a single concept. Thus, my perspective on the explanatory gap will conclusively be: *What would the consequences be of looking to the results of the behavioural and brain sciences for an understanding of mind and consciousness from an integrated Wissenschaftliges perspective?* Can we view qualia and meaning as coming from the culturally embodied distributed linguistic mind and understand it in a grander scientific, evolutionary and ecological view?

This is where I think only a Peircean biosemiotics can answer 'yes'. A realistic and pragmatic conceptualization of sign processes in all their variations could be seen as the unitary phenomenon that connects all living natural systems with human cultures and furthermore distinguishes them from inanimate nature. It could serve as the framework that provides the human, social, engineering, business, life and natural sciences with a common theoretical basis for empirical research. Peirce's realism is, among other things, based on his belief in secondness, or the unexplainable random fact. There are immediate differences and resistances between phenomena or different things (haecceities). Peirce adopts Duns Scotus' term *haecceity* to designate the arbitrary here-and-now-ness of existence, a person's or object's 'this-ness', that is, the brutal facts based on relations. *Peirce* identified this haecceity as 'pure secondness'. Peirce's view of haecceities as being unexplainable as singular events is close to the modern understanding of quantum events. It is interesting that quantum physics has realized that it cannot explain the singular event either; it can only make a probability model from thousands of them, describing the thirdness of the phenomena. There is an undetermined spontaneity of the single event that is not explainable in itself from a scientific point of view (Stapp 2007).

So how does the mind collect all these haecceities to one quale experience? One way of formulating this question is in the form of *the binding problem*, widely discussed in brain and consciousness studies (Chalmers 1996). It asks how the unity of conscious perception is created in the neurological processes that make up the central nervous system. Thus, two unsolved aspects of the phenomenon of conscious awareness are the mechanisms and laws that produce the *unity of conscious perception*. Physiologically we can ask, how do we create a unified percept from the input from many separate neuronal systems? But phenomenologically we must also ask how does the unity of conscious self appear, as it seems to be the background for our judgement of singular experiences, not produced as the sum of them.

Some researchers see this as only a neurophysiological question, but in fact it is a question that demands types of answers that extend beyond the realm of physical science alone, since it concerns meaningful subjective and intersubjective experiences that point beyond physical explanations. Searle defends the view 'that consciousness consists of unified, qualitative subjectivity, caused by brain processes and realized in the brain' (2007: 102). In that case, how do we integrate all those different perceptual inputs from inside and outside the body into a lifeworld or a conscious horizon, with ourselves in the centre? The question from science should be, *How can we systematically work with any reality beyond the physical?* It is a foundational philosophical problem prior to any empirical science.

Peirce's whole semiotic philosophy of science is an answer to this question, as he believed that nominalism and derivatives of it like sensationalism, phenomenalism, individualism and materialism all based solely on secondness were a great threat to the advancement of science and civilization. His semiotics was a nuanced realism in which he distinguished reality from existence in a way that allowed him to admit general and abstract entities, which he conceptualized as belonging to thirdness, as reals. He did that without attributing to them direct physically efficient causal powers, but these non-existent reals could influence the course of events by means of final causation.

It is crucial to Peirce's semiotic realism that thirds are as real as firsts and seconds. They are connected through the semiosis that carries scientific knowing. Thus, the argument does not need to lead to the introduction of elements or worlds outside nature in the way in which Cartesian dualism, for instance, can be interpreted to do in its postulation of a *res cogitans* (i.e. a thinking substance). Signs are relations. The ontological idea is not placing consciousness and the world of thought outside nature in a special mental world. The idea, rather, is to expand our ontological views of living nature to a biosemiotic-based interdependent thinking of lived sense making (Cowley et al. 2010).

Husserl's work and Gadamer's hermeneutical philosophy (Gadamer 1989) are attempts to give another more comprehensive model for reality, including the sciences as well as a theory of understanding, communication and history of culture. Gadamer's theory of interpretation and understanding goes through pre-understanding and the process of the hermeneutical circle in order to integrate parts of interpretation, as well as the subjects' and the objects'[11] horizons. His view is that truth does not spring automatically from using one type of method and naming it 'scientific' or 'mathematical-logical' or 'empirical' or a combination thereof. One has to reflect on the horizon from which one produces knowledge. This is done in order to create understanding in the form of fusing knowledge and experiential horizons (Heelan 1983, 1987) for all living beings with conscious awareness. Thus, consciousness in the form of awareness and the ability to have sense experiences need to be conceptualized within an understanding of a natural reality bigger than physics, unless one wants to deny that animals have sense experience and deny that our own animal body is a prerequisite for self-consciousness. We will therefore assume that consciousness, matter and signs are coexisting in, or comprise, nature as well as culture.

To go one step further, we might add the work of David Chalmers. Chalmers (1995: 201–202, 1996) is well-known for defining what he calls *the easy and the hard problems of consciousness*. The easy problem has to do with the inner workings of consciousness, such as the ability to discriminate, categorize and react to environmental stimuli; to be able to report mental states by accessing internal states; and

[11] Which can be another subject's mind, an artefact, a piece of art or a text.

to focus attention, deliberately control behaviour and distinguish between mental states. *The hard problem*, which is the one we are speaking about here, has to do with solving the problem of how sense experiences and their different qualia—such as pleasure and pain, sweet and sour, colours, and mental images—emerge from physical brain and body matter. That is the problem we are dealing with here in a naturalistic and therefore also evolutionary framework. Thus, our question now can, align with Chalmer's, be stated as: How can the ability to experience emerge from, what science presumes to be a material world?

This very question is asked by Colin McGinn (2000) in his famous book on consciousness: *The Mysterious Flame: Conscious Minds in a Material World.* McGinn is sceptical towards our ability to explain the phenomenon of consciousness, at least with our present vocabulary. How it is possible in a natural world, which we so far have defined as 'material', to 'feel like someone' in the way it is framed in Nagel's famous article, 'What is it Like to be a Bat?' (Nagel 1974), or to experience the sight qualities of, say, red or blue? The problem of explaining and modelling in a scientific way the ability to experience qualitative differences in sense experiences is formulated as the question of qualia (Jackson 1982).[12] How do nervous systems produce sense experiences? But opposing the importance of qualia are functionalistic philosophers. They argue that in understanding the function of a system, it is not its materiality or its experiential quality that matters. There is no reason to give causal powers to experience. This often leads to the assumption that computers have minds (Harman 1990). But it is important to note that this functionalist view of mind is then not the experiential mind I speak about herein.

Another handle on the problem of the limitations of computers for our theories of experiential consciousness is Roger Penrose's work (1989, 1994, 1997) in which he shows that even in mathematics, human minds are capable of non-computable or non-algorithmic processes that go beyond the present capabilities of computers. Based on this observation, my position in this chapter will be that only *aspects* of mind processes can be simulated by computers or algorithms, since most researchers presently agree that computers—as we presently know them—cannot compute awareness, qualia and meaning.

Based on Peircean biosemiotics (Brier 2008b), I side with Searle (1980) and Penrose (1994, 1997) against the view of hard AI that symbol manipulation in itself is the core of intentionality. I fail to see how automatic symbol manipulation in computers has anything to do with the production of intentionality and qualia. Jackendoff (1987) has very precisely framed the problem in the form of the concept of *the mind-mind problem.* I agree with him, when he formulates the gap problem as the relationship between *the computational* and the *phenomenological mind*! As the philosopher Nagel (1986: 259) also points out:

[12] The question of what 'it' is denied by Bennett and Hacker (2007) as a wrong type of question in their Wittgensteinian-inspired pragmatic linguistic theory of mind. But I side with Searle (2007) on this problem that we cannot define the ontological dimension of this problem away.

> If we try to understand experience from an objective viewpoint that is distinct from that of the subject of the experience, then even if we continue to credit its perspectival nature, we will not be able to grasp its most specific qualities unless we can imagine them subjectively.... Since this is so, no objective conception of the mental world can include it all.

Thus, if we do not believe that the brain is just a computer and that informational computation is what creates consciousness in the human body, then it must be something else. Searle (1980, 1989, 1997 and 2007) argues that it has something to do with our biology. Consciousness and intentionality must be biological products. The secret of consciousness is also the secret of life, one could say.

The tragedy is that biology so far has only been able to give functional definitions of life. Searle (1980) believes that the brain's production of intentionality is like chlorophyll's production of carbohydrates through photosynthesis. Boden (1990) in a critique points out rightly that experience is a qualitatively different product than carbohydrates. We can describe and measure carbohydrates scientifically, but this is not the case with the quality of experience. As far as we know today, only living bodies can produce the awareness necessary for having experience. To live is to experience! *But the living, experiencing flesh is still a mystery to the physico-chemical sciences as well as to the life sciences in their present non-semiotic form,* as Merleau-Ponty (1962, 1963, 2003) thoroughly argued from the philosophy of embodied phenomenology. As experience is a prerequisite for science, science may not be able to explain it.

Still we must conclude that consciousness has an inescapable biological component. Consciousness is (also) a feature of the brain. But as Favareau (2010: vi) points out, if this is the case, then what we considered the *one* central problem is rather a triplet: 'What is the relation between mental experience, biological organization, and the law-like processes of inanimate matter?' This is at least how biosemiotics, which analyses the processes of life from a semiotic viewpoint in addition to the physico-chemical view, sees it. Scientific biology in the form of physics, chemistry and physiology is unable to describe important aspects of the processes of living systems. The suggestion here is that we supplement our physico-chemical knowledge with a semiotic view.

As a mode of inquiry into the psychological activities of the human brain, semiotics has always sought to investigate and develop models of how the mind extracts meaning from physical forms through interaction, as well as the way in which such forms can stand for something else. Biosemiotics, including human and cultural semiotics, can be defined as the study of how meanings are created in living systems between signs and the information they encode in the perceptual and cognitive apparatus (Hoffmeyer 2010).

The realization that the embodied cognitive apparatus in humans is developed in evolution has given rise to biosemiotics as the field investigating how different species transform sense experience into perceptual schemas through species-specific semiosis. As a consequence, it has become evermore obvious that sign study cannot avoid biological considerations. As one of the contributors to biosemiotics, I find that, especially in its stringent Peircean formulation (Brier 2008b) with its triadic phaneroscopic categories, the field represents a promising way out of dualism, monistic eliminative materialism and other sorts of physicalism and informationalism, as well as radical forms of constructivism.

Favareau's way of formulating the gap problem is, interestingly, a bit broader than asking how brains produce minds, as it broadens the field from specifically *human* physiology to evolutionary and ecological semiotics and the (comparative) psychology of all living systems having the ability to experience and communicate aspects of their environment.

Such a paradigm was originally formulated as *Umweltlehre* by Jacob von Uexküll (1982, 1934) and later, inspired by him, as *ethology* by Konrad Lorenz (1970–1971) and Niko Tinbergen (1973) (see Brier 1999, 2000a, b, 2001). Connected to these questions is also the problem of how living systems perceive sense experiences and communicate in the frame of *meaning* and why and how they seem to have intentionality. Furthermore, it is a scientific enigma how signs and the grammatically ordered symbols of language can evoke feelings, qualia and images from the body. How can individual emotional purpose such as a love through a poem enter the nervous system of another human and create semiotic interpretations in the form of feelings? What is the physical causality? How can free will have causal influence on, for instance, the movement of our bodies, when physics believes that causality is primarily based in initial conditions and universal mathematical laws (Penrose 1997)?

In the world of matter, energy and objective information—as the natural scientific paradigms presently see the basic ontology of nature—no meaning as such is supposed to be found. But then how can the life sciences, of which biology is the most prominent, avoid working with the reality of emotions, intentionality and meaning? This is a problem Konrad Lorenz struggled with over 30 years (Brier 2008a; Lorenz 1970–1971) and could not solve within the natural scientific paradigm. As Hinde (1970) argues, biology is not able to encompass the psychological 'level of existence' or, to be more Wittgensteinian, 'description'.

The point is, again, that if biology is to encompass the felt experience of animals, its foundation has to differ from that of physics and chemistry. Current biology is therefore not enough. As Hoffmeyer (2008) writes, 'scientific description in gene-fixed reductionistic biology, exclusively deals with phenomena that may be described in the language of third-person phenomena, and thus … excludes this science from arriving at a theoretical understanding of the human biosystem as a first-person being' (Hoffmeyer 2008: 333–334).

Thus, we need a Wissenschaft, which includes a theory of signification and meaning, which is exactly what biosemiotics attempts to do. Emmeche (1998, 2004: 118) writes, 'The semiotic approach means that cells and organisms are not primarily seen as complex assembles of molecules, as far as these molecules – rightly described by chemistry and molecular biology – are sign vehicles for informational and interpretation processes, briefly, sign processes or *semiosis*'.

But this view is not a possibility for energetic, molecular or even informationally founded biology. Kull (2009) discusses what this kind of Wissenschaft biosemiotics could and should be and suggests a qualitative modelling science he calls

Sigma-science after Vihalemm (2007). In the humanities there are dominant paradigms designed to analyse human qualitative and intentional consciousness, culture and language. These include phenomenology, hermeneutics, linguistics, rhetoric, discourse and cultural analyses and semiology. The humanities deal with the world of meaning as produced by humans in society through language, art and social interactive practice. But if you ask contemporary researchers in the humanities what the *ontology* of meaning is, they usually answer, 'it is just a social and cultural construction', as if that was not real and not also biologically based! But on the other hand, most do agree that the social world, held together by communication, power and institutions, is the dominant reality we live in.

The reality of social phenomena is surely something other than physical reality, but the social world of meaning and values is real, and interactions in it can be described systematically, as Max Weber showed in his research method of ideal types, exemplified most famously in *The Protestant Ethic and the Spirit of Capitalism* (Weber 1920). Social constructivists can only give answers within the historical time frame of hundreds and up to thousands of years. Biological evolution is not part of their paradigmatic framework, since in the biological evolutionary viewpoint, meaning has a history of millions of years in the development of embodied living systems. This is the story biosemiotics attempts to tell, since the sciences are not conceptually equipped to do it (Emmeche 2004). Thus, we should encompass the social as well as the individual experiential reality and their history in nature. But how are we going to connect them? Where to put the brain in experience?

Chalmers' *The Conscious Mind: In Search of a Fundamental Theory* (1996) collects nearly all the material in science and philosophy we had on the subject at that time, except Peirce's semiotic philosophy. His suggestion of a solution is a type of double-aspect theory, where the experiential is the inside of information in the brain. But viewing objectively defined information and experiential meaning as two aspects of 'the same' does not solve the deep troublesome problem lying in the obvious observation, that I am not my brain and that emotions like jealousy can make a person murder the one he/she loves. The murderer is not his/her brain but him/her. One should not commit the mereological fallacy to contribute to the part that which only makes sense when attributed to the whole. It is not the brain that experiences; it is embodied human persons in a culture with a language (Bennet and Hacker 2007; Cowley et al. 2010). But the person seems to be a biological, psychological as well as a social and linguistic product—a wholeness not reducible to the brain.

My brain is part of me. So who or what is phenomenological me? Am I the nonmaterial linguistically informed product of my brain? Is it then possible that conscious awareness and experience are something we are missing in our scientific explanations of living systems such as perception, cognition and communication as we know them? For instance, dark matter and energy were missing in early cosmological descriptions of the universe's evolution. They were concepts later introduced because we were lacking something to harmonize what we observed astronomically with the physical laws we had developed. What we saw and measured did not fit with the laws we believed were universal. After introducing the new aspects of physical reality christened 'dark energy' and 'dark matter',[13] what we before had considered being the whole of material reality, now showed to be 3–4% of the whole (Bertone 2010). Thus, a revolutionary new cosmology was created by introducing new ontological elements.

The parallel I am arguing for is that it might turn out that what we now consider the material reality of biological systems is just a small percentage of the whole of living system because we missed something vital for the functioning of living systems! Namely, signs and sign functions.

In the context of the social sciences, we know that we are consciously experiencing a world through processes that are unconscious for us. We do not know what we do when we see, feel, intend and act accordingly. But most cultures and societies hold their citizens responsible for the actions they take from their interpretation of sense experience. Materialistically based evolutionary and ecological theory forces the question that if culture comes out of nature, *how do experiential subjects emerge from an objective world*? Here, I am not thinking about research, which accepts the experiential aspect of life in the living and therefore describes how it has developed through evolution like Donald (1991, 2001). He describes the evolution of consciousness and its forms from a biopsychological platform. Sonesson (2009) bases his work on phenomenology, Piaget and aspects of Peircean semiotics. The work of Zlatev (2009a, b) uses aspects of Peircean semiotic terminology, but not his ontological foundation, in an evolutionary framework. Nor am I thinking of Deacon (1997) or his later articles (2007, 2008), which stray away from a Peircean foundation. None of these works attempt to solve the hard problem.

Thus, in my view, a pure materialistic and scientistic theory cannot answer the question I am asking, because it cannot describe the feeling of being aware or the phenomena of experiencing qualia, will and intentionality. Such theories can only describe physiological and behavioural consequences. Thus, the philosophy of ontological reflection going beyond physics and scientific knowledge in general seems to be required because the unity of conscious experience—in spite of the numerous neurophysiological systems—that underpins it does not really have a physical scientific meaning. It can have a social meaning, since we talk about it, based on our interpretation of others' behaviour in the belief that they have inner mental states with causal powers over their behaviour.

---

[13] Wikipedia writes, 'Dark matter came to the attention of astrophysicists due to discrepancies between the mass of large astronomical objects determined from their gravitational effects, and mass calculated from the "luminous matter" they contain; such as stars, gas and dust. It was first postulated by Jan Oort in 1932 to account for the orbital velocities of stars in the Milky Way and Fritz Zwicky in 1933 to account for evidence of "missing mass" in the orbital velocities of galaxies in clusters.... According to consensus among cosmologists, dark matter is believed to be composed primarily of a new, not yet characterized, type of subatomic particle'.

5 The Idea of Cybersemiotics

The transdisciplinary frame for information, cognition and communication science called Cybersemiotics (Brier 2008a, b, c, d; 2010a, b) is an attempt to show, using Peircean Biosemiotics, how to combine knowledge produced in the natural, life and social sciences and the humanities, as each describes an aspect of consciousness.

But first we have to deal with the incompatibility between the two transdisciplinary paradigms attempting to create a theory of consciousness. With an expression from Kuhn's (1970) paradigm theory, the two paradigmatic theories on thinking and communication suffer from incommensurability. The first paradigm is cybernetic information theory and cognitive science, which is actually a technologically oriented paradigm that has a background in a scientific, materialistic and mathematics or logic, as a more abstract and general part of nature, metaphysics.

Many members of this world view have the deep problem that they usually do not consider their views to be founded on metaphysical postulates at all, but only common-sense reality. Therefore, they do not want to be drawn into 'metaphysical speculation' or philosophy. Many people have the misconception that modern physics deals with the world as we know it in our daily life. Nothing can be further from the truth. Quantum field theory and the special and general theories of relativity, super string theory and black holes, dark matter and the like are totally outside of our common experience. If you ask people to interpret everyday physical processes, most of them give explanations close to Aristotelian physics. Thus, the majority of human beings have not even moved into a Newtonian paradigm, let alone Einstein's, Bohr's, Feynman's or Hawking's. Modern physics has no direct bearing on our awareness, meaning or common sense. Still to this physicalistic world view, many researchers of the World War II era inspired by cybernetics attempted to add information and computation to explain the emergence of conscious awareness.

Cyberneticists built an expanded new world view by adding the concept of information to energy, space, time and force and imagining that all natural processes including consciousness and emotion could be fruitfully described and understood in a grand theory of natural computation (Dodig-Crnkovic 2010; Dodig-Crnkovic and Müller 2011). This pan-computational/pan-informational project is an interesting scientific endeavour as such, but I fail to see how it will ever be able to solve the experiential and qualitative aspects of conscious feeling and experience as it lacks the experiential aspect of reality. As mentioned above, Chalmers (1995) attempts to solve this problem with a double-aspect ontology in such a way that he can keep the mathematical foundation of information theory and still get the experiential aspect at the same time. But I do not think he has any good arguments for how this should work, and he misses the meaning process dynamics, which is inherent in Peirce's semiotics. Thus, like Peirce, I want to expand our wissenschaftliges concept of reality. I do talk about not only that aspect of it that can be described by physics (often reified as the physical world, turning an epistemological concept into an ontological one and reifying it) but also what can be described by the life sciences, communication sciences and psychology. Thus, reality includes at least a material environment, a living body, a lifeworld of experience and a social communicative world all necessary to produce experiential knowing. Science is based on intersubjectively well-functioning communication in a field of meaning, coordinating knowledge and practice in the real world. I am therefore asking what kind of transdisciplinary ontology and epistemology we need in order to construct the theory of a evolution of meaning and conscious lived experience that is coherent with the natural, life and social sciences.

6 Phenomenology and the Lifeworld

What is then the rational basis of my insistence that the physical aspect of the world is not the paramount foundation of reality? It is basically acceptance of the main point of the whole phenomenological movement, the history of which Spiegelberg (1965) has made a highly recognized exposition of, including Peirce. We will not go into that grand history here, but many researchers take their departure from the work of the father of modern European phenomenology, Husserl (1970, 1997, 1999), and the father of the American variant called phaneroscopy, namely, C. S. Peirce (1931–1934), who is also the father of the pragmatic, triadic transdisciplinary semiotics, upon which much of biosemiotics is being built.

Husserlian phenomenology claims that the so-called *lifeworld* is a unit of reality before science splits the world into subjects and objects or interior and exterior. The dualism of subject and object is really not essentially relevant for the phenomenological paradigms, which, like hermeneutics, claim to deal with the cognitive processes that are prerequisites for the invention of science in our cultures. This is the area where the philosophical grounding for the natural, life and social sciences becomes relevant for the analysis.

Thus, in phenomenology the percept is a primary reality, *before* scientists try to explain the origin of sense perception and its information and meaning from a combination of interior physiological processes and exterior physical information disturbing the sense organs, or biology tries to explain the function of the sense organs and the nervous system from evolutionary and eco-physiological theories.

Phenomenologically, we must accept that biology cannot explain why and how we see and hear and smell the world (Edelmann 2000: 222). It can only model the physiological way the organs work, *but it has nothing to say about how they produce experience.* This is a choking fact for a neuro- and behavioural scientist studying the philosophy of science. But it is only a problem for those scientists who take philosophy of science seriously—and they are fairly few. Many empirical researchers do not see the problem and believe that more empirical research will solve any problem. And science concurs! I am arguing for a different, more philosophical, reflective view here.

In phenomenology, the knower, the known and knowing are viewed as one living whole in *the lifeworld*. The knowing consciousness contains the known objects (Drummon 2003: 65). Thus, phenomenology considers the lifeworld experiential first-person awareness to be producing knowledge more foundational than that produced by the natural and social sciences.

The phenomenologist argument that knowledge starts in the non-dual lifeworld is one of the clearest arguments for the necessity of philosophy when determining how to evaluate and use the knowledge from the natural as well as the social sciences. It is especially Husserlian phenomenology upon which Merleau-Ponty draws, which figures the lifeworld as more fundamental than natural as well as social scientific knowledge and therefore claims that there is no scientific explanation for consciousness as it is the primary given. Consciousness in itself is not viewed as a product of the brain or of culture and language in Husserl (1997, 1999); only the content of consciousness and way of that content are expressed. But, on the other hand, Merleau-Ponty does not privilege the body over the mind—the body *is* the mind and vice versa, in that they are one whole synthesis. The phenomenological 'I' is a universal, natural, human sense-perceiving 'I' that brings things into existence for oneself through one's intentionality; this includes 'the other'. Merleau-Ponty writes (1962: xi):

> Perception is not a science of the world, it is not even an act, a deliberate taking up of a position; it is the background from which all acts stand out, and is presupposed by them.

It is through being in the world and experiencing the world that we have consciousness, but that world is not ontologically the same as the 'physical world' as it also includes the subjective and intersubjective world of living and communicating with other living, embodied conscious linguistic beings. Thus, the physicalistic and/or computational brain science, on the one hand, and phenomenology, on the other, operate in two different worlds that each sees the other as only describing a small part of reality that is not so important for the big picture. Both claim to be the most fundamental description of reality. They each have their map of the world on which the other almost does not exist or at least is not represented in a way they will themselves accept.

One of the deepest conundrums for the sciences is the undeniable fact of our own ability to undergo qualitatively varied sense experiences, such as internal drives and urges, as well as states of feelings and will that alter body processes. These lead to the ability to make our body carry out goal-directed movements which, in turn, fulfil goals, some of which can be bodily and psychological desires. Furthermore, this poses a very general problem for the sciences because this experiential aspect of reality is not just a matter of the special category of human consciousness—*all living beings have these abilities to varying degrees.* This is one of the reasons why biosemiotics is a necessary supplement to ordinary scientific biology as well as cultural semiotics.

One can try to avoid the problem, of course, by claiming that our experience of making conscious decisions on the basis of analysis of our qualitative experiences is an illusion or folk psychology (Churchland 2004a, b, and Dennett 1991, 2007) and that consciousness has no causal effect in the world as we know it. But I refuse to take eliminative materialism seriously, as I consider it to be a self-defeating paradigm, since by its elimination, it denies the fact that science has sense experience and the ability to think and create and communicate meaningful theories, plus the ability to make purposeful experiments as a prerequisite. As Gadamer (1989) shows in his hermeneutics, science also has meaning and interpretation, based on a cultural historical horizon as a prerequisite, because it is dependent on the ability to create linguistic concepts and interpret them through one of many natural languages produced by cultures and their world views. That is very much the insight that Kuhn's paradigm theory (Kuhn 1970) builds on. Put simply, science is a cultural product.

7 Evolution and Teleonomy

I argue here that knowledge needs an experiential component added to the functional because sense experiences and awareness are usually not part of the biological story of the development of life and knowing. Thus, structural couplings in autopoiesis theory, affordances à la Gibson and Uexkull's tones are all important parts of a pragmatic evolutionary understanding of cognition, but it is not enough to make a theory of the emergence of the experiential mind in evolution.

Surviving entities in the course of evolution are those wherein the heritable structures of their DNA molecules contributed to solving survival problems. But how exactly this should happen as a mechanical process, we do not know. But the general idea is that starting from random noise, the autopoietic functions of the cell make it possible to selectively filtrate for useful functionality. As such, researchers often say that this process gradually builds knowledge of the world into the DNA sequence. But how, and what kind of knowledge?

Barbieri, in the further development of his code semiotics (2001), sees a parallel between the problem of the emergence of life from the physico-chemical world and the emergence of experience from the self-organization of living systems. To Barbieri, the production of new codes can solve both. Life is built out of new artificial molecular assembles by the DNA, RNA and ribosomal apparatus that combine amino acids in new, inventive ways. The solution to how the capacity to experience emerges from the brain of mammals is the production of new neural codes, which generate the brain's capacity for sense experience, emotions and imaginary abilities. Barbieri (2011) in his most interesting grand theory of code semiotics writes:

> The idea of a deep parallel between life and mind leads in this way to a parallel between proteins and feelings, and in particular to a parallel between the processes that generate them. We already know that the assembly of proteins does not take place spontaneously because no spontaneous process can produce an unlimited number of identical sequences of amino acids. The Code model of mind is the idea that the same is true in the case of feelings, i.e., that feelings are not the spontaneous result of lower level brain processes. They can be generated only by a neural apparatus that assembles them from components according to the rules of a code. According to the Code model, in short, feelings are brain-artifacts that are manufactured by a codemaker according to the rules of the neural code.
>
> In the case of feelings, the codemaker is the intermediate brain of an animal, the system that receives information from the sense organs and delivers orders to the motor organs. The

> sense organs provide all the information that an animal is ever going to have about the world, and represent therefore in an animal what the genotype is in a cell. In a similar way, the motor organs allow a body to act in the world, and have in an animal the role that the phenotype has in a cell. Finally, the intermediate brain is a processing and a manufacturing system, an apparatus that is in an animal what the ribotype is in a cell.
>
> The parallel between life and mind, in conclusion, involves three distinct parallels: one between proteins and feelings, one between genetic code and neural code, and one between cell and animal codemaking systems. The categories that we find in the cell, in other words, are also found in animals, because at both levels we have information, code and codemaker. The details are different, and yet there is the same logic at work, the same strategy of bringing absolute novelties into existence by organic coding.
>
> (Barbieri (2011: 380))

Thus, one can say that Barbieri offers a solution to Searle's problem of how biological processes allow the brain to produce qualitative consciousness. A later section in the article shows that Barbieri thinks of sense experience as modelling. It certainly is, but seen from my phenomenologically informed view, the problem is that it is a qualitatively unique kind of modelling. Barbieri (2011) writes:

> The results of brain processing are what we normally call feelings, sensations, emotions, perceptions, mental images and so on, but it is useful to have also a more general term that applies to all of them. Here we follow the convention that all products of brain processing can be referred to as brain *models*. The intermediate brain, in other words, uses the signals from the sense organs to generate distinct *models* of the world. A visual image, for example is a model of the information delivered by the retina, and a feeling of hunger is a model obtained by processing the signals sent by the sense detectors of the digestive apparatus.
>
> (Barbieri (2011: 388))

Barbieri uses the modelling idea from Lotman developed further by Sebeok and Danesi (2000). It is a good *functionalist approach* that catches some important practical aspects of reality. But when I make a model of the route I have to follow to get home from a new place in town, I actually visualize the streets. I see them and thereby experience them. I make the images for my 'inner eye' and draw on my lifetime's experiential memory of this town, in which I have lived my whole life. It is not just a logical map that directs my way home. It is embodied and experiential. I claim that it is qualitatively different from what such a map is to a robot, not least because I have the free will to choose not to follow it and to instead change the route. I am not in any way automatically determined to follow it. Clayton (2004: 601) also argues that the emergence into the quality of experience is different from other emergence theories. I agree though with Barbieri when he writes:

> The evolution from single cells to animals was a true macroevolution because it created absolute novelties such as feeling and instincts (the first modelling system). Later on, another major transition allowed some animals to evolve a second modeling system that gave them the ability to *interpret* the world. That macroevolution gave origin to a new type of semiosis that can be referred to as *interpretive* semiosis, or, with equivalent names, as *abductive* or *Peircean* semiosis.
>
> (Barbieri (2011: 391))

As many before him, Barbieri wants to use Peirce's triadic semiotic theory, but refuses his triadic metaphysics of firstness, secondness and thirdness—his synechism, hylozoism and tychism.[14] But this is the foundation of Peirce's general paradigm. Denying the ontological, epistemological and methodological foundation, he then tries to solve the problem that Peirce's pragmaticist triadism attempts to solve in the framework of what current scientific thinking is on the mammalian brain. From this foundation, he wants to explain the brain's production of mind through code-sign processes, introducing the triadic sign process including interpretation on this level as a result of the emergence of experience now explained from the code-semiotic paradigm. A semiotic system is here defined as a triadic set of processes and objects linked by a code. But this is not triadic in the Peircean sense, since the metaphysics does not entail his three categories as they emerge as indestructibles in the phaneroscopic analysis. Peirce combines phenomenology, mathematics and empirical data in his pragmaticism. Code semiotics is not able to integrate a phenomenological view in its paradigmatic foundation—neither ontological nor epistemological. To establish the genuine interpretative sign function, it has to be Peircean 'all the way down' to power the basic categories, which makes the sign triad function as a meaning-generating process (Ketner 2009). I challenge Barbieri to produce an alternative framework than can compete with Peirce's instead of introducing Peircean semiotics at the level of the brain on an implicit materialistic ontology wherein molecules assume agency and become code makers. The central question unexplained by Barbieri is how the macromolecules resume agency and make codes suddenly in an unspecified materialistic ontology.

Peircean biosemiotics suggest that what are transferred in and between living systems are signs, not objective information. Signs have to be interpreted, and it has to happen on three levels. On the most basic level, we have the basic coordination between the bodies as a dance of black boxes to allow for meaningful exchange. This goes on at the next level of instinctual sign plays of drive and emotionally based communication about meaningful things in life like mating, hunting, dominating, food and territory seeking. Barbieri (2011) distinguishes between a cybernetic and instinctive aspect of the brain function and argues that the emotions emerge from the instinctual brain. I agree with this, but cannot see that he solves the problem Konrad Lorenz (who saw the same two aspects) could

[14] Peirce writes that tychism is '… absolute chance – pure tychism…' (CP 6.322, c. 1909). So tychism is connected to firstness as real objective chance in the universe. But it has to be integrated with the secondness of resistance, facts and individuality to create thirdness to mediate connections between the two in synechism. This is connected to his pragmatism: 'It is that synthesis of tychism and of pragmatism for which I long ago proposed the name, Synechism' (CP 4.584, 1906). Synechism is '…that tendency of philosophical thought which insists upon the idea of continuity as of prime importance in philosophy and, in particular, upon the necessity of hypotheses involving true continuity' (CP 6.169, 1902). This deep continuity between everything, including mind and matter as well as the three categories, is synechism:'…I chiefly insist upon continuity, or Thirdness,…and that Firstness, or chance, and Secondness, or Brute reaction, are other elements, without the independence of which Thirdness would not have anything upon which to operate' (CP 6.202, 1898).

not in his creation of the ethological paradigm (Brier 2008a, b, c, d). Based on these two aspects or levels, a new third level of meaning is created that the socio-communicative system can modulate to conscious linguistic meaning.

Today, it is widely recognized that what we call a human being is a conscious social being, living in language. Terrance Deacon, in his book *The Symbolic Species* (1997), sees our language-processing capacity as a major selective force for the human brain in the early stages of human evolution. We speak language, but we are also spoken by language. To a great extent, language carries our cultures as well as our theories of the world and of ourselves. As individuals, we are programmed with language—to learn a language is to learn a culture. As such, prelinguistic children are only potentially human beings, as they have to be linguistically programmed in order to become the linguistic animal cyborgs we call human. However, getting behind language as such is difficult without creating a broader platform beyond linguistics. Peircean semiotics and its modern evolution to a biosemiotics is such an attempt for a doctrine of cognition and communication and therefore the creation of knowledge in the widest sense.

I do not see quantum physics, general relativity theory or non-equilibrium thermodynamics as being of any particular help concerning this problem, although they may be helpful in explaining the physical aspect of consciousness (Penrose 1994, 1997). This is my argument why a bottom-up, empirically based physicalism or pan-computationalism is inadequate to solve the gap problem. Here is where Peirce's theory of the tendency to take habits[15]—what he calls thirdness—brings the physical and the mental together in that he sees the tendency to take habits in both nature and mind. Here is one of those deep Peircean quotations arguing with the mechanical view of natural law:

> The law of habit exhibits a striking contrast to all physical laws in the character of its commands. A physical law is absolute. What it requires is an exact relation. Thus, a physical force introduces into a motion a component motion to be combined with the rest by the parallelogram of forces; but the component motion must actually take place exactly as required by the law of force. On the other hand, no exact conformity is required by the mental law. Nay, exact conformity would be in downright conflict with the law; since it would instantly crystallize thought and prevent all further formation of habit. The law of mind only makes a given feeling more likely to arise. It thus resembles the 'non-conservative' forces of physics, such as viscosity and the like, which are due to statistical uniformities in the chance encounters of trillions of molecules.
>
> (Peirce 1892)

This is why thirdness is so important in Peirce's categories and at the same time it is critical to remember that thirdness includes secondness and firstness.

The Cybersemiotic transdisciplinary theory accepts Peirce's view and sees scientific explanations as going from our present state of sociolinguistically based conscious semiosis in self-organized autopoietic systems towards a better understanding of the prerequisites of language and the self-conscious being. Science gives a good economically and practically useful understanding of certain

[15] As Peirce calls it.

processes, often in a way that allows prediction with a wanted precision within certain circumstances. However, it does not give universal explanations of the construction of reality, energy, information, life, meaning, mind and consciousness. Natural science deals only with the outer material aspect of the world and our body, not with experiential consciousness, qualia, meaning and human understanding in its embodiment (Edelmann 2000: 220–222).

Nicolescu (2002: 65–66)—who is also a quantum physicist—promotes, like Peirce does, the theory that consciousness is a vital and active part of the wholeness of the universe. The subjective and the objective sides of nature make up the whole of reality to an integrated whole based in what Nicolescu calls trans-nature or the zone of nonresistance. As such, he is close to Peirce's evolutionary concept of hylozoism.[16] We are the systems developed in and by the universe that are most highly developed to make the universe look at itself. As the universe in its fundamental quantum level is still partly undetermined, it is in an ongoing rearranging process of building itself (even all the way back to the Big Bang) (Rugh and Zinkernagel 2009). Nicolescu explains this further when he writes: 'Nature seems more like a book in the process of being written: the book of Nature is therefore not so much to be read as experienced, as if we are participating in the writing of it' (Nicolescu 2002: 65). That also seems to be Wheeler's (1994, 1998)(Davies 2004) view, as well as Peirce's. New foundational theories of agency and the quality necessary to be an observer have appeared (Sharov 2010; Arrabales et al. 2010). That problem cannot be solved here, but seems to be related to C.S. Peirce's idea of semiosis—the ability to make signs and interpret them meaningfully—as not only being limited to humans but including all living systems with a fuzzy border to the precursor systems of life, making thinking something that goes on in an ecological systemic context, as Bateson (1973) also views it (Brier 2008c).

8 Conclusion

Let us return to the Kant quotation on nature and free will and expand on it a bit further. Kant writes about the contradiction between free will and a lawful view of nature:

[16] In philosophy 'hyle' refers to matter or stuff; the material causes underlying change in Aristotelian philosophy. It is what remains the same in spite of the changes in form. In opposition to Democritus' atomic ontology, hyle in Aristotle's ontology is a plenum or a sort of field. Aristotle's world is an uncreated eternal cosmos, but Peirce used the term in an evolutionary philosophy of a world that has an end and a beginning. Hylozoism—in this context—is the philosophical conjecture that all material things possess life, very much like Whitehead's (1978) panexperientialism. It is not a form of animism either, as the latter tends to view life as taking the form of discrete spirits. Scientific hylozoism is a protest against a mechanical view of the world as dead, but, at the same time through synechism, upholds the idea of a unity of organic and inorganic nature and derives all actions of both types of matter from natural causes.

> It is an indispensable problem of speculative philosophy to show that its illusion respecting the contradiction rests on this, that we think of man in a different sense and relation when we call him free, and when we think of him as subject to the laws of nature.... It must therefore show that not only can both of these very well co-exist, but that both must be thought of *as necessary united* in the same subject.
>
> Kant (1909: 76)

I think this is what we have done in our work *towards a Wissenschaft of consciousness* that should be able to include mental events in an absolute naturalism.

But to make such a shift, one needs to develop an ontology that can encompass the ontologies of all the four views in a transdisciplinary setting. I have suggested to take our point of departure in C.S. Peirce's pragmatistic, evolutionary semiotic process philosophy, where semiotic social interactions between embodied more or less free minds in nature are viewed as the central process of knowledge production, which is also behind the self-same 'sciences' that attempt to explain the meaning of production and consciousness. Thus, we return to a partly Aristotelian view adding evolution plus phaneroscopy and biology in the form of a biosemiotics.

References

Arrabales, R., Ledezma, A., & Sanchis, A. (2010). ConsScale: A pragmatic scale for measuring the level of consciousness in artificial agents. *Journal of Consciousness Studies, 17*(3–4), 131–164.

Barbieri, M. (2001). *The organic codes: The birth of semantic biology.* Ancona: PeQuod. (Republished in 2003 as *The organic codes. An introduction to semantic biology*. Cambridge: Cambridge University Press).

Barbieri, M. (2011). Origin and evolution of the brain. *Biosemiotics, 4*, 369.

Barrow, J. D. (2007). *New theories of everything*. Oxford: Oxford University Press.

Barrow, J. D., Davies, P. C. W., & Harper, C., Jr. (Eds.). (2004). *Science and ultimate reality. Quantum theory, cosmology, and complexity*. Cambridge: Cambridge University Press.

Bateson, G. (1973). *Steps to an ecology of mind: Collected essays in anthropology, psychiatry, evolution and epistemology*. St. Albans: Paladin.

Bennet, M., & Hacker, P. (2007). The philosophical foundation of neuroscience. In M. Bennet, D. Dennet, P. Hacker, & J. Searle (Eds.), *Neuroscience and philosophy: Brain, mind and language*. New York: Columbia University Press.

Bennet, M., Dennet, D., Hacker, P., & Searle, J. (2007). *Neuroscience and philosophy: Brain, mind and language*. New York: Columbia University Press.

Bertone, G. (2010). *Particle dark matter: Observations, models and searches*. Cambridge: Cambridge University Press.

Blackmore, S. (1999). *The meme machine*. Oxford: Oxford University Press.

Boden, M. A. (1990). Escaping from the Chinese room. In M. A. Boden (Ed.), *The philosophy of artificial intelligence*. Oxford: Oxford University Press.

Brier, S. (1999). Biosemiotics and the foundation of cybersemiotics. Reconceptualizing the insights of ethology, second order cybernetics and Peirce's semiotics in biosemiotics to create a non-Cartesian information science. *Semiotica, 127*(1/4), 169–198.

Brier, S. (2000a). Biosemiotic as a possible bridge between embodiment in cognitive semantics and the motivation concept of animal cognition in ethology'. *Cybernetics & Human Knowing, 7*(1), 57–75.

Brier, S. (2000b). Transdisciplinary frameworks of knowledge. *Systems Research and Behavioral Science, 17*(5), 433–458.

Brier, S. (2001). Cybersemiotics and Umweltslehre'. *Semiotica, 134–1*(4), 779–814.
Brier, S. (2008a). *Cybersemiotics: Why information is not enough*. Toronto: University of Toronto. New edition 2010.
Brier, S. (2008b). The paradigm of Peircean biosemiotics. *Signs, 2008*, 30–81.
Brier, S. (2008c). Bateson and Peirce on the pattern that connects and the sacred. Chapter 12. In J. Hoffmeyer (Ed.), *A legacy for living systems: Gregory Bateson as a precursor for biosemiotic thinking, biosemiotics 2* (pp. 229–255). London: Springer.
Brier, S. (2008d). A Peircean panentheist scientific mysticism. *International Journal of Transpersonal Studies, 27*, 20–45.
Brier, S. (2009a). Cybersemiotic pragmaticism and constructivism. *Constructivist Foundations, 5*(1), 19–38.
Brier, S. (2010a). Cybersemiotics and the question of knowledge. Chapter 1. In G. Dodig-Crnkovic & M. Burgin (Eds.), *Information and computation*. Singapore: World Scientific Publishing Co.
Brier, S. (2010b). Cybersemiotics: An evolutionary world view going beyond entropy and information into the question of meaning'. *Entropy, 2010*, 12.
Cartwright, N. (1997). Why physics? Chapter 5. In R. Penrose (Ed.).
Chalmers, D. (1995). Facing the problem of consciousness. *Journal of Consciousness Studies, 2*(3), 200–219.
Chalmers, D. (1996). *The conscious mind: In search of a fundamental theory*. New York: Oxford University Press.
Churchland, P. (2004). Eliminative materialism and the propositional attitudes. In J. Heil (Ed.), *Philosophy of mind: A guide and anthology* (pp. 382–400). Oxford: Oxford University Press.
Clayton, P. D. (2004). Emergence: Us from it. In J. D. Barrow, P. C. W. Davies, & C. Harper Jr. (Eds.), *Science and ultimate reality. Quantum theory, cosmology, and complexity* (pp. 577–606). Cambridge: Cambridge University Press.
Colling, F. (2003). *Konstruktivisme*. Frederiksberg: Roskilde Universitetsforlag.
Cowley, S. J., Major, J. C., Steffensen, S. V., & Dinis, A. (2010). *Signifying bodies, biosemiosis, interaction and health*. Braga: The Faculty of Philosophy of Braga Portuguese Catholic University.
Davies, P. C. (2004). John Archibald Wheeler and the clash of ideas. In J. D. Barrow, P. C. W. Davies, & C. Harper Jr. (Eds.), *Science and ultimate reality. Quantum theory, cosmology, and complexity* (pp. 3–23). Cambridge: Cambridge University Press.
Dawkins, R. (1989). *The selfish gene*. Oxford: Oxford University Press.
Deacon, T. W. (1997). *The symbolic species: The co-evolution of language and the brain*. New York: Norton.
Deacon, T. W. (2007). Shannon – Boltzmann – Darwin: Redefining information (Part I). *Cognitive Semiotics, 1*, 123–148.
Deacon, T. W. (2008). Shannon – Boltzmann – Darwin: Redefining information (Part II). *Cognitive Semiotics., 2*, 169–196.
Dennett, D. C. (1991). *Consciousness explained*. Boston: Back Bay Books.
Dennett, D. C. (2007). Philosophy as naïve anthropology. In M. Bennet, D. Dennet, P. Hacker, & J. Searle (Eds.), *Neuroscience and philosophy: Brain, mind and language*. New York: Columbia University Press.
Dodig-Crnkovic, G. (2010). The cybersemiotics and info-computationalist research programmes as platforms for knowledge production in organisms and machines. *Entropy, 12*(4), 878–901.
Dodig-Crnkovic, G., & Müller, V. (2011). A dialogue concerning two world systems: Info-computational vs. mechanistic. In G. Dodig-Crnkovic & M. Burgin (Eds.), *Information and computation* (Series in information studies). Singapore: World Scientific Publishing Co.
Donald, M. (1991). *Origins of the modern mind: Three stages in the evolution of culture and cognition*. Cambridge, MA: Harvard University Press.
Donald, M. (2001). *A mind so rare: The evolution of human evolution*. New York/London: W.W. Norton & Co.
Drummon, J. J. (2003). The structure of intentionality. In D. Welton (Ed.), *The new Husserl: A critical reader* (pp. 65–92). Bloomington: Indiana University Press.

Edelmann, G. M. (2000). *A universe of consciousness: How matter becomes imagination*. New York: Basic Books.
Ellis, G. F. R. (2004). True complexity and its associated ontology. In J. D. Barrow, P. C. W. Davies, & C. Harper Jr. (Eds.), *Science and ultimate reality. Quantum theory, cosmology, and complexity* (pp. 607–636). Cambridge: Cambridge University Press.
Emmeche, C. (1998). Defining life as a semiotic phenomenon. *Cybernetics & Human Knowing*, 5(1), 33–42.
Emmeche, C. (2004). A-life, organism and body: The semiotics of emergent levels. In M Bedeau, P Husbands, T Hutton, S Kumar, & H Suzuki (Eds.), *Workshop and tutorial proceedings. Ninth international conference on the simulation and synthesis of living systems (Alife IX)* (pp. 117–124), Boston, MA.
Esposito, J. L. (1980). *Evolutionary metaphysics: The development of Peirce's theory of the categories*. Athens: Ohio University Press.
Favareau, D. (Ed.). (2010). *Essential readings in biosemiotics: Anthology and commentary*. Berlin/New York: Springer.
Feyerabend, P. (1975). *Against method*. London: NLB.
Gadamer, H.-G. (1989). *Truth and method* (2nd rev. ed., J. Weinsheimer & D. G. Marshall, Trans.). New York: Crossroad.
Harman, G. (1990). The intrinsic quality of experience. *Philosophical Perspective, 4*, 31–52.
Heelan, P. A. (1983). *Space-perception and the philosophy of science*. Berkeley: University of California Press.
Heelan, P. A. (1987). Husserl's later philosophy of natural science. *Philosophy of Science, 1987*(53), 368–390.
Hinde, R. (1970). *Animal behaviour: A synthesis of ethology and comparative behavior* (International student edition). Tokyo: McGraw-Hill.
Hoffmeyer, J. (2008). *Biosemiotics*. Scranton: University of Scranton Press.
Hoffmeyer, J. (2010). A biosemiotic approach to health. In S. J. Cowley, J. C. Major, S. V. Steffensen, & A. Dinis (Eds.), *Signifying bodies, biosemiosis, interaction and health* (pp. 21–41). Braga: The Faculty of Philosophy of Braga Portuguese Catholic University.
Hofstadter, D. (2007). *I am a strange loop*. New York: Basic books.
Husserl, E. (1970). *The crisis of European science and transcendental phenomenology* (D. Carr, Trans.). Evanston: Northwestern University Press.
Husserl, E. (1997). *Fænomonologiens ide*. København: Hans Reitzels forlag (Die Idee der Phenomenologie).
Husserl, E. (1999). *Cartesianske meditationer*. København: Hans Reitzels forlag (Cartesianische Meditationen).
Jackson, F. (1982). Epiphenomenal qualia. *Philosophy Quarterly, 32*, 127–136.
Kant, E. (1909). *Fundamental principle of the metaphysics of morals* (T. K. Abbott, Trans.). London: Forgotten Books, 1938.
Ketner, K. L. (2009). Charles Sanders Peirce: Interdisciplinary scientist. In E. Bisanz (Ed.), *Charles S. Peirce: The logic of interdisciplinarity* (pp. 35–57). Berlin: Akademie Verlag.
Kuhn, T. (1970). *The structure of scientific revolutions* (2nd enlarged ed.). Chicago: The University of Chicago Press.
Latour, B. (1993). *We have never been modern* (C. Porter, Trans.). Cambridge, MA: Harvard University Press.
Latour, B. (2004). *Politics of nature: How to bring the sciences into democracy*. New York: Harvard University Press.
Latour, B. (2007). *Reassembling the social: An introduction to actor network theory*. New York: Oxford University Press.
Levine, J. (1983). Materialism and the qualia: The explanatory gap. *Pacific Philosophy Quarterly, 64*, 1983.
Lorenz, K. (1970–1971). *Studies in animal and human behaviour I and II*. Cambridge, MA: Harvard University Press.

McGinn, C. (2000). *The mysterious flame: Conscious minds in a material world.* London: Basic Books.

Merleau-Ponty, M. (1962). *Phenomenology of perception* (C. Smith, Trans.). London: Routledge & Kegan Paul, 2002. (Originally published as *Phenomenologie de la Perception.* Paris: Callimard, 1945, English 1962).

Merleau-Ponty, M. (1963/2008). *The structure of behavior.* Pittsburgh: Duquesne University Press.

Merleau-Ponty, M. (2003). *Nature: Course notes from the Collège de France.* Evanston: North Weston University Press.

Nagel, T. (1974). What is it like to be a bat? *Philosophical Review, 83*, 435–450.

Nagel, T. (1986). *The view from nowhere.* New York: Oxford University Press.

Nicolescu, B. (2002). *Manifesto of transdisciplinarity.* Albany: State of New York University Press.

Peirce, C. S. (1931–1935). *The collected papers of Charles Sanders Peirce.* Intelex CD-ROM edition (1994), reproducing Vols. I–VI, C. Hartshorne, & P. Weiss (Eds.). Cambridge, MA: Harvard University Press, 1931–1935; Vols. VII–VIII, A.W. Burks (Ed.); same publisher, 1958. Citations give volume and paragraph number, separated by a period like (Peirce CP 5. 89).

Peirce, C. S. (1980). *New elements of mathematics.* Amsterdam: Walter De Gruyter Inc.

Penrose, R. (1989). *The Emperor's new mind: Concerning computers, minds, and the laws of physics.* Oxford: Oxford University Press.

Penrose, R. (1994). *Shadows of the mind: A search for the missing science of consciousness.* London: Oxford University Press.

Penrose, R. (1997). *The large, the small and the human mind.* Cambridge: Cambridge University Press.

Popper, K. R. (1972). *Objective knowledge: An evolutionary approach.* Oxford: Clarendon.

Schrödinger, E. (1967/2006). *What is life and mind and matter.* Cambridge: Cambridge University Press.

Searle, J. (1980). Minds, brains, and programs. *The Behavioral and Brain Sciences, 3*(3), 417–457.

Searle, J. (1989). *Minds, brains and science.* London: Penguin.

Searle, J. (1997). *The mystery of consciousness.* New York: New York Review of Books.

Searle, J. (2007). Putting consciousness back in the brain. In M. Bennet, D. Dennet, P. Hacker, & J. Searle (Eds.), *Neuroscience and philosophy: Brain, mind and language.* New York: Columbia University Press.

Sebeok, T. A., & Danesi, M. (2000). *The forms of meaning: Modeling systems theory and semiotic analysis.* Berlin: Walter de Gruyter.

Sharov, A. A. (2010). Functional information: Towards synthesis of biosemiotics and cybernetics. *Entropy, 12*(5), 1050–1070.

Sonesson, G. (2009). New considerations on the proper study of Man – And, marginally, some other animals. *Cognitive Semiotics, 2009*(4), 34–169.

Spiegelberg, H. (1965). *The phenomenological movement: A historical introduction* (2 Vols., p. 765). The Hague: Martinus Nijhoff.

Stapp, H. P. (2007). *The mindful universe.* New York: Springer.

Steffensen, S. V., & Cowley, S. (2010). Signifying bodies and health: A non-local aftermath. In S. J. Cowley, J. C. Major, S. V. Steffensen, & A. Dinis (Eds.), *Signifying bodies, biosemiosis, interaction and health* (pp. 331–355). Braga: The Faculty of Philosophy of Braga Portuguese Catholic University.

Thompson, E. (Ed.). (2003). *The problem of consciousness: New essays in the phenomenological philosophy of mind.* Alberta: University of Calgary Press.

Tinbergen, N. (1973). *The animal in its world* (pp. 136–196). London: Allan & Unwin.

Vihalemm, R. (2007). Philosophy of chemistry and the image of science. *Foundations of Science, 12*(3), 223–234.

von Uexküll, J. (1982). The theory of meaning. *Semiotica, 42*(1), 25–82.

von Uexküll, J. (1934). A stroll through the worlds of animals and men. A picture book of invisible worlds. In C. H. Schiller (Ed.) (1957), *Instinctive behavior. The development of a modern concept* (pp. 5–80). New York: International Universities Press, Inc.
Weber, M. (1920). *The protestant ethic and "The Spirit of Capitalism"* (S. Kalberg, Trans.) (2002). Los Angeles: Roxbury Publishing Company.
Wheeler, J. A. (1994). *At home in the universe*. New York: American Institute of Physics.
Wheeler, J. A. (1998). *Geons, black holes & quantum foam: A life in physics*. New York: W. W. Norton & Company.
Whitehead, A. N. (1978). *Process and reality: An essay in cosmology*. New York: The Free Press.
Wilson, E. O. (1999). *Consilience. The unity of knowledge*. New York: Vintage Books, Division of Random House, Inc.
Zlatev, J. (2009a). The semiotic hierarchy: Life, consciousness, signs and language. *Cognitive Semiotics, 2009*(4), 170–185.
Zlatev, J. (2009b). Levels of meaning, embodiment, and communication. *Cybernetics & Human Knowing, 16*(3–4), 149–174.

Part II
Mental Representation

The Emergence of Empathy in the Context of Cross-Species Mind Reading

John Sarnecki

Abstract Evolutionary accounts of the origins of mind reading and empathy have emphasized the reproductive and social value of understanding other human minds. On this view, selective pressures within human communities contributed to our capacity to imagine ourselves in the spatiotemporal and cognitive place of other individuals. I argue that these social accounts of empathy neglect the phenomenon of mind reading between humans and other species. In particular, I argue that the cognitive demands on early human hunters privileged the ability to take on the perspective of potential prey in tracking. These selective pressures on mind reading not only have serious consequences for how we view empathy but may also have had substantive consequences for how we read other human minds.

1 Evolutionary Explanations of Our Empathetic Capacities

Humans are avid mind readers. As toddlers, we begin to explain the actions of others in terms of their mental states and point of view, and we feel comfortable ascribing mental states and agency to other animals and inanimate objects, as well as unseen (and often nonphysical) forces in the spiritual world.[1] In this chapter, I will examine the evolutionary origins of these empathetic capacities, with particular regard to their emergence in Pleistocene hunter-gatherers.

The predominant view of the origins of empathy suggests that our characteristically human empathetic abilities emerged from selective pressures associated with social interactions *within* prehistoric communities. While the details of theories vary,

[1] See, for example, Boyer (2001) and Currie (2011).

J. Sarnecki (✉)
University of Toledo, Toledo, OH, USA
e-mail: john.sarnecki@utoledo.edu

L. Swan (ed.), *Origins of Mind*, Biosemiotics 8, DOI 10.1007/978-94-007-5419-5_6,

these accounts suggest that empathy arose to help individuals navigate increasingly complex social environments associated with the growing size of human communities (Dunbar 2000) or as part of a cognitive arms race within the human species (Humphrey 1976). However, I will argue that social models provide an incomplete account of the origins of human empathetic capacities and that cross-species mind reading played a significant evolutionary role in early human development. Using prehistoric endurance or persistence hunting as a case study, I argue that how we empathize has been shaped by our need to understand the perspectives of animals as a means to predict their behaviors. In sketching this alternative account of the evolution of empathy and its consequences for human cognition, I will argue that empathy across species not only provided our early human ancestors with valuable insight into the minds of animals but also played a major role in shaping how we read the minds of other humans.

A univocal definition of empathy has been notoriously difficult to pin down, but scholars typically use the term to describe shared affective or cognitive states, compassion or feeling for the distress of others, and/or imagining one's self in another's place in order to understand his or her cognitive, volitional, and affective states. In this chapter, I will focus on this latter process of perspective taking. On this account, empathy allows humans to take on or imagine another individual's experiences, thoughts, drives, as well as his or her emotional responses and commitments. In adopting this conception of empathy, I hope to avoid, at least for this forum, the current debate between advocates of simulationist (e.g., Goldman 2006) and theory-theory (e.g., Carruthers 1996) conceptions of mind reading. While this debate is not wholly independent of the issues I will discuss, it is, nevertheless, outside the immediate scope of this chapter.

Explanations of the origins of human empathy and mind reading emphasize two fundamental forces, pressures from nurturing and cooperative relationships between parents, caregivers, and children and social pressures emerging from within-group sexual or resource competition.

Nurture. Frans De Waal and Stephanie Preston have argued, for example, that increased sensitivity of mothers to their children's psychological states has positive adaptive consequences (Preston and De Waal 2002). Sarah Hrdy locates the origins of empathy not in the sensitivity to the cognitive lives of one's own offspring so much as the shared responsibility for children in cooperative breeding or "alloparenting". Humans are unique in the animal world in recruiting caretakers and sharing resources in raising offspring (Hrdy 2009). Mind reading emerges in consequence of the need for individuals to be aware of and sensitive to the needs of infants and children with whom that there has been little previous interaction. Conversely, the capacity to enlist both biological and nonbiological parental assistance requires an increased sensitivity to the cognitive states of potential caregivers for infants and children.

Competition. Others emphasize the importance of empathy in predicting the behavior of adversaries for in-group resources (e.g., food or shelters) or potential sexual competitors (Humphrey 1976). This form of empathy is often rolled into a broader conception of intelligence that is thought to emerge from a kind of cognitive arms race within human communities. This "Machiavellian intelligence" would advantage individuals within particular groups in competition for sexual partners or communal resources. On this model, the ability to empathize or understand the perspective of others would impart potentially valuable clues about the cognitive lives (and hence future behaviors) of potential rivals or prospective mates.

In each of these socially based models, empathy arises in response to the complex social demands of communal living. Within these communities, mind reading helps individuals navigate and manipulate elements of their personal relationships in the aid of socially and evolutionarily valuable life skills. Being better able to predict and explain the behavior of others affords greater opportunities for social bonding and mating, as well as entering into beneficial resource-sharing relationships.

Socially based models of empathy are committed to its emergence within human communities; accordingly, each model locates the selective pressures responsible for the origin of empathy in interactions between humans. According to this view, my ability to enter into the thoughts of others depends importantly on the fact that the others are like me. In fact, empathy studies have been dominated by the view that our ability to empathize with others is proportional to the cognitive similarities shared between mind readers and their subjects. Cultural similarities further facilitate empathy. Hence, a Western Canadian would be both more inclined to empathize with and more reliably predict the behavior of other Western Canadians, as opposed to those from Toronto, for example.[2] I argue that the origins of empathy do not fundamentally depend on shared background or morphology. Instead, I will argue that the cognitive mechanisms necessary for empathy and mind reading were forged in the context of what Mary-Catherine Harrison has called "empathy across difference" (Harrison 2011). While Harrison emphasizes sociological differences like race and class, however, I argue that differences between species may have played a particularly significant role in the evolutionary development of human empathic capacities.

Finally, two caveats. Establishing any claim in evolutionary psychology is fraught with difficulty, especially when drawing largely from behavioral evidence and comparisons with modern hunter-gatherers. Hence, the arguments developed below are meant to be more suggestive than demonstrative. The purpose of this chapter is to loosen our conviction in the view that the cradle of human empathy was limited to contexts in which mind readers shared fundamentally the same backgrounds or social group. The sources of evidence here differ little from those offered in support of social accounts of empathy. However, in many of these cases, the data are presented solely within a framework that locates the value of empathy strictly within social interactions. This chapter seeks to challenge that assumption.

Second, in developing this view, I will often be casting it in stark opposition to the current orthodoxy. To some extent, this opposition is contrived. Very little in

[2] Goldman (2006) calls this assumption the "resemblance to self" thesis. See also Trout (2009, 23–25) for a typically uncritical example of this view.

my argument turns on the idea that social models are wholly false or that our mind-reading capacities arose entirely in isolation from social forces. However, in presupposing this opposition, we may perhaps get a clearer picture of the advantages of empathy outside of contexts of social similarity and in-group selection. Ultimately, how these forces can augment each other is the province of a different paper.

2 Persistence Hunting and the Evolution of Empathy

Empathy with animals might have proven beneficial to early humans in many different ways. For example, many forms of hunting depend on a rich understanding of how animals see the world relative to their interests and desires. Hence, hunting techniques often involve the manipulation of environmental signals or conditions to trick unsuspecting animals into close range, while others involve obscuring pitfall traps or snares that may tip off animals to the hunter's presence. Each of these techniques centrally depends on manipulating the visual environment of the animals. In each case, the hunter makes suppositions about what the animal can or cannot see and which elements of its perspective are visible in particular conditions. It is significant that hunters do not merely anthropomorphize animal perspectives by grafting onto them human-like visual capacities, for example, but instead they tailor their readings of what the animals can or cannot see to the prey being pursued. Hence, visual signals that may be unlikely to mislead a human onlooker may nevertheless be employed to successfully mislead particular animals. For example, it is the discernment of the unwanted pests that guides the structure and detail of the figures employed in fields to deter encroachment into the farmer's crops. Hence, it is not surprising that scarecrows do not typically deceive humans even when they are effective in fulfilling their intended purpose.[3]

While evidence for the prehistoric camouflaging and obscuring of traps would be nearly impossible to produce, Holliday (1998) cites evidence from bone assemblages and ecological context that supports the view that Pleistocene hunters used traps on small animals, such as foxes and hares. In this use of camouflage, Pleistocene hunters would be no different from modern hunters, who employ a wide variety of camouflage techniques to both attract animals and obscure traps.

We also have no reason to suppose that such techniques would have been limited to strictly visual environmental cues. Modern hunters disguise scents or use prevailing winds to avoid tipping off prospective prey. Bird and animal calls can be used to draw animals into ambush. These techniques not only presuppose what the animal can or cannot see but also implicitly commit the hunter to a theory about what attracts or repels the approach of targeted animals. That is, the hunter develops an account of the cognitive life of the animal—in terms of their interests and

[3] In this section, I am indebted to Robert Lurz for a helpful discussion on the environmental and cognitive conditions on what he calls *allocentric spatial perspective taking.*

desires—in order to manipulate and predict its behavior. This suggests that it is a short step from strictly visual or perceptual manipulation of environmental signals to more complex theories of the animal's cognitive states.

Consider, for example, how we might develop theories of the animal's desires and interests. Speculating about what a particular animal, in a particular part of the day, during a particular season, during its lifecycle, etc., may believe or desire will inform decisions regarding the placement of traps or the site of blinds or ambush locations. For example, supposing that a quadruped may face increased thirst late on a summer day may contribute to a mid-afternoon hunting strategy that emphasizes staking out potential sources of water. Hence, hunting techniques that attend to the cognitive states of particular creatures may routinely go beyond the immediate perceptual environment of the creature under pursuit. In sum, there is potential value in a richly textured understanding of the animal's psychological states in particular circumstances.

There is perhaps no better demonstration of this complexity than the case of what has been called *persistence* or *endurance hunting*. This ancient hunting technique appears to place special emphasis on tracking the full complement of an animal's mental states over the course of a single hunt. Despite this, it is among the least celebrated of ancient human hunting techniques.[4] It relies little on strength or stealth. Hunters merely engage the animal (usually a large quadruped) in a prolonged chase, often over days, until exhaustion renders the animal defenseless.

Evolutionary accounts of persistence hunting have emphasized the physiological adaptations that allowed humans to pursue animals that are stronger and faster than we are. The upright gait of early modern humans limited direct skin surface exposure to sunlight, which diminishes water loss from sweating and panting.[5] Similarly, the relative absence of hair on humans allows for faster cooling than animals with fur coats (Carrier 1984). These features allow humans to stay cool and hydrated compared to larger animals they were pursuing. Bipedal locomotion permits enormous economy in traveling long distances that are not shared by quadrupeds. Horses consume nearly the same amount of energy crossing a given distance whether they run or walk; a human consumes roughly half when walking instead of running. This metabolic difference suggests that endurance hunting is an evolutionary outgrowth of the increased efficiency of slow speed human locomotion (see, e.g., Carrier 1984).

These accounts may imply that the success of endurance techniques relies primarily on adaptations in human physiology rather than cognitive capacities. But these physical traits would provide little benefit if endurance hunters typically lost the trail of their prey, which would have been quite common. The quadrupeds most widely pursued are capable of outrunning any human over relatively short distances

[4] At least compared with the traditional imagery of Pleistocene hunting—like driving animals over cliffs or raining boulders down on mammoths.

[5] See Bramble and Lieberman (2004) for a discussion of the adaptive significance of endurance running (and its physiological basis) in humans.

(Bramble and Lieberman 2004). It was the ability of early human hunters to *regain* the trail of their quarry (often many times) over the course of the hunt that enabled them to cut off the animal's access to water or sources of food. In order to do this, hunters needed cognitive strategies that would allow them to predict animal's movements when they could no longer be seen.

Grover Krantz (1968) has argued that changes in human brain size from early Australopithecines to Homo erectus might be accounted for, at least in part, in terms of the cognitive demands of endurance hunting. Persistence hunting requires the hunter to keep a particular goal or strategy in mind, often for days at a time, while simultaneously anticipating and planning for a wide variety of contingencies. While Krantz emphasizes the computational complexity of the tracking problem, my concern is more with its content. That is, the contingencies predicted by a persistence hunter are at once environmental and psychological. Since the prey is often lost or unseen, the hunter must consistently attempt to predict not only the environment in which the animal will move but also the kinds of decisions it will make.

Peter Carruthers (2002) has argued that the origins of scientific inquiry may be traced to the cognitive capacities required by prehistoric hunting and tracking. However, similar claims have not been made for the role of hunting and tracking in the development of human empathetic capacities. Even so, recent descriptions of modern persistence hunting have emphasized the tracker's keen awareness of how things appear from his prey's point of view. Following Louis Liebenberg (1990), Carruthers writes, "[i]n predicting what an animal will do in given circumstances a hunter will rely, in part, on his folk-psychology – reasoning out what someone with a given set of needs and attitudes would be likely to do in those circumstances" (2002, 89). Studies of modern hunter-gatherer communities suggest that hunters do not rely solely on physical indicators of the animal's whereabouts—their tracks or other markings; rather, these hunters try to predict what escape routes and defensive strategies animals would employ based on a reading of their point of view. In short, the hunter tries to imagine himself in the shoes (or hooves) of the animals he is pursuing.[6]

In this section, I have emphasized the value of cross-species empathy in terms of different forms of prehistoric hunting. However, some of these same survival benefits might be associated with being quarry ourselves. Predator avoidance strategies would also benefit from a keen conception of how hunting animals think. Hence, paths to be avoided or strategic methods for protection in choosing shelters or forage might be informed by empathetic identification with animals that pose specific dangers. Attending to how lions or other predators view the world might

[6] Of course, using modern hunter-gatherers to draw these comparisons is not unproblematic. However, in these cases the comparison seems less fraught than usual. The hunts themselves cover similar landscapes and involve tools that are little different from those employed in prehistoric times. In addition, the fact that persistence hunting has been observed in many unrelated modern hunter-gatherer societies suggests that the strategy was likely common in prehistory as well (see Liebenberg 2006 for a survey of persistence hunting in Australia, the American Southwest, Mexico, several different African locations, and South America).

help in adopting defensive measures designed to thwart their advances. Empathy, in this sense, cannot only benefit human hunting but also lessen the likelihood of falling prey to animal hunters.[7]

3 How Hunting Pays: The Evolutionary Advantage of Cross-Species Empathy

The advantages of these predictive strategies should be clear. Insofar as empathy across species renders accurate predictions of future actions, a hunter who can get a better sense of the perspective of his/her prey will fare better than one who lacks or misapplies this kind of empathetic imagination.[8] Successful hunters will be better able to provide themselves and their families, as well as members of their community.[9] Moreover, since meat could not be easily preserved, an informal economy of food sharing might simultaneously prevent waste while developing relationships with other successful hunters to insure a food supply in harder times. These contributions are significant for more than nutrition. As Carruthers (2002) points out, the fitness benefits of hunting could also be defined in terms of sexual success. Because virtually all hunter-gather communities share meat equally between group members, advocates of the costly signaling theory (Zahavi and Zahavi 1997) have argued that the selectional benefits of being a good hunter would come less from being better fed than acquiring a higher status within the group.[10] On this view, successful hunting signals qualities that make for a superior mate or a more formidable competitor (and, conversely, a worthy ally). Eric Alden notes that despite a relative paucity of studies in this area, this view has collected some significant empirical support (2004, 354). Together, these considerations suggest a strong evolutionary pressure for tracking the cognitive lives of animals.

As we shall see, none of what I have argued above is designed to rule out the view that empathy between humans has played an important role in conditioning our capacity to read and understand the perspectives of others. Rather, my claim is

[7] This is not intended to suppose that no other cognitive strategies would prove useful in this context. Early humans would likely have availed themselves of many distinct predation strategies for animals based on the full complement of their observable behaviors.

[8] This is, of course, a legitimate empirical question. It may turn out that perspective taking with animals may not prove to be an effective hunting strategy. Hunter gatherers may be inclined to use these techniques, even if they do not increase the likelihood of hunting success.

[9] This is supported by modern anthropological observations that "[m]en who fail to hunt, or fail to help in cooperative hunts are generally not invited to participate on future forest treks" (Gurven and Hill 2009). This suggests that effective hunting has significant consequences for one's ability to provide for one's family. In a meta-analysis, Gurven notes "Good hunters have been shown to display higher reproductive success almost everywhere the relationship has been investigated" (2006, 81).

[10] An early version of this theory was called "the show-off hypothesis" (Hawkes 1991).

that we should be open to the possibility that empathy with animals played a pivotal role in cognitive development. From this perspective, we can observe that many arguments for the social intelligence hypothesis are surprisingly equivocal about the source of these capacities. For example, advocates of the social intelligence hypothesis have speculated that the massive increase in brain size in human evolution is a function of the cognitive capacities necessary to anticipate and respond to the increase in size of human communities (Dunbar 1992, 1993). However, recent studies of predatory animals suggest that increases in brain size typically correspond more closely with increases in the brain size of one's prey than they do with increases in sociality or group size. Kay Holekamp (2007) notes that this is problematic for the social intelligence hypothesis because the correlation in brain size holds both for social animals and those that live relatively solitary existences (e.g., bears). Hunting may therefore be a better predictor of increases in brain size than sociality.

Although it is difficult to gather direct behavioral evidence of cross-species empathy in prehistory, we can gather some intriguing evidence from prehistoric cave paintings. These images are remarkable not only for their realistic depictions of large animals but also for the virtual absence of images that depict everyday life or social interactions beyond those associated with hunting or sex. Moreover, where there are images of humans, they are starkly simplistic and lacking the detail given to images of animals. They are also often merged with images of animals themselves. Gregory Curtis writes, "when they [the cave painters] did paint or engrave pictures of humans they did so with little care or effort; most of such pictures are stylized stick figures or simple line drawings of crude faces that look like cartoons or caricatures" (Curtis 2006, 20). The close attention to animal forms coupled with the relative paucity of human ones suggests a strong identification not with other people but rather with the animals themselves.[11]

4 How Empathy Pays

Discussions of the social value of empathetic capacities often emphasize the selective advantages of empathy. On this view, empathy can play a central role in being better able to predict behaviors of conspecifics, either in the competition for resources or sexual partners or to facilitate interpersonal interactions in cooperative or nurturing relationships. However, the success of this enterprise will depend in large

[11] Both the subject matter of these images and those created from templates of body parts suggest a preponderance of male artists (Guthrie 2004), a point that is relevant to the discussion that follows. Nicholas Humphrey argues that the similarity of cave painting to artistic works by autistic children—which also show more attention to animals than other humans—demonstrates dramatic cognitive difference between the cave painters and modern humans (Humphrey 1998). Given the strong connection between autism and deficits in mind reading of other humans, Humphrey's theory does not appear to be incompatible with my hypothesis.

part on how successful mind-reading strategies are for predicting future behaviors. Failures of empathy are certainly common, and attempts to read minds can often result in costly miscalculations of the responses of others. Consider, for example, how frequently we misconstrue the behavioral signals of prospective mates or even close friends. Updating and correcting our conception of others' psychological states is almost a constant feature of our social interactions. It is for these reasons that despite the evident ubiquity and naturalness of empathetic mind reading, we may, nevertheless, regard our own intuitions about other minds with some suspicion.

This would seem especially true in cases where my intuitive mind reading conflicts with other sources of evidence about a person's behavior. That is, we can imagine cases wherein a preponderance of observational evidence of an individual's behavior may suggest preferring predictive strategies that do not rely on perspective taking. Put simply, in the relatively small social world of prehistoric human hunter-gather societies, individuals could develop a large database on each group member's past behavioral propensities from which future behavior may be more reliably inferred. Hence, we may be more likely to emphasize someone's past behavior in predicting his or her future behavior than use a calculus that factors in a reading of that person's current cognitive states.

Consider, for example, a common academic experience for university faculty. We may have a student who has failed to make any of several deadlines for assignments or take-home exams over the course of a semester. In adopting a mind-reading perspective, we may entertain several different considerations of the individual's current psychological states. Hence, we can attribute and appropriately weigh the desire to pass the course, coupled with the extra time to make up late assignments, etc., and on this basis reckon that the student will make good on an extended, but final, deadline. At the same time, based on the student's earlier behavior, we may also hypothesize that they will not make the deadline (as the student has not in the past). It is unsurprising that in my informal survey, most faculties consider the predictions based on the mind-reading option to be less reliable than that generated by past experience. Perhaps it is because the mind reading position is often more general and therefore less tailored to the individual for whom the prediction is made. Whatever the explanation, cases like these are easy to multiply, especially with individuals for whom there is a great deal of past interaction. Mind reading may be valuable in aiding our social interrelations, but it is eminently defeasible. Once we have something more to go on, we can easily overthrow generic readings of an individual's behavior in favor of a more nuanced and more informed predictive matrix. Psychologically, mind reading may be our first option, but epistemologically it can be our last resort.

This seems especially clear when we consider the role that language plays in communicating our own psychological states. In understanding another person's behavior, I am not limited to observations that are specifically nonverbal. Instead, I often rely on explicit testimony regarding mental states (even if I may be wary of always taking these on face value). These reports may aid future perspective taking, but they may also obviate the need for mind reading in the first place. Announced intentions are likely to be more accurate than surmised ones.

Animal interactions, however, are less informed by daily and intimate interactions with specific individuals. We have a much smaller inductive basis for generating tailored predictions about behavior. Hence, relying on past experience may prove less valuable here than in cases where there is far more past experience to go on. In such cases, the cognitive work of perspective taking may take on greater relative predictive value in our interactions with animals.

5 Gender Differences in Hunting and Empathy

If human empathetic capacities evolved both within social communities and in the context of early hunting strategies, then it is likely that these environmental and cognitive pressures were experienced differently by different segments of the human population. Most importantly, evidence from modern hunter-gatherer communities suggests that persistence hunting is almost entirely the province of male hunters.[12] Hence, if we were to suppose that hunting produces significant evolutionary pressures for empathy, these pressures would be largely limited to males within the community. Moreover, women could face different pressures for mind reading as a function of their social and economic roles within the group. This dynamic suggests a tempting line of inquiry, one that distinguishes empathetic capacities by the cues that elicit them. On this hypothesis, males are more likely to have developed empathetic capacities for reading the minds of animals, while female group members might have benefited more by tuning their empathetic capacities toward other humans. On this differential model of empathy, socially based empathy might be more verbally mediated and attentive to human facial expression, while the less anthropocentric model I am suggesting may place more emphasis on nonverbal behavior or situational cues (like concrete physical or spatiotemporal relations). These differences suggest that miscommunication between genders might arise not from distinct interest in sexual or reproductive strategies so much as developing distinct mind-reading perspectives.

Although at first glance this differential model of empathy might seem simplistic (and make no mistake, it is), there is evidence to suggest that empathetic abilities are in fact significantly gendered. Simon Baron-Cohen has argued that there is strong empirical support for the view that men and women vary markedly in their empathetic capacities.[13] Consider, for example, how these differences are born out in language. Studies of language use in children suggest that girls and boys exhibit marked differences in the style and content of their speech. As Baron-Cohen notes, "girls' speech has been described as more cooperative, more reciprocal and more

[12] Liebenberg (2006) notes that there have been no observed cases of persistence hunting involving women and more generally, in a meta-analysis of hunting among hunter-gatherers, men alone hunt in 166 of the 179 societies examined.

[13] Space limits me to only one example, but Baron-Cohen develops his case relative to many different lines of inquiry. See Baron-Cohen (2003).

collaborative," whereas men typically engage in speech patterns that are more attuned to group activities and status enhancement (2009, 48–49). Girls also learn language earlier than boys and tend to be more proficient with it once it has been acquired. None of this would be surprising on the hunting account. Social pressures for empathy were likely to be, at least on some level, linguistically mediated, both in terms of reading the behavior of others and interacting with them. Game hunters, insofar as they hunted in groups, may have faced similar pressures, but in this there is less selective pressure for mind reading and perspective taking. Put simply, on this model the currency of male empathy makes fewer linguistic demands on men than that required by women. Different levels of empathy, or more precisely different kinds of empathy, would develop to suit different kinds of evolutionary pressures.

While I describe this view as tempting, it is not one that I would explicitly endorse. These arguments may be suggestive, but there is currently not enough evidence to suppose that we can draw firm conclusions about the source of gender differences in empathetic capacities. Moreover, it is important to distinguish this position from the thesis being developed in the rest of the chapter. Whether perspective taking developed in the context of prehistoric hunting or across distinct social or cognitive groups may be supported by observations of gender differences in mind-reading capacities, it does not depend on the view that our empathetic capacities are significantly gendered.

6 Modeling Mind Reading

Perhaps the most famous philosophical account of prehistoric mind reading comes from Wilfrid Sellars in his discussion of the Myth of Jones (1956). Shaun Nichols (forthcoming) offers a capsule version:

> [I]n our distant past, our ancestors never spoke of internal mental states like beliefs and desires. Rather, these "Rylean" ancestors only spoke of publicly observable phenomena like behavior and dispositions to behave...Then one day Jones, a great genius, arose from this group. Jones recognized that positing inner states like *thoughts* as theoretical entities provides a powerful basis for explaining the verbal behavior of his peers, and Jones developed a *theory* according to which such behavior is indeed the expression of internal thoughts. Jones then taught his peers how to use the theory to interpret the behavior of others.

Sellars has often been viewed as the source of the theory model of mental explanation. Jones develops an account of how we understand the psychological states of others, which envisages them as part of a theory that is ultimately passed on to others through explicit instruction. The main difference between Sellars and the social accounts considered above is the source of the theory used to describe others. What Sellars views as a social construction, social accounts of empathy usually ascribe to evolutionary pressures. Even so, in orientating his theory construction to the reading of individuals of the same tribe, Sellars' account is no less committed to the social origins of mind reading than the evolutionary psychologists who advocate the social intelligence hypothesis.

But it would be easy to imagine a different version of the myth. Suppose that Jones is less fond of campfire conversation and more inclined to pursue big game. Rather than produce a parlor game of human explanation, he hones his skills by imagining what it would be like to be the animal that he is hunting. He may, as a consequence, posit a host of tightly related internal states that determine the animal's future decisions. He might find that he can train his own responses to mimic that of his prey. On this version of the tale, any success would meet with immediate and tangible rewards. Perhaps, like the original Jones, he realizes that he can use this same capacity to predict the behavior of his peers.

Sellars' myth culminates with the realization that the same theory could be applied to oneself. My account is no different. In understanding my own actions and cognitive states, I adhere to the same model of cognition that I apply to other people. These empathetic projections provide a means of not merely understanding others, but also myself. In this way, we may be informed, not merely by our understanding of the behavior and attitudes of other humans but also by our historic readings of different species of game animals. To the extent that we employ these in our understanding of others and ourselves, *we are reading the minds of other humans according to a script written by animals hunted in our prehistoric past.*

This is of course a simple version of a complicated story. I do not expect that cross-species empathy can explain the full complement of human empathetic capacities nor does the success of this account depend on it. Instead, cross-species empathy and, more generally, empathy with individuals from widely divergent social and biological backgrounds may be located within a complex framework of evolutionary and social pressures that are connected to the predictive value of attending to the causal structure of minds themselves. Attributing minds (and with them perspectives, desires, beliefs, and other intentional phenomena) can thus allow individuals not merely to interpret and predict the behavior of others like themselves, but may be employed to bridge the gulf between individuals with very different backgrounds.[14]

References

Alden, E. (2004). Why do good hunters have higher reproductive success? *Human Nature, 15*(4), 343–364.

Baron-Cohen, S. (2003). *The essential difference: Men, women and the extreme male brain.* London: Penguin.

[14] This chapter has benefited from countless comments from both readers and audience members. I would like to thank Mary-Catherine Harrison for introducing me to the idea of empathy that is the foundation of this chapter and also Justine Kingsbury for a close and detailed reading of a late draft of the text and Robert Lurz for an enlightening set of conference comments. I would also like to thank John Baker, Peter Carruthers, Alan Gibbard, Nythamar de Oliveira, Ellen Fridland, Lila Hart, and many others for their comments and questions at presentations of this work. I am also grateful to Liz Swann for her valuable suggestions and careful editing of this text.

Baron-Cohen, S. (2009). Why so few women in math and science? In C. Sommers (Ed.), *The science on women and science* (pp. 7–23). Washington, DC: American Enterprise Institute for Public Policy Research.
Boyer, P. (2001). *Religion explained: The evolutionary origins of religious thought*. London/New York: Random House/Basic Books.
Bramble, D., & Lieberman, D. (2004). Endurance running and the evolution of homo-erectus. *Nature, 432*(7015), 345–352.
Carrier, D. (1984). The energetic paradox of human running and hominid evolution. *Current Anthropology, 25*(4), 483–495.
Carruthers, P. (1996). Simulation and self-knowledge: A defence of the theory-theory. In P. Carruthers & P. K. Smith (Eds.), *Theories of theories of mind*. Cambridge: Cambridge University Press.
Carruthers, P. (2002). The roots of scientific reasoning: Infancy, modularity and the art of tracking. In P. Carruthers, S. Stich, & M. Siegal (Eds.), *The cognitive basis of science* (pp. 73–96). Cambridge: Cambridge University Press.
Currie, G. (2011). Empathy for objects. In A. Coplan & P. Goldie (Eds.), *Empathy: Philosophical and psychological perspectives* (pp. 82–98). Oxford: Oxford University Press.
Curtis, G. (2006). *The cave painters: Probing the mysteries of the world's first artists*. New York: Knopf.
Dunbar, R. (1992). Neocortex size as a constraint on group size in primates. *Journal of Human Evolution, 22*(6), 469–493.
Dunbar, R. (1993). Coevolution of neocortical size, group size and language in humans. *Behavioral and Brain Sciences, 16*(4), 681–735.
Dunbar, R. (2000). The origin of the human mind. In P. Carruthers & A. Chamberlain (Eds.), *Evolution and the human mind* (pp. 238–253). Cambridge: Cambridge University Press.
Goldman, A. (2006). *Simulating minds: The philosophy, psychology and neuroscience of mind-reading*. New York: Oxford University Press.
Gurven, M., & Hill, K. (2009). Why do men hunt? A re-evaluation of "Man the Hunter" and the sexual division of labor. *Current Anthropology, 50*(1), 51–74.
Gurven, M., & von Rueden, C. (2006). Hunting, social status and biological fitness. *Biodemography and Social Biology, 53*(1), 81–99.
Guthrie, R. Dale (2004). *The Nature of Paleolithic Art*. Chicago, IL: The University of Chicago Press.
Harrison, M. C. (2011). How narrative relationships overcome empathic bias: Elizabeth Gaskell's empathy across difference. *Poetics Today, 32*(2), 255–288.
Hawkes, K. (1991). Showing off: Tests of an hypothesis about men's foraging goals. *Ethology and Sociobiology, 12*, 29–54.
Holekamp, K. E. (2007). Questioning the social intelligence hypothesis. *Trends in Cognitive Science., 11*, 65–69.
Holliday, T. (1998). The ecological context of trapping among recent hunter-gatherers: Implications for subsistence in terminal Pleistocene Europe. *Current Anthropology, 39*, 711–719.
Hrdy, S. (2009). *Mothers and others, the evolutionary origins of mutual understanding*. Cambridge, MA: Harvard University Press.
Humphrey, N. K. (1976). The social function of intellect. In P. P. G. Bateson & R. A. Hinde (Eds.), *Growing points in ethology* (pp. 303–317). Cambridge: Cambridge University Press.
Humphrey, N. K. (1998). Cave art, autism, and the evolution of the human mind. *Cambridge Archaeological Journal, 8*(2), 165–191.
Krantz, G. (1968). Brain size and hunting ability in earliest man. *Current Anthropology, 9*(5), 450–451.
Liebenberg, L. W. (1990). *The art of tracking: The origin of science*. Cape Town: David Philip.
Liebenberg, L. W. (2006). Persistence hunting by modern hunter gatherers. *Current Anthropology, 47*(6), 1017–1025.

Nichols, S. (forthcoming). Mindreading and the philosophy of mind. In J. Prinz (Ed.), *The Oxford handbook on philosophy of psychology.* New York: Oxford University Press.
Preston, S., & De Waal, F. (2002). Empathy: Its ultimate and proximate bases. *Behavioral and Brain Sciences, 25*(1), 1–20.
Sellars, W. (1956). Empiricism and the philosophy of mind. *Minnesota Studies in the Philosophy of Science, 1*, 253–329.
Trout, J. D. (2009). *The empathy gap: Building bridges to the good life and the good society*. New York: Viking/Penguin.
Zahavi, A., & Zahavi, A. (1997). *The handicap principle: A missing piece of Darwin's puzzle*. New York: Oxford University Press.

The Evolution of Scenario Visualization and the Early Hominin Mind

Robert Arp

Abstract In this chapter, I argue that *scenario visualization*—viz., a mental activity whereby visual images are selected, integrated, and then transformed and projected into visual scenarios for the purposes of solving problems in the environments one inhabits—emerged in our hominin past and accounts for certain kinds of vision-related creativity. The kinds of problems with which our hominin ancestors were confronted most likely were of the spatial relation and depth relation types related to basic survival—such as judging the distance between an object and oneself, determining the size of an approaching object, matching an object to any number of associated memories, and anticipating the need for a particular kind of tool to accomplish a task—and so the capacity to scenario visualize would have been useful for their survival. Thus, scenario visualization has been and continues to be relevant for *vision-related* forms of creative problem-solving.

Keywords Bissociation • Cognitive fluidity • Creative problem-solving • Evolutionary psychology • Hominin • Mithen • Scenario visualization • Visual imagery

1 Introduction

The construction of novel tools and pieces of art, as much as language, would seem to characterize our apparent human uniqueness among species in the animal kingdom. Humans not only manufacture products; they manufacture products *to manufacture other* products, synthesize disparate ideas, successfully negotiate environments, invent, innovate, imagine, improvise, and solve all kinds of problems in creative ways. In field observations and in controlled laboratory experiments, we witness

R. Arp (✉)
Independent Scholar
e-mail: robertarp320@gmail.com

L. Swan (ed.), *Origins of Mind*, Biosemiotics 8, DOI 10.1007/978-94-007-5419-5_7,

and document chimps, orangutans, dolphins, elephants, octopi, crows, and other animals engaging in fairly sophisticated forms of problem-solving; however, even the most advanced animals—such as chimps (Whiten 2010; Whiten et al. 1999; Call and Tomasello 1994; Lonsdorf et al. 2010; Watanabe and Huber 2006; Pearce 2008; von Bayern et al. 2009)—achieve the problem-solving capacities of a normal human 3- or 4-year old, and no animal to date has been able to solve the kinds of problems that even our earliest hominin ancestors apparently were able to solve, like the problem of how to kill a mammoth without getting yourself killed which was solved simply by placing a flake on the end of a stick to produce a projectile such as the spear.

Following Mayer (1995), we can distinguish between *routine problem-solving* and *nonroutine creative problem-solving* (NCPS) (also see the papers in Smith et al. 1995). In routine problem-solving, an animal recognizes many possible solutions to a problem if they have worked in the past. Animals constantly perform routine problem-solving activities that are concrete and basic to their survival, an example of which is to pursue a short-term goal that has been established in memory or immediate perceptual association.

The *human* animal performs routine problem-solving activities too, but also can engage in activities that are more abstract and creative, such as inventing new tools based on mental blueprints; synthesizing concepts that, at first glance, seem wholly disparate or unrelated; devising novel solutions to problems; and producing sublime works of art. If a person decided to pursue a *wholly new way* to solve a problem by, say, inventing some kind of tool, then we would have an instance of NCPS.

In this chapter, I present the ideas and arguments put forward by archeologist Steven Mithen (1996, 1999, 2001, 2005), according to which the human mind evolved an ability to use mental images creatively so as to generate novel pieces of artwork, invent tools, and solve *nonroutine* problems. Several evolutionary psychologists believe that these complex cognitive abilities are the result of specified Swiss Army knife-like mental modules (there are numerous versions of this idea) having evolved in our early hominin Pleistocene past to deal with the various and sundry problems a human might have experienced. Mithen shows the deficiency in this position and argues that creativity is possible because the mind has evolved what he calls *cognitive fluidity*, an ability to exchange information flexibly between mental modules—or, to use a term from Koestler (1964), an ability to *bissociate*. In fact, according to Mithen, cognitive fluidity *is* conscious reasoning, our uniquely human mental ability. This is a plausible view that has been well received in the literature concerning the evolution of consciousness, imagination, and creativity (e.g., Ruse 2006; Calvin 2004; Gregory 2004; Goguen and Harrell 2004; Arp 2005a, b, 2006, 2008; cf. Fodor 1998).

Mithen's view cannot be the full story, however. My claim in this chapter is that what I call *scenario visualization* emerged as a mental property to act as a kind of metacognitive process that selects and integrates relevant visual information from psychological modules in order to perform vision-related, NCPS tasks in environments. However, if this kind of mental activity were *merely* free flow of information—as suggested by Mithen—there would be no mental coherency; the information would be chaotic and directionless and not really *informative* at all. Data need to be segregated and integrated so that they can become informative for the cognizer; in fact, selecting and integrating visual information from mental modules are the function of scenario visualization.[1]

2 Evolutionary Psychology and the Swiss Army Knife Modular Mind

According to many evolutionary psychologists, the mind is like a Swiss Army knife loaded with specific mental *tools* that evolved in the Pleistocene epoch (which began some 1.8 million years ago and lasted almost one million years) to solve specific problems of survival, such as face recognition, mental maps, intuitive mechanics, intuitive biology, kinship, language acquisition, mate selection, and cheating detection. The list of mental tools could be longer or shorter, and there are many variations of the Swiss Army knife model, with the human mind having evolved one larger all-purpose tool to complement the more-specified tools or several dual-purpose tools coexisting with several more-specified tools or any combination thereof (Cosmides and Tooby 1987, 1994; Buss 2009; Gardner 1993; Palmer and Palmer 2002; Confer et al. 2010; Hampton 2010).

Evolutionary psychologists speak of these mental modules as domains of specificity. What this means is that any given module handles only one kind of adaptive problem to the exclusion of others. Modules are encapsulated in this sense and do not share information with one another. For example, one's cheater-detection module evolved under a certain set of circumstances and has no direct connection to one's fear-of-snakes module, which evolved under a different set of circumstances. This kind of encapsulation works best for environments where the responses need to be quick and routine; such developments enabled these organisms to respond efficiently and effectively in their regular or accustomed environments.

[1] I have argued for my scenario visualization view in the past (Arp 2005a, b, 2006, 2008), and not only has it been applauded as "innovative and interesting," and even "ambitious" (Downes 2008; Jarman 2009; Thomas 2010; O'Connor et al. 2010), it also has been utilized by numerous philosophical psychologists, cognitive scientists, A.I. researchers, and others (Sloman and Chappell 2005; Gomila and Calvo 2008; Weichart 2009; Sugu and Chatterjee 2010; Arrabales et al. 2008, 2010; Rivera 2010; Bullot 2011; Langland-Hassan 2009; Boeckx and Uriagereka 2011). Thus, the view likely has *at least* initial plausibility. Still, I have critics (Kaufman and Kaufman 2009; Picciuto and Carruthers 2008), and I welcome the continued dialogue concerning the evolution of the human mind. Although I desire to explain the specific ways various researchers have utilized my scenario visualization view, as well as offer numerous responses to my critics, given space limitations here—coupled with the nature of this book--I will stick to the basic plan of explaining and arguing for scenario visualization as a plausible hypothesis associated with the evolution of our mental architecture.

3 A Problem for the Swiss Army Knife Modular Mind

There seems to be a fundamental flaw, however, in the evolutionary psychologist's reasoning. If mental modules are encapsulated and are designed to perform certain *routine* functions, how can this modularity account for *novel* circumstances? When routine perceptual and knowledge structures fail or when atypical environments present themselves, it is *then* that we need to be innovative in dealing with this novelty. Imagine the Pleistocene epoch. The climate shift in Africa from jungle life to desert savanna life forced our early hominins to come out of the trees and survive in totally new environments. Given a fortuitous genetic code, some hominins re-adapted to the new African landscape, some migrated elsewhere to places like Europe and Asia, and most died out. This environmental shift had a dramatic effect on modularity, since now the specific content of the information from the environment in a particular module was no longer relevant. *The information that was formerly suited for jungle life could no longer be relied upon in the new environment of the savanna*. Appeal to modularity alone would have led to certain death and extinction for our hominin ancestors.

The successful progression from the typical jungle life to the atypical and novel savanna life of our early hominin ancestors would have required some other kind of mental capacity to emerge that could creatively handle the new environment. But how is it that we can be creative?

4 Mithen and Cognitive Fluidity

Steven Mithen advanced the evolutionary psychologists' modular mind by introducing *cognitive fluidity*, which enables one to respond creatively to novel environments. Mithen sees the evolving hominin mind as going through a three-step process beginning prior to 6 mya when the primate mind was dominated by what he calls a *general intelligence*. This general intelligence was similar to chimpanzee mindedness in that it consisted of an all-purpose, trial-and-error learning mechanism that was devoted to multiple tasks where all behaviors were imitated, associative learning was slow, and there were frequent errors made.

The second step coincides with the evolution of the *Australopithecine* line and continues all the way through the *Homo* lineage to *H. neanderthalensis*. In this second step, multiple *specialized intelligences*, or modules, emerge alongside general intelligence. Associative learning within these modules was faster, so more complex activities could be performed. Compiling data from fossilized skulls, tools, foods, and habitats, Mithen concludes that *H. habilis* probably had a general intelligence as well as modules devoted to social intelligence (because they lived in groups), natural history intelligence (because they lived off the land), and technical intelligence (because they made tools). *Neanderthals* and *H. heidelbergensis* would have had all of these modules, including a primitive language module, because their skulls exhibit bigger frontal and temporal areas—areas that in the modern human brain are engaged in language functioning. According to Mithen (1996, 1999, 2001, 2005), the *Neanderthals* and *H. heidelbergensis* had the Swiss Army knife mind that the standard evolutionary psychology account describes.

Now, a problem arises of which Mithen, too, is aware: it cannot be the case that the emergence of distinct mental modules that evolutionary psychologists today postulate as accounting for learning, negotiating, and problem-solving took place *during the Pleistocene*. The potential variety of problems encountered in generations subsequent to the Pleistocene is too vast for a limited Swiss Army knife mental repertoire; there are too many hypothetical situations for which *nonroutine creative problem-solving* would have been needed in order to survive and dominate the earth. There are potentially an *infinite number* of problems confronting animals constantly as they negotiate environments. That we negotiate environments so well shows that we have some capacity to handle the various and sundry *potential nonroutine* problems that arise in our environments.

Here is where the third step in Mithen's evolution of the mind, known as *cognitive fluidity,* comes into play. In this final step—which coincides with the emergence of modern humans—the various mental modules are working together with a flow of knowledge and ideas between them. The modules can now influence one another, resulting in an almost limitless capacity for imagination, learning, and problem-solving. The working together of the various mental modules as a result of this cognitive fluidity *is* consciousness for Mithen and represents the most advanced form of mental activity (Mithen 1996, 1999, 2001, 2005).

5 Cognitive Fluidity and Creativity

Mithen notes that his model of cognitive fluidity accounts for human creativity in terms of problem-solving, art, ingenuity, and technology. His idea has initial plausibility, since it is arguable that humans would not exist today if they had not evolved consciousness to deal with novelty. No wonder, then, Crick (1994) maintains that "without consciousness, you can deal only with familiar, rather routine situations or respond to very limited information in new situations" (p. 20). Also, as Searle (1992) observes, "one of the evolutionary advantages conferred on us by consciousness is the much greater flexibility, sensitivity, and creativity we derive from being conscious" (p. 109).

Mithen's idea resonates with what researchers refer to as *bissociative creativity* and creative problem-solving. Scientists have documented chimps looking pretty creative in their problem-solving by trying a couple of different ways to get at fruit in a tree—like jumping at it from different angles or jumping at it off tree limbs—before finally using a stick to knock it down. Scientists also document young chimps watching older chimps do the same thing what same thing? (Whiten 2010; Lonsdorf et al. 2010). In fact, several observations have been made of various kinds of animals engaged in imitative behaviors that look like creative problem-solving (Norris and Papini 2010).

However, the number of possible solutions is limited in these examples of routine problem-solving because the mental repertoire of these animals is environmentally fixed and their tool usage (if they have this capacity) is limited. In fact, all attempts to get chimpanzees and other primates to imitate the basic knapping method utilized by *Homo habilis* (2.33–1.4 mya), for example—where essentially a stone tool is used to knap (strike and chip) to make another stone tool—have failed (Merchant and McGrew 2005; De Beaune et al. 2009; Whiten 2010; Lonsdorf et al. 2010).

Unlike routine problem-solving, which deals with associative connections within familiar perspectives, nonroutine creative problem-solving entails an innovative ability to make connections between *wholly unrelated* perspectives or ideas. A human seems to be the only kind of being who can solve nonroutine problems *on her or his own, without imitation or help*. Koestler (1964) referred to this quality of the creative mind as a *bissociation of matrices*. When a human bissociates, that person puts together ideas, memories, representations, stimuli, and the like, in wholly new and unfamiliar ways *for that person*. Echoing Koestler, Boden (1990) calls this an ability to "juxtapose formerly unrelated ideas" (p. 5). Thus, Dominowski (1995) claims that "overcoming convention and generating a new understanding of a situation is considered to be an important component of creativity" (p. 77; also see the papers in Smith et al. 1995).

Humans *bissociate* and are able to ignore normal associations, trying out *novel* ideas and approaches in solving problems. Bissociation also has been pointed to as an aid in accounting for the ability to laugh, the hypothesis-formation, the art, the technological advances, and the proverbial "ah-hah," creative insight eureka moments humans experience when they come up with a new idea, insight, or tool.

So, when we ask how it is that humans can be creative, part of what we are asking is how they bissociate, viz., *juxtapose formerly unrelated ideas in wholly new and unfamiliar ways for that person*. To put it colloquially, humans can take some visual perception, concept, or idea found "way over here in the left field" of the mind and make some coherent connection with some other wholly disparate and unrelated visual perception, concept, or idea found "way over here in the right field" of the mind. And humans seem to be the only species that can engage in this kind of mental activity.

6 Scenario Visualization: Advancing Mithen's View

Mithen's account of cognitive fluidity allows for the free movement of information between modules (Koestler's bissociation). I believe this is important as a *precondition for* mental activities, such as imagination, that require the simultaneous utilization of several modules. So, for example, Mithen would think that totemic anthropomorphism associated with animals in, say, a totem pole made up of part-human and part-animal figures derives from the free flow of information between a natural history module dealing specifically with animals and their characteristics and a social module dealing specifically with people and their characteristics. A totem carved out of wood is the *material* result of the free flow of information between the natural history and social modules that occurred in the mind of the artist.

Mithen's model is unsatisfactory, however, because he makes consciousness out to be a passive phenomenon. On his account, consciousness is just a flexible fluidity, and this does not seem to me to be the full account of consciousness. When we are engaged in conscious activity, we are *doing* something. The fundamental insight, derived from Kant (1929) and reiterated by numerous philosophers, psychologists, and neuroscientists, is that consciousness is an active process (e.g., Crick and Koch 2003; Singer 2000; Arp 2005a, b, 2006, 2008).

Kandel et al. (2000) bolster Kant's insight when they claim that perception "organizes an object's essential properties well enough to let us handle the object" (p. 412). Drawing directly on Kant's insights, they claim further that our perceptions "are constructed internally according to constraints imposed by the architecture of the nervous system and its functional abilities" (p. 412). I am proffering Kant's fundamental insight and suggesting that mental activities associated with the selecting and integrating of visual information from mental modules for the purposes of negotiating environments are essential to creative problem-solving *and* that Mithen's account of cognitive fluidity acts as a precondition for the possibility of the information contained in these modules to intermix. So on one hand, Mithen is correct about the possibility of information between and among mental modules as intermixing, and he is correct that cognitive fluidity probably is a better description of our mental architecture, given the early hominin ability to survive in the ever-changing Pleistocene environments. On the other hand, I am transforming and adding to Mithen's account by arguing that possible intermixing of modular information is not the full story concerning vision-related, creative problem-solving.

I am arguing for a view I call *scenario visualization*, which is:

> a mental process that entails selecting pieces of visual information from a wide range of possibilities, forming a coherent and organized visual cognition, and then projecting that visual cognition into some suitable imagined scenario, for the purpose of solving some problem posed by the environment which one inhabits.

In my example of totemism (above), the images utilized had to be *selected from* other visual images as relevant. In the totem, visual information from both the social and the natural history modules is *synthesized*, allowing for something sublimated or innovative *to emerge anew* as a result of the process. While speaking about Mithen's idea of cognitive fluidity, Fodor (1998) expresses a similar claim about integration: "Even if early man had modules for 'natural intelligence' and 'technical intelligence,' he couldn't have become modern man just by adding what he knew about fires to what he knew about cows. The trick is in thinking out what happens when you put the two together; you get *steak au poivre* by *integrating* (italics ours) knowledge bases, not by merely summing them" (p. 159).

It is important to mention that other thinkers acknowledge the fact that the mind's architecture is made up of flexible interacting modules, and, similarly, have put forward mechanisms of integration to account for mental coherency. Damasio

(2000), Singer (2000), Velmans (1992), and Tononi and Edelman (1998) each have put forward a view of consciousness as entailing an integrating mechanism. Fauconnier and Turner (2002) use the concept of "conceptual blending" or "conceptual integration" to account for "making human beings what they are, for better or worse," as language bearers and creative problem-solvers. Also, Goguen and Harrell (2004) lay out a view of conceptual blending that utilizes mathematical algorithms and computational implementations to generate narrative and metaphor.

I think that scenario visualization comes to light most clearly when humans engage in vision-related forms of problem-solving. I am not suggesting that people *always* visualize or *never* use semantic forms of reasoning or other forms of reasoning when solving nonroutine problems. Whether one utilizes scenario visualization most likely will depend upon the type of problem with which one is confronted. There are some problems—for example, certain mathematical problems—that can be solved without the use of scenario visualization. Other problems, like spatial relation or depth perception problems, may require scenario visualization. The kinds of problems with which our hominin ancestors were confronted most likely were of the spatial relation and depth relation types related to basic survival—such as judging the distance between an object and oneself, determining the size of an approaching object, matching an object to any number of associated memories, and anticipating the need for a particular kind of tool to accomplish a task—and so the capacity to scenario visualize would have been useful for their survival. Thus, scenario visualization has been and continues to be relevant for *vision-related* forms of creative problem-solving.

7 Scenario Visualization and Toolmaking

It is generally agreed upon by biologists, anthropologists, archeologists, and other researchers that a variety of factors contributed to the evolution of the modern human brain including bipedalism, diversified habitats, social systems, protein from large animals, higher amounts of starch, delayed consumption of food, food sharing, language, and toolmaking (Aiello 1997; Donald 1997; Calvin 2004; Dawkins 2005). It is not possible to get a complete picture of the evolution of the brain without looking at all of these factors, since brain development is involved in a complex coevolution with physiology, environment, and social circumstances. The emergence of language in our species clearly occupies a central place with respect to our ability to flourish and dominate the earth (Tallerman 2005). However, I wish to focus on toolmaking as essential in the evolution of the brain and visual system, and I do this for four reasons:

1. First, toolmaking is the mark of intelligence that distinguishes the *Australopithecine* genus from the *Homo* genus in our evolutionary past. *Homo habilis* was the first toolmaker, as meaning the Latin name, "handyman," denotes.
2. Second, tools offer us indirect, but compelling, evidence that psychological states emerged from brain states. In trying to simulate ancient toolmaking techniques,

archeologists have discovered that certain tools can only be made according to *mental* templates, as Pelegrin (1993), Isaac (1986), Wynn (1993), and De Beaune et al. (2009) have demonstrated.
3. Third, as I mentioned above, our hominin ancestors were not solving math problems, they were concerned with recognizing and discerning prey, predator, friend, and/or foe (and other basic survival activities), and so the capacity to scenario visualize with respect to toolmaking would have been useful for their survival.
4. Finally, as I attempt to show, the evolution of toolmaking parallels the evolution of visual processing in terms of scenario visualization.

The breakthrough in tool technology that is central to my scenario visualization theory was the Mousterian industry that arrived on the scene with the *H. neanderthalensis* lineage, near the end of the *H. heidelbergensis* lineage, around 300,000 ya. Mousterian techniques involved a more complex three-stage process of constructing (a) the basic core stone, (b) the rough blank, and (c) the refined finalized tool. Such a process enabled various kinds of tools to be created, since the rough blank could follow a pattern that ultimately could become cutting tools, serrated tools, flake blades, scrapers, or lances. Further, these tools had wider application as they were being used with other material components to form handles and spears and were being used to make other tools, such as wood and bone artifacts. Consistent with the increase in complexity of toolmaking, the brain of *H. heidelbergensis* and *H. neanderthalensis* increased to 1,200 and 1,500 cm in volume, respectively, up 300–600 cm from *H. erectus*.

By 40,000 ya, some 60,000 years after anatomically modern *H. sapiens* evolved, we find instances of human art in the forms of beads, tooth necklaces, cave paintings, stone carvings, and figurines. This period in tool manufacture is known as the Upper Paleolithic, and it ranges from 40,000 ya to the advent of agriculture around 12,000 ya. Sewing needles and fish hooks made of bone and antlers first appeared, along with flaked stones for arrows and spears, burins (chisel-like stones for working bone and ivory), multi-barbed harpoon points, and spear throwers made of wood, bone, and antler (Pelegrin 1993; Mithen 1996; Isaac 1986; McHenry 1998; De Beaune et al. 2009).

I suggest that scenario visualization emerged as a natural consequence of the development of a complex nervous system in association with environmental pressures that occasioned its evolution. In attempts to recreate early hominin tools from the later Mousterian and Upper Paleolithic industries, archeologists such as Mithen (1996) and Wynn (1993) have shown that the construction of such tools would require several mental visualizations, as well as numerous revisions of the material, so as to attain optimal performance of such tools. Such visualizations likely included the abilities to, at least, identify horizontal or vertical lines, select an image from several possible choices, distinguish a target figure embedded in a complex background, construct an image of a future scenario, project an image onto that future scenario, as well as recall from memory the particular goal of the project. If an advanced form of toolmaking acts as a mark of the most advanced mind, given complex and changing Pleistocene environments, as well as the scenario visualization that is necessary to produce tools so as to survive these environments, what I am suggesting is that visual processing most likely was the primary way in which this advanced mind emerged on the evolutionary scene.

8 Evolution of the Javelin

In what follows, I trace the development of the multipurposed javelin from its meager beginnings as a stick, through its modifications into the spear, and finally its specialization into a javelin equipped with a launcher. We need an example that illustrates the emergence of scenario visualization in our evolutionary past, and the development of this tool gives us concrete evidence. The following story is meant to be presented as a plausible account of how it is that scenario visualization would have emerged in our early hominin past, and, like most evolutionary stories, it is not meant to be an account for which we have *decisive* evidence.

8.1 Step 1: The Stick

We can take present-day chimpanzee activities to be representative of early hominin life, and we can see that chimps in their native jungle environments do indeed use tools. The chimps use rocks, leaves, and sticks to crack open nuts, carry items, fish for termites, and hit in self-defense or in attack. This is probably what our early hominins did while in the jungles, savannas, and grasslands of Africa.

As previously mentioned, chimps engage in trial-and-error and imitative learning. Baby chimps try to imitate the actions of older chimps, including the tool usage. Researchers have tried to get chimps to use tools to make other tools with cobbles and stones (the way early *H. habilis* is likely to have done) by flaking and edging, but they cannot do it (McGrew 2004; Byrne 2001). So, it seems that chimps can form and recall visual images from memory when using tools. But they clearly do not have the capacity to produce tools like those found in the Upper Paleolithic industry; their tool usage merely is imitative and wholly lacking in innovation.

When the climate changed and early hominins moved from the jungles to forage food on African savannas, they constructed javelins they could throw from a distance in order to kill prey (Ambrose 2001; Churchill and Rhodes 2009). One could continue to hit prey with a stick until it dies, as was done in jungle environments. This may work for some prey, but what about the ones that are much bigger than you? Imagine being stuck on the savanna with a stick as your only tool of defense against woolly mammoths and saber-toothed tigers. Stated simply, you would need to become more creative in your toolmaking just to survive. Calvin (2004) asks a simple question related to the survival of our early hominins: "Could they innovate?" (p. 25). If the answer is *no*, then such hominins ultimately went the way of the dodo.

The progression from stick to javelin went through its own evolution that is indicative of the advance from visual processing to scenario visualization. The kind of toolmaking that our early *Homo* ancestors engaged in was likely to be little more than trial-and-error or imitative learning that was passed on from generation to generation. Flakes were constructed. So too, sticks were constructed. Apparently, however, it never occurred to members of these species to place one of their flaked stones on the end of a stick.

8.2 Step 2: The Spear

By the end of the Mousterian industry, archaic *H. heidelbergensis* and *H. neanderthalensis* had adopted a three-step stone-forming process, which allowed for the construction of a variety of tools. Also, stone flakes were placed on the ends of sticks as spears. The most basic step in constructing a stone tool has to do with simply striking a flake from a cobble.

When we consider that our early hominin ancestors not only had to select certain materials that were appropriate to solve some problem in a particular environment but also utilized a diverse set of stone working techniques involving a number of steps that resulted in a variety of tool types, it becomes apparent that a fairly advanced form of mental activity had to occur. Striking a sequence of flakes (knapping) in such a way that each one aids in the removal of others demands much more control of the brain, as well as a hand equipped with a variety of grips. The various steps in the process must be evaluated, and it may be the case that previous steps are seen in light of future steps. Wynn (1993) claims that tool behavior "entails problem solving, the ability to adjust behavior to a specific task at hand, and, for this, rote sequences are not enough" (pp. 396–397). This mental complexity has caused McNabb and Ashton (1995) to refer to our hominin toolmaking ancestors as "thoughtful flakers."

It is safe to say that the variety of tools constructed is evidence that these hominins were visualizing future scenarios in which these tools could be used; otherwise, *what would be the point of constructing a variety of tools in the first place*? Chimps use the same medium of sticks or rocks to hit, throw, or smash. However, the construction of a variety of tools indicates that the tools have a variety of purposes. What is the purpose in this context other than the formation of a visual image, the projection of that visual image onto some future scenario, and the intent to act on said visualization? The variety of tools is the material result of purposive scenario visualization. Following Wynn, Mithen (1996) notes that a mind with an ability to "think about hypothetical objects and events is absolutely essential for the manufacture of a stone tool like the handaxe. One must form a mental image of what the finished tool is to look like before starting to remove flakes from the stone nodule. Each strike follows from a hypothesis as to its effect on the shape of the tool" (p. 36).

8.3 *Step 3: The Javelin*

Around 40,000 ya, 60,000 years after the emergence of modern humans, we find evidence of a variety of types of javelins, spears, and javelin launchers. Archeologists such as Mithen (1996) and Wynn (1993) have shown that the construction of a javelin would require several mental visualizations, as well as numerous revisions of the material, so as to attain optimal performance of such a tool (also see Ambrose 2001; Churchill and Rhodes 2009). Such visualizations likely included the abilities to (a) identify horizontal or vertical lines, (b) select an image from several possible choices, (c) distinguish a target figure embedded in a complex background, (d) construct an image of a future scenario, (e) project an image onto that future scenario, and (f) recall from memory the particular goal of the project in the first place.

Different types of javelins with different shaped heads and shafts were constructed, depending upon the kind of kill or defense anticipated. If our early hominin ancestors tried simply to walk up to and hit a large animal, they likely would have been killed. In fact, this is probably what happened on more than one occasion to the early hominin working out of the environmental framework of the jungle in this totally new environmental framework of the savanna (Ambrose 2001; Churchill and Rhodes 2009). Eventually our ancestors, such as *H. neanderthalensis*, developed the spear; however, the evidence suggests that they could only develop spears, and not javelins (Mithen 1996; Ambrose 2001; Churchill and Rhodes 2009). *H. sapiens sapiens* developed javelins, equipped with launchers, that could be used in creative ways not only to throw from a distance but also to spear at close range, hack, and cut (Mithen 1996; Ambrose 2001; Churchill and Rhodes 2009).

Our hominin ancestors were living in social groups, watching and learning from each other. I am not suggesting that scenario visualization occurs in some solipsistic vacuum. Just as with other primates, our ancestors would have learned a lot from trial and error and other forms of mimetic expression in their social groups. At the same time, we can think of the proverbial "mad scientist" who might lock himself or herself away to work on some problem into which they have some insight. There are always innovators present in every social group. My suggestion is that, by 40,000 ya, the brains of our hominin ancestors were fortunate enough, through genetic variability, to have the right connections in their neural hardware so as to allow for the possibility of scenario visualization. With these *neural* connections already in place, all that was needed was some environmental cue to prompt the *psychological* connections, inferences, and insights to be made. All it takes is some psychologically creative "good trick" (to use the words of Dennett 1995)—implemented, possibly, by even one hominin—to get the creative juices flowing, so to speak, and prompt scenario visualization in our hominin ancestry. I would imagine that there would have been a complex interplay of trial-and-error and creative learning and implementation occurring in our hominin lineage with respect to negotiating environments, just as there is today.

Through the fortunes of genetic variability and natural selection, the brains of our hominin ancestors would have needed all the right neural connections in place to allow for scenario visualization. The hominins were living in social groups, learning from each other, and implementing behaviors through trial and error. This good trick is just that, a *useful device* for handling certain vision-related problems encountered by our ancestors, and the ones who could utilize it survived so as to pass their genes and memes (trial-and-error kinds as well as more innovative kinds) on to the next generation (Dawkins 1976; Blackmore 1999). Those of us in our species living today still retain this capacity.

9 The Harpoon

Below is a diagram that has to do with the construction of a harpoon. This schematization is supposed to represent the slower, intelligent processes associated with one of our early hominin ancestor's abilities to scenario visualize.

The diagram (Fig. 1) is based upon information gathered from Mithen (1996) and regarding the Angmagssalik hunters of Greenland and their construction of harpoons utilized to hunt seals. Their harpoons are fairly complex, having a spearhead equipped with a line attached to a flotation device, as well as several other parts designed to make the harpoon sturdy, accurate, and easy to throw. These hunters are an interesting case because it is likely that their harpoon technology has not changed much in thousands of years; thus, their technology can be studied to get a sense of what early hominin toolmaking may have been like.

In the schematization, I ask you to imagine that the problem to be solved has to do with throwing a projectile at a seal from a distance, for the purposes of killing it, skinning it, and using its body parts for food and warmth during the approaching winter months. I also ask you to imagine that this is the *very first instance* of some hominin coming up with the idea of the harpoon. At first, this particular hominin has no prior knowledge of the harpoon, but through the process of scenario visualization, he or she eventually "puts two and two together" and devises the mental blueprint for the harpoon. In other words, this is supposed to be a schematization of bissociative, nonroutine creative problem-solving at work in the early hominin mind.

In the first step, the hunter has separate visual images associated with the seal characteristics, the properties of objects in water, the manufacture of the bifaced hand ax, and the projectiles moving through the air. Consistent with Mithen's idea of cognitive fluidity, the visual information among these mental spheres has the potential to intermix and is represented by the dotted-line bubbles. Further, consistent with the data presented by developmental and evolutionary psychologists, there are several mental modules (dotted-line bubbles) that make up a person's mind. In the second step, scenario visualization is beginning as the animal biological, technological, and intuitive physics modules are bracketed off or segregated from the other mental modules. In the third step, the process of visualization is continuing because the hominin is manipulating, inverting, and transforming the images as they

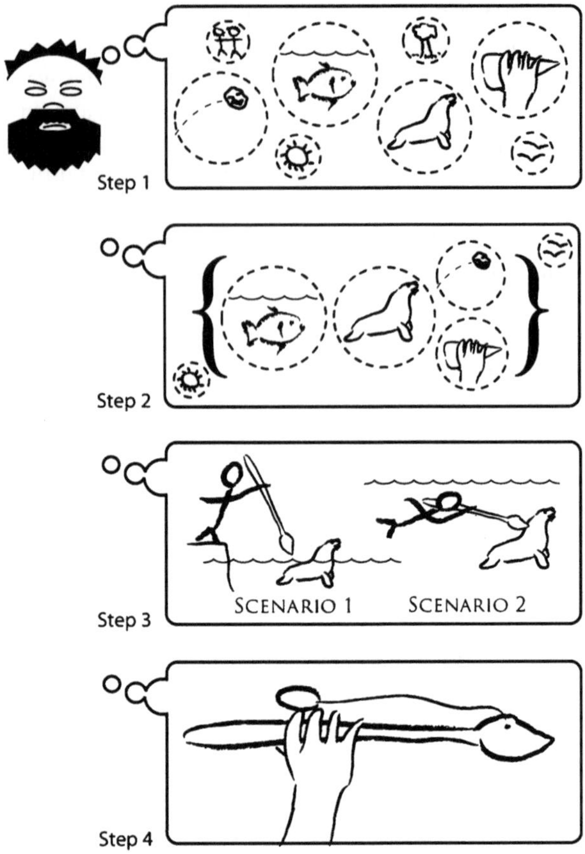

Fig. 1 The construction of a harpoon

are projected into a future imagined scenario. In the fourth step, these modules are actively integrated so that a wholly new image is formed that can become implemented in the actual production of the harpoon.

10 Conclusion

In this chapter, I presented the ideas and arguments put forward by evolutionary psychologists that our hominin ancestors evolved certain capacities to solve nonroutine, vision-related problems creatively. I argued that cognitive fluidity as well as what I call scenario visualization—viz., a mental activity whereby visual images are selected, integrated, and then transformed and projected into visual scenarios for the purposes of solving problems in the environments in which one inhabits—emerged in our hominin past and accounts for vision-related creativity. I hope that I have given a plausible account concerning a certain aspect of our mental architecture.

References

Aiello, L. (1997). Brain and guts in human evolution: The expensive tissue hypothesis. *Brazilian Journal of Genetics, 20*, 141–148.

Ambrose, S. (2001). Paleolithic technology and human evolution. *Science, 291*, 1748–1753.

Arp, R. (2005a). Scenario visualization: One explanation of creative problem solving. *Journal of Consciousness Studies, 12*, 31–60.

Arp, R. (2005b). Selectivity, integration, and the psycho-neuro-biological continuum. *Journal of Mind and Behavior, 6&7*, 35–64.

Arp, R. (2006). The environments of our Hominin ancestors, tool usage, and scenario visualization. *Biology and Philosophy, 21*, 95–117.

Arp, R. (2008). *Scenario visualization: An evolutionary account of creative problem solving*. Cambridge, MA: MIT Press.

Arrabales, R., Ledezma, A., & Sanchis, A. (2008). Criteria for consciousness in artificial intelligent agents. In *Proceedings of the autonomous agents and multiagent systems conference, 2008, Estoril, Portugal* (pp. 1187–1192).

Arrabales, R., Ledezma, A., & Sanchis, A. (2010). ConsScale: A pragmatic scale for measuring the level of consciousness in artificial agents. *Journal of Consciousness Studies, 17*, 131–164.

Blackmore, S. (1999). *The meme machine*. Oxford: Oxford University Press.

Boden, M. (1990). *The creative mind: Myths and mechanisms*. New York: Basic Books.

Boeckx, C., & Uriagereka, J. (2011). Biolinguistics and information. In G. Terzis & R. Arp (Eds.), *Information and living systems: Philosophical and scientific perspectives* (pp. 353–370). Cambridge, MA: MIT Press.

Bullot, N. (2011). Attention, information, and epistemic perception. In G. Terzis & R. Arp (Eds.), *Information and living systems: Philosophical and scientific perspectives* (pp. 309–352). Cambridge, MA: MIT Press.

Buss, D. (2009). The great struggles of life: Darwin and the emergence of evolutionary psychology. *The American Psychologist, 64*, 140–148.

Byrne, R. (2001). Social and technical forms of primate intelligence. In F. DeWaal (Ed.), *Tree of origin: What primate behavior can tell us about human social evolution* (pp. 145–172). Cambridge, MA: Harvard University Press.

Call, J., & Tomasello, M. (1994). The social learning of tool use by orangutans (*Pan pygmaeus*). *Human Evolution, 9*, 297–313.

Calvin, W. (2004). *A brief history of the mind: From apes to intellect and beyond*. Oxford: Oxford University Press.

Churchill, S., & Rhodes, J. (2009). *The evolution of the human capacity for "killing at a distance": The human fossil evidence for the evolution of projectile weaponry* (Vertebrate paleobiology and paleoanthropology: Special issue on the evolution of Hominin diets, pp. 201–210). Dordrecht: Springer.

Confer, J., Easton, J., Fleischman, D., Goetz, C., Lewis, D., Perilloux, C., & Buss, D. (2010). Evolutionary psychology: Controversies, questions, prospects, and limitations. *The American Psychologist, 65*, 110–126.

Cosmides, L., & Tooby, J. (1987). From evolution to behavior: Evolutionary psychology as the missing link. In J. Dupre (Ed.), *The latest on the best: Essays on evolution and optimality* (pp. 27–36). Cambridge, MA: Cambridge University Press.

Cosmides, L., & Tooby, J. (1994). Origins of domain specificity: The evolution of functional organization. In L. Hirschfeld & S. Gelman (Eds.), *Mapping the mind: Domain specificity in cognition and culture* (pp. 71–97). Cambridge, MA: Cambridge University Press.

Crick, F. (1994). *The astonishing hypothesis*. New York: Simon & Schuster.

Crick, F., & Koch, C. (2003). A new framework for consciousness. *Nature Reviews Neuroscience, 6*, 119–126.

Damasio, A. (2000). A neurobiology for consciousness. In T. Metzinger (Ed.), *Neural correlates of consciousness* (pp. 111–120). Cambridge, MA: MIT Press.

Dawkins, R. (1976). *The selfish gene*. Oxford: Oxford University Press.
Dawkins, R. (2005). *The ancestor's tale: A pilgrimage to the dawn of evolution*. New York: Mariner Books.
De Beaune, S., Coolidge, F., & Wynn, T. (2009). *Cognitive archeology and human evolution*. Cambridge: Cambridge University Press.
Dennett, D. (1995). *Darwin's dangerous idea: Evolution and the meanings of life*. New York: Simon & Schuster.
Dominowski, R. (1995). Productive problem solving. In S. Smith, T. Ward, & R. Finke (Eds.), *The creative cognition approach* (pp. 73–96). Cambridge, MA: MIT Press.
Donald, M. (1997). The mind considered from a historical perspective. In D. Johnson & C. Erneling (Eds.), *The future of the cognitive revolution* (pp. 355–365). New York: Oxford University Press.
Downes, S. (2008). Evolutionary psychology. In *Stanford encyclopedia of philosophy*. Retrieved from http://plato.stanford.edu/entries/evolutionary-psychology/
Fauconnier, G., & Turner, M. (2002). *The way we think: Conceptual blending and the mind's hidden complexities*. New York: Basic Books.
Fodor, J. (1998). *In critical condition: Polemical essays on cognitive science and the philosophy of mind*. Cambridge, MA: MIT Press.
Gardner, H. (1993). *Multiple intelligences: The theory in practice*. New York: Basic Books.
Goguen, J., & Harrell, D. (2004). Style as a choice of blending principles. In S. Argamon, S. Dubnov, & J. Jupp (Eds.), *Style and meaning in language, art, music and design* (pp. 49–56). New York: American Association for Artificial Intelligence Press.
Gomila, T., & Calvo, P. (2008). Directions for an embodied cognitive science: Toward an integrated approach. In P. Calvo & T. Gomila (Eds.), *Handbook of cognitive science: An embodied approach* (pp. 1–26). Oxford: Elsevier.
Gregory, R. (Ed.). (2004). *The Oxford companion to the mind*. Oxford: Oxford University Press.
Hampton, S. (2010). *Essential evolutionary psychology*. Thousand Oaks: SAGE Publishers.
Isaac, G. (1986). Foundation stones: Early artifacts as indicators of activities and abilities. In G. Bailey & P. Callow (Eds.), *Stone age prehistory* (pp. 221–241). Cambridge: Cambridge University Press.
Jarman, R. (2009). Review of scenario visualization: An evolutionary account of creative problem solving. *Journal of Consciousness Studies, 16*, 199–208.
Kandel, E., Schwartz, J., & Jessell, T. (Eds.). (2000). *Principles of neural science*. New York: McGraw-Hill.
Kant, I. (1929). Critique of Pure Reason, Norman Kemp Smith, trans. New York: St. Martin's Press.
Kaufman, A., & Kaufman, J. (2009). Review of scenario visualization: An evolutionary account of creative problem solving. *American Journal of Human Biology, 21*, 199–208.
Koestler, A. (1964). *The act of creation*. New York: Dell.
Langland-Hassan, P. (2009). A puzzle about visualization. *Phenomenology and the Cognitive Sciences, 10*, 145–173.
Lonsdorf, E., Ross, S., & Matsuzawa, T. (Eds.). (2010). *The mind of the chimpanzee: Ecological and experimental perspectives*. Chicago: University of Chicago Press.
Mayer, R. (1995). The search for insight: Grappling with Gestalt psychology's unanswered questions. In R. Sternberg & J. Davidson (Eds.), *The nature of insight* (pp. 3–32). Cambridge, MA: MIT Press.
McGrew, W. (2004). *The cultured chimpanzee: Reflections on cultural primatology*. Cambridge: Cambridge University Press.
McHenry, H. (1998). Body proportions in *A. afarensis* and *A. africanus* and the origin of the genus *Homo*. *Journal of Human Evolution, 35*, 1–22.
McNabb, J., & Ashton, N. (1995). Thoughtful flakers. *Cambridge Archeological Journal, 5*, 289–301.
Merchant, L., & McGrew, W. (2005). Percussive technology: Chimpanzee baobab smashing and the evolutionary modeling of hominid knapping. In V. Roux & B. Bril (Eds.), *Stone knapping: The necessary conditions of a uniquely hominid behaviour* (McDonald Institute monograph series, pp. 339–348). Cambridge: McDonald Institute for Archaeological Research.
Mithen, S. (1996). *The prehistory of the mind: The cognitive origins of art, religion and science*. London: Thames and Hudson.

Mithen, S. (1999). Handaxes and ice age carvings: Hard evidence for the evolution of consciousness. In S. Hameroff, A. Kaszniak, & D. Chalmers (Eds.), *Toward a science of consciousness: The third Tucson discussions and debates* (pp. 281–296). Cambridge, MA: MIT Press.
Mithen, S. (2001). Archeological theory and theories of cognitive evolution. In I. Hodder (Ed.), *Archeological theory today* (pp. 98–121). Cambridge: Polity Press.
Mithen, S. (2005). *The singing Neanderthals: The origins of music, language, mind and body*. London: Weidenfeld and Nicolson.
Norris, J., & Papini, M. (2010). Comparative psychology. In I. Weiner & W. Craighead (Eds.), *The Corsini encyclopedia of psychology* (pp. 507–520). Malden: Wiley-Blackwell.
O'Connor, M., Fauri, D., & Netting, F. (2010). How data emerge as information: A review of scenario visualization. *The American Journal of Psychology, 123*, 371–373.
Palmer, J., & Palmer, A. (2002). *Evolutionary psychology: The ultimate origins of human behavior*. Needham Heights: Allyn and Bacon.
Pearce, J. (2008). *Animal learning and cognition: An introduction*. New York: Psychology Press.
Pelegrin, J. (1993). A framework for analyzing stone tool manufacture and a tentative application to some early stone industries. In A. Berthelet & J. Chavaillon (Eds.), *The use of tools by human and non-human primates* (pp. 302–314). Oxford: Clarendon.
Picciuto, E., & Carruthers, P. (2008). Creativity explained? Review of scenario visualization: An evolutionary account of creative problem solving. *Evolutionary Psychology, 6*, 427–431.
Rivera, F. (2010). *Toward a visually-oriented school mathematics curriculum: Research, theory, practice, and issues*. London: Springer.
Ruse, M. (2006). *Darwinism and its discontents*. Cambridge: Cambridge University Press.
Searle, J. (1992). *The rediscovery of the mind*. Cambridge, MA: MIT Press.
Singer, W. (2000). Phenomenal awareness and consciousness from a neurobiological perspective. In T. Metzinger (Ed.), *Neural correlates of consciousness* (pp. 121–138). Cambridge, MA: MIT Press.
Sloman, A., & Chappell, J. (2005). The altricial-precocial spectrum for robots. In *Proceedings of the international joint conferences on artificial intelligence: 2005, Edinburgh, Scotland* (pp. 1–8).
Smith, S., Ward, T., & Finke, R. (Eds.). (1995). *The creative cognition approach*. Cambridge, MA: MIT Press.
Sugu, D., & Chatterjee, A. (2010). Flashback: Reshuffling emotions. *Cognitive Computation, 1*, 109–133.
Tallerman, M. (Ed.). (2005). *Language origins: Perspectives on evolution*. New York: Oxford University Press.
Thomas, N. (2010). Mental imagery. In *Stanford encyclopedia of philosophy*. Retrieved from http://plato.stanford.edu/entries/mental-imagery/
Tononi, G., & Edelman, G. (1998). Consciousness and integration of information in the brain. *Advances in Neurology, 77*, 245–279.
Velmans, M. (1992). Is consciousness integrated? *The Behavioral and Brain Sciences, 15*, 229–230.
von Bayern, A., Heathcote, R., Rutz, C., & Kacelnik, A. (2009). The role of experience in problem solving and innovative tool use in crows. *Current Biology, 19*, 1965–1968.
Watanabe, S., & Huber, L. (2006). Animal logics: Decisions in the absence of human language. *Animal Cognition, 9*, 235–245.
Weichart, A. (2009). Sub-symbols and icons. *International Journal on Humanistic Ideology, 1*, 342–347.
Whiten, A. (2010). A coming of age for cultural panthropology. In E. Lonsdorf, S. Ross, & T. Matsuzawa (Eds.), *The mind of the chimpanzee: Ecological and experimental perspectives* (pp. 87–100). Chicago: University of Chicago Press.
Whiten, A., Goodall, J., McGrew, W., Nishida, T., Reynolds, V., Sugiyama, Y., Tutin, C. E. G., Wrangham, R. W., & Boesch, C. (1999). Cultures in chimpanzees. *Nature, 399*, 682–685.
Wynn, T. (1993). Layers of thinking in tool behavior. In K. Gibson & T. Ingold (Eds.), *Tools, language and cognition in human evolution* (pp. 389–406). Cambridge, MA: Cambridge University Press.

Representation in Biological Systems: Teleofunction, Etiology, and Structural Preservation

Michael Nair-Collins

Abstract In this chapter I propose a novel thesis about the nature of representation in biological systems. I argue that what makes something a representation is distinct from what determines representational content. As such, it is useful to conceptualize *what it is to be* a representation in terms of fundamental concepts from biology, particularly the concept of a biological function (or teleofunction). By contrast, representational *content* is best understood as a structured relation involving two parts, and the explanation of how states of biological systems have content involves the preservation of internal structural relations and causal history.

I review recent literature on the neurophysiologic mechanisms underlying a sensory discrimination task, in which neurons use a variety of mechanisms for encoding, storing, and comparing information about vibrotactile stimuli. These mechanisms include a one-to-one burst code, a temporal code in which periodicity is the operative mechanism, and a variety of rate codes, some with opposite slopes, and some reflecting neither the base nor comparison stimuli, but rather their quantitative difference. In motor cortex, a binary behavioral outcome is reflected in a sigmoidal shape of firing patterns. A theory of biological representation, if it is to be empirically useful, ought to be able to unify these various encoding mechanisms under an overarching conceptual framework that explains what biological representation is and how representational content is determined, from a general standpoint, and I suggest that the theory on offer takes significant steps toward this aim.

M. Nair-Collins (✉)
Medical Humanities and Social Sciences, Florida State University College of Medicine, Tallahassee, FL, USA
e-mail: michael.nair-collins@med.fsu.edu

L. Swan (ed.), *Origins of Mind*, Biosemiotics 8, DOI 10.1007/978-94-007-5419-5_8,

1 Introduction

Representation is a foundational concept. At its core, it is simply *aboutness*, pointing-to, or standing-in-for. For example, my belief that I have a cup of coffee on my desk is *about* the cup of coffee. As I consider what would happen were I to turn it upside down, the processes involved in counterfactual reasoning and visual imagery involve states that "stand in for" or represent the cup in different positions, the likely outcomes such as coffee spilling on my desk, and so on. This basic concept of "aboutness" is appealed to routinely – in various incarnations – in the cognitive sciences, the neurosciences, our commonsense psychology, and in the philosophy of mind and language. It is used to explain many aspects of neurological and cognitive functioning as well as adaptive (and maladaptive) behavior. Indeed, we might reasonably consider the concept of representation to be the *single* foundation upon which our understanding of mind rests. Yet it is widely agreed that we lack an adequate naturalistic understanding of representation and its place in the physical world. There is something deeply mysterious about how physical systems have states that bear this sense of "directedness," particularly given that such systems can make errors and can represent counterfactual scenarios, both of which seem to imply a relation between the representation and a nonexistent state of affairs. My purpose in this chapter is to propose a thesis about the nature of representation in biological systems.

As mentioned above, the concept of representation gets imported into a number of distinct theoretical approaches to understanding the mind/brain and behavior, from neurophysiology, to cognitive psychology, to our commonsense belief-desire psychology. My purpose in this chapter however is only to address the primitive or basic representational states instantiated in the nervous systems of living biological organisms, from which more complicated states presumably arise.

2 Representation: Accuracy, Error, and Logical Structure

Not everything in the universe is a representation. This is obvious, surely, but the question then arises as to what differentiates things that do, from things that do not, bear representational content. One of the most prominent responses given among philosophers of mind is that representations are states that are truth or satisfaction-evaluable, meaning that they are states that can be evaluated as to whether they are accurate or inaccurate, satisfied or not. For example, my belief that there is a cup of coffee on my desk is truth-evaluable; it might be accurate, or the belief might be inaccurate. Suppose I have an intention to pick up the cup; that intention might be satisfied (i.e., I might actually pick up the cup) or not. Indeed, the problem of incorporating an understanding of *misrepresentation* into an account of representation is perhaps the single most discussed problem in the last 30 years of work on mental representation in analytic philosophy.

This is a key conceptual point that bears emphasis. It is common in neuroscience to assume something like an implicit causal theory, wherein neurons or ensembles of neurons that are differentially responsive to certain forms of energy at the periphery (e.g., edge detectors in primary visual cortex) are taken to represent what typically causes them to fire (e.g., bars of light at a particular orientation relative to the retina; cf. Bechtel 2001, for discussion). Farther downstream, other neurons are assumed to take the information encapsulated in the firing of edge detectors with affinities for specific orientations and generate progressively more complex and abstract representations of visually encoded objects.[1] However, simply differentially responding to (i.e., being caused by) specific kinds and levels of energy is not sufficient for something's having representational content.[2] A tropical storm system, for example, is differentially responsive to specific causal factors involving atmospheric pressure and temperature, wind speed and direction, and so forth. But the states of that system do not bear representational content, and there is no sense in assigning to them semantic properties such as accuracy or error. If a state of a system is not truth- or satisfaction-evaluable, then there is no distinction between its simply having a causal history or playing some causal role (which everything does) and its being a representation, being an encoding, bearing representational content, etc. Representations, of course, play causal roles as well, but they are also semantically evaluable; indeed, this is what generates the mystery in the first instance.

It also bears emphasis that the concept of truth-evaluability is not specific to human languages or linguistically expressed beliefs and desires. Presumably, the honeybee's dance represents the location of nectar to its conspecifics, with variables on the dance structure such as tempo and the angle of its long axis corresponding to variables on nectar location such as distance from the hive and direction relative to the sun (cf. Millikan 1984, chapters 2 and 6). Such dances are semantically evaluable: The dancing bee can send its conspecifics directly to the nectar by accurately representing its location, or it can send them in the wrong direction by misrepresenting the location of nectar.

Similar comments can be made regarding early perceptual and discriminatory processes: Rats are able to use their whiskers to discriminate the size of an aperture in order to select one of two options that will lead to their acquisition of a food pellet in a laboratory task (Nicolelis and Ribeiro 2006; cf. Swan and Goldberg 2010). Supposing the animal's trained task is to press the left button when the aperture is narrow (vs. wide), the animal might be in error by pressing the right button instead. In this case, its behavioral signal is incorrect; this might be a result of any number of

[1] The classic "hierarchical processing" view of visual representation adumbrated above is of course complicated by the fact that feedback modulation occurs at every hierarchical level, even prior to primary visual cortex (V1) in the lateral geniculate nucleus of the thalamus. But that does not alter the basic conceptualization of the representational capacity of early sensory neurons as being grounded in a specific causal etiology.

[2] This is not to say that edge detectors are, or are not, representational; rather, it is to say that if they are, it is not solely in virtue of their affinity for firing in response to certain types of energy impinging on the periphery.

factors. Its early perceptual encoding and discriminatory processes might be in error by encoding the width as wide when it is in fact narrow; its short-term memory might be in error by losing the informational content from early perceptual processes as it transforms sensory and mnemonic information into motor plans; its long-term memory might be inaccurate by reversing the task instructions (e.g., recalling the task instructions as to press the right button for narrow rather than the left); and its motor command processing might generate a motor output different than what the system had intended (e.g., pressing the right rather than left button). Each of these would lead to the behavioral manifestation of task error in a given trial. But each of these states, from perceptual discrimination to short- and long-term memory, to motor plans, to behavioral output, is semantically evaluable in the sense that it can be accurate or inaccurate (for sensory and mnemonic representations as well as behavioral signals of the choice made), or satisfied or not (for motor plans). By contrast, the states of a tropical weather system, though such systems are nearly as causally complex, are not amenable to such interpretation and are not representational. *Representational content demands the possibility of accuracy or error*, and this has a significant consequence for a theory of representation in biological systems.

For a state to bear representational content and hence be truth- or satisfaction-evaluable, it must be logically structured. A linguistic example is instructive here. The sentence, "Johnny has green hair," let us assume, is true. It is true in virtue of (1) the subject term "Johnny" refers to, or points to, Johnny, thereby rendering the sentence itself as referring to Johnny, and (2) the predicate term "has green hair" predicates the property of *having green hair* of whatever thing the sentence refers to. In this case, that thing is Johnny. Furthermore (we are assuming), Johnny does indeed instantiate the property of having green hair, and therefore, the sentence is true; if he did not, then the sentence would be false. This basic linguistic distinction between subjects and predicates maps onto the ontologically basic distinction between individuals and the properties that they bear, with subjects referring to individuals and predicates applying to properties. I'll henceforth refer to the relation between subject and object as *reference* and the relation between predicate term and property as *predication*. It is crucial to recognize that subjects, or referential terms, in and of themselves, are not truth-evaluable, and neither are individual predicates truth-evaluable. The term "Johnny" is neither true nor false, and the term "has green hair" is neither true nor false. It is only their concatenation, or joining together in a unified semantic construct, that renders truth- or satisfaction-evaluability, and hence accuracy or error, possible. Thus, neither reference nor predication alone, in the absence of their logical concatenation, suffices to generate representational content. Representational content demands the possibility of accuracy or error, as discussed above, and accuracy and error do not occur except in the context of a representation that bears logical structure.

This concatenation, or logical structure, need not imply physico-mechanical or symbolic structure. In natural languages, the logical structure of sentences supervenes on its syntactic structure, itself realized by orthographic or phonetic structural properties (for written and spoken utterances, respectively). However, even apparently (physically) unstructured entities can bear logical structure. By "logical structure" I mean simply that the vehicle of representation both refers to a thing and predicates a property of that thing. An example that Devitt and Sterelny use in discussing whether representations can be simple is the yellow flag once hung on a ship's mast to signify to other passing ships that the ship has yellow fever (Devitt and Sterelny 1999, 139). This seems like a simple, nonstructured vehicle of representation, but it isn't, at least not in the sense that I'm using the term. The fact that the flag *is yellow* signifies that whatever ship is flying it has yellow fever. But it is not the yellowness of the flag that signifies *which ship* has yellow fever. The fact that the flag is attached to *this* ship's flagpole is what determines the referent of the predicate, "has yellow fever," as this particular ship. Thus, different aspects of a vehicle of representation can determine different aspects of its representational content; logical structure can, but need not, map onto physical or symbolic structure.[3]

Building on this background, I'll next briefly outline a proposed theory of representation in biological systems, followed by an illustrative example appealing to recent work on the neurophysiological mechanisms involved in Macaque (and by extension human) vibrotactile discrimination.

3 A Theory of Representation: Teleofunction, Etiology, and Structural Preservation

There is a conceptual distinction between what makes something a representation and, given that a thing is a representation, what determines its content. The former involves the metaphysics of what it is to be a representation, and the latter, the

[3] The argument I'm building here is that the fundamental semantically evaluable units are themselves truth-evaluable; hence, those units bear logical structure in the sense I'm using the term here. A different possibility is that the basic semantically evaluable units are not themselves truth-evaluable, but are instead something like subsentential units that concatenate to form larger sentence-like, truth-evaluable complexes. These fundamental units are like words in a language of thought, admitting of syntactic rearrangement which generates the productivity and systematicity of the language of thought, itself responsible for the productivity and systematicity of natural languages (Fodor 1975, 2008). This is the (or at least one of the) standard view(s) in classical cognitive science. However, the key step is the concatenation of numerically distinct, neurologically instantiated symbols: How does it work? How and why do those two neurologically instantiated symbols "come together" in that particular thought, and not some others? In virtue of what is this complex well-formed in its neurological syntax? In virtue of what are these symbols "joined together"? The appeal to concatenating neurologically instantiated symbols at the lowest level introduces a new binding problem: How and why do those particular symbols join together, excluding others, and in what does this joining consist? Just like the more familiar binding problem of explaining how different aspects of an experience (e.g., bluishness and squareness) join together in the brain to form a coherent, unified percept (e.g., as of a blue square), the *syntactic binding problem* demands an explanation for how distinct symbols join together to form a unified meaningful mental representation. If, however, the fundamental semantic units are, as I suggest, themselves logically structured and hence truth-evaluable, then the syntactic binding problem is avoided for those units. Furthermore, many suppose that even the lowest-level sensory states can *accurately* or *inaccurately* reflect peripheral energy states. If that is the case, it follows that the sensory states must have logical structure because neither accuracy nor inaccuracy is possible without it, as argued in the text. There is of course a great deal more to be said on this issue, but I will leave further discussion for a different venue.

semantics of representational content. To compare, consider the difference between what makes something money and, given that a thing is money, what determines its particular value (Michael Levin proposed this analogy in conversation). Although the conditions that determine each are closely related (involving complex relations and interactions among social agents), there is nonetheless a conceptual distinction between a thing's being money and, given that it is money, what its particular value is. For example, the value of a dollar, understood in terms of its relative purchasing power either locally or globally in exchange for foreign units of currency, fluctuates. But its status as *being money* (at all) does not; therefore, they are conceptually distinct.

This distinction is helpful in the present context as follows. Representations are states of biological organisms. As such, it is useful to conceptualize *what it is to be* a representation in terms of fundamental concepts from biology, particularly the concept of a biological function (or teleofunction). Just as hearts have the function of circulating oxygenated blood, but can fail to do so, representational states of the nervous system also have biological functions to play (but can fail to do so). Living, mobile organisms have the capacity to selectively respond to labile environmental conditions – in ways that reflect those changing conditions – which enables them to maintain physiologic stability, to avoid predation, or to reproduce. The behavioral flexibility that manifests as appropriate responses to changing environmental conditions is rooted in the organism's capacity to represent both internal and external conditions; more specifically, *what it is to be* a representation is to have the biological function of bearing certain correspondence relations, as follows.

Some things have the biological function of corresponding to environmental conditions in such a way that other states, the *users* or *consumers* of the first, use the state of the first in reacting appropriately to changing internal or external conditions. Other things have the biological function of producing or helping to produce the states to which they correspond. The former are indicative or sensory representations, and the latter are procedural representations or motor plans (cf. Millikan 1984, 1989, 2004).

For example, the nematode *C. elegans* performs chemotaxis, or oriented movement in response to a chemical stimulus, to locate its primary food source of bacteria. The chemotaxis circuit includes four pairs of chemosensory neurons, four pairs of interneurons, and five pairs of motor neurons (Bargmann and Horvitz 1991). *C. elegans* neurons exhibit graded voltage potentials (rather than action potentials); the voltage of the chemosensory neurons at the tip of its nose bears specific correspondence relations to the concentration of chemoattractant in the environment, whereby increases in chemical concentration correspond to proportional increases in voltage. By comparing the scalar value of the current chemical environment to its first derivative (i.e., the change in chemical concentration as the nematode moves), the sensory and interneurons generate a signal to the motor neurons, which then generate a motor output signal to the neck muscles, enabling the animal to orient itself up the chemical gradient and toward food (Ferree and Lockery 1999; cf. Mandik et al. 2007, for computer simulations of evolved neural network control of chemotaxis). In this example, the sensory neurons have the teleofunction of bearing specific correspondence relations to the concentration of chemoattractant at the periphery of the organism. In virtue of the sensory neurons realizing this correspondence relation, the interneurons and motor neurons are able to use that information to generate output signals appropriate to the local environment by comparing the present concentration to the change in concentration in order to determine in which direction the gradient increases. Thus, the changing voltages of the chemosensory neurons are sensory or indicative representations. The motor neurons evince similar proportional changes in voltage relative to the degree of extension of specific muscles in the neck which determine the neck's turning angle, and the activity of the motor neurons is causally relevant to producing those specific turning angles. Thus, these neurons have the function of producing (or causing) the muscle states to which they correspond, and should be considered procedural representations or motor plans. What it is to be a representation, therefore, is to have the biological function of bearing specific correspondence relations which enable adaptive behavior of the organism of which those states are a part.

However, as discussed in the previous section, representational *content* demands the possibility of accuracy or error, which in turn requires logical structure. Having the biological function to bear specific correspondence relations to environmental or muscle states is insufficient for generating logical structure, and thus is insufficient for generating representational content. In order to explain what determines representational content (as opposed to what makes a thing a representation at all), some analogue of predication and reference must be built in to the theory. I emphasize again that these concepts are not specific to language, but instead map onto the basic ontological distinction between properties and the bearers of properties. Even the representational states of worms – if they are to bear representational content and thus admit of accuracy and error – must both refer to a thing and predicate some property of that thing. The states of the chemosensory neurons of *C. elegans*, for example, might predicate *having concentration X of chemoattractant* (a property) of the immediate environment located at the tip of its nose (a *thing* of which the property is predicated). Of course, the worm does not use words like "concentration," "chemoattractant," or "local environment" to *express* such representational contents, but this does not imply that its neural states do not thereby *have* that representational content.

I propose that what determines representational content is a combination of causal etiology and isomorphism. As discussed above, it is common in neuroscience to implicitly presume some version of a causal theory of representation, whereby states of the nervous system are taken to represent what typically causes them, or what they typically cause. Although this is an insufficient condition on being a representation, it is nonetheless a key component of a theory of representational content. However, it is also well understood from the philosophy of mind literature how profoundly difficult it is to make sense of the possibility of error, given a purely causal theory of representational content.

There are two kinds of causal theories: causal history (or etiology) and counterfactual covariation. Causal history theories state that representations represent whatever caused them. In this circumstance, it should be obvious that error is impossible: Representation R represents precisely its causal antecedent; therefore, no sense can be had in stating that the representation is in error. The frog that snaps after a passing bit of darkly colored leaf blowing erratically in the wind, which resembles a fly, cannot be said to have misrepresented the leaf as a fly. Instead, it must be said that the frog correctly represented the leaf, but then it is difficult to make sense of why the frog snapped at it. To deal with such problems, the concept of counterfactual covariation was introduced, in which representational states are taken to represent whatever they counterfactually causally covary with, perhaps under ideal circumstances, or ideal circumstances in the environment of evolutionary origin. But a different set of problems then arise, the most significant of which is that attempting to discern the item or property of maximal counterfactual covariance inevitably leads to a disjunction of such things and thereby, again, the impossibility of error. For example, the states of the frog's nervous system which are typically taken to represent the fly as food do not maximally covary with flies, but rather with the disjunctive property *fly-or-passing-leaf*. In this circumstance, error is again impossible because the frog correctly represents the passing leaf as *fly-or-passing-leaf*, but it seems clear that we should say that the frog has mistaken the leaf for a fly. That's why the frog snapped at it.

However, there is also wisdom in causal theories, which (I suspect) is why they are implicitly presumed in the neuroscience literature and why so much energy has been expended in the philosophical literature to attempt to correct their serious deficiencies. To appreciate why causal etiology is relevant, consider the parallels between reference and causation. The basic problem with causal theories is that a causal relation either obtains or does not, and if it does, it becomes very puzzling to say why in some circumstances, but not others, this causal relation should determine representational content. But reference (alone), like causation, either obtains or does not. There is no such thing as "mis-reference"[4]; semantically evaluable units must either succeed or fail in referring (to anything). Thus, while we cannot reduce representational content to causal etiology because of the impossibility of error, we can reduce *reference* to causal etiology, without needing the possibility of "error." Referring expressions are neither true nor false; rather, they either refer or they don't. In explaining reference in terms of causal etiology, however, it should be understood that causal history determines the object or thing that the representation

[4] We'll need to be careful here: If I "refer" to my dog Mac as "that cat," it might seem that I've mis-referred, but I haven't. Rather, the ostensive act referred to an individual, and I predicated the property *catness* of it. The reference relation obtained, whereas I misapplied a predicate of that to which I referred. On the other hand, there are tricky issues regarding reference to nonexistents; can I refer to Sherlock Holmes or unicorns? These are larger issues in the philosophy of language which will not be addressed here; better to understand the simpler kinds of representation first. If you like, consider my claim that there is no mis-reference as both axiomatic and using the word "reference" to mean something like, only the most fundamental kind of reference. The argument for accepting any axiom is, of course, dependent on how well the theory constructed from that axiom works.

is about, but does not determine the property that the representation predicates of that thing.

A different and much older idea says that representation is a picturing or resemblance relation, where the vehicle of representation bears structural similarities to, or shares properties with, that which it represents. The guiding idea here is that there is a kind of resemblance or "mirroring" between representation and represented in virtue of which the representation relation obtains. The strength of this view is its intuitive appeal: A realistic portrait of President Obama represents President Obama himself, due to the structural similarities, or the resemblance, between the two. However, due to a number of problems with a simple resemblance view, among them that resemblance is symmetric while representation is not, resemblance was abandoned long ago as a viable theory of representation. It has lately been revived, however, by appealing to a more sophisticated form of resemblance, namely, an isomorphism among a *system* of representations and a *system* of states of affairs, rather than a structural similarity between the token vehicle of representation and whatever it represents.

On this latter theory, the guiding motivation is the same: The preservation of internal structural relations between representation and represented is of the essence of representation. However, the structural similarity obtains between a set of items and relations on that set, and another set of items and relations on it. By appealing to systems of states of representational vehicles and transformations over them, a more abstract kind of resemblance can obtain, which need not respect any first-order structural similarity between a token vehicle of representation and its content. This is important because for the most part, a first-order picturing or mirroring relation does not hold between brain states and world states (e.g., the chemosensory neurons of *C. elegans* do not share first-order structural similarities with the changing chemical concentration at the tip of its nose, in the same way that a realistic portrait of President Obama shares a first-order structural similarity with President Obama himself).

While the system isomorphism approach is in many ways an improvement over its ancestor, it still faces many of the same problems. The most important of these is the problem of multiple isomorphisms. If isomorphism is the sole determinant of content, then it seems to follow that representations are about or represent far too many things. For example, given any relational system (i.e., a set with relations on it), there exist infinitely many relational systems to which it is isomorphic; furthermore, given two isomorphic systems, there exist numerous if not infinitely many distinct mappings between those two systems that preserve isomorphism equally well. Apparently, this would seem to preclude the possibility of false representations since a representation may be true under one mapping but false under another, and if there is no principled means of selecting among the numerous mappings, then there seems no way to account for error.

Consider, however, the parallels between predication and isomorphism. Unlike causation, and unlike reference, predication is not specific. The predicate "has green hair" applies to all and only the things that have green hair; predicates are multiply applicable because properties are multiply instantiated, unlike individuals which are not. Unless concatenated with a referential expression, a predicate does not apply to any specific individual. But notice that this is precisely the problem with isomorphism-based theories: They are not specific. The multiplicity of isomorphisms, and the multiplicity of things to which predicates apply (due to the multiple instantiability of properties), suggests that isomorphism or something like it is the element responsible for predication in basic representations.

More specifically, states of individual neurons or ensembles of neurons admit of certain transformations that realize an ordering relation over those states, resulting in empirical relational systems. Firing rate, for example, admits of transformations by increasing or decreasing how quickly action potentials fire; the set of firing rates ordered by the greater-firing-rate relation constitutes an empirical relational system. Similar remarks apply to neurons that admit of graded voltage potentials, ordered by the greater-voltage relation. Furthermore, transducible energy states impinging on the periphery of an organism can be ordered according to transformations in similar fashion, resulting in relational systems composed of distinct energy states and transformations over them. For example, the set of concentrations of chemoattractant in the local environment can be ordered by the greater-chemoattractant relation, resulting in a relational system. The idea is that representations are not found in biological organisms as punctate atoms, but rather there are *systems* of representations, the members of which are organized in such a way that those systems are isomorphic to different organized systems of representeds. A mapping, or mathematical function, from the elements of one system to the elements of the other maps states of one system (say, a particular firing rate) to states of the other (say, a particular frequency of vibration at the skin) so that that particular firing rate predicates the property of vibrating at that particular rate. This mapping just is the specific correspondence relation mentioned above, which these representational states have the teleofunction of bearing.

Furthermore, there is no need to constrain this idea to the activity of single neurons. Populations of neurons can be described using vectors and relations on them, and multivalued functions between higher-order relational systems and other relational systems describing energy states can define isomorphisms between systems. On the represented side, anything can be a member of a relational system, not just parametric energy states at the periphery of the organism. Thus, in addition to mechanical, electromagnetic, thermal, and other forms of energy, relational systems may include things like predator, food source, conspecific, shelter, etc. There is also no reason to suppose that the mappings between relational systems must involve linear or even monotonic relations.[5] They can be sigmoidal, quadratic, or anything at all. Finally, for marshaling the concept for use in a theory of biological representation, there seems no reason to maintain the relatively strict technical requirements imposed by

[5] Akins (1996), for example, argues that the "traditional naturalist" project of Dretske (1981, 1988), Fodor (1987, 1990), Millikan (1984, 2004), and others rests on a mistaken view of the senses, which is that they must be "veridical." Akins argues instead that sensory systems are not veridical but are what she calls "narcissistic." That is, they do not "dispassionately" report what is going on out in the world, but instead are highly dependent on local context (as in, "what does this mean for

the mathematical construct of isomorphism. There are numerous ways to extend or relax these technical constraints while maintaining the fundamental aspect of the preservation of internal relational structure (for some examples, see Swoyer 1991). I use the term *structural preservation* to refer to the class of structure-preserving relations between relational systems that includes isomorphism, homomorphism, and several others, which are weakened versions of these constructs.

Before delving into a detailed example to illustrate the theory, I'll summarize the main ideas. Not everything is a representation; what differentiates things that are, from things that are not representations, is semantic evaluability, which requires the possibility of accuracy or error. This applies to even the simplest biological organisms, not just language-using humans. Furthermore, the possibility of accuracy or error requires logical structure, or a concatenation of some analogue of reference and predication, where reference maps subjects to objects (or things) and predication maps predicates onto properties. However, logical structure need not imply physico-mechanical or symbolic structure; rather, different aspects of a vehicle of representation might be responsible for the different aspects of representation.

There is a conceptual distinction between what makes something a representation and what determines representational content; a theory of representation must explain both. I've suggested that what makes a thing a representation (at all) is its having the teleofunction of bearing certain correspondence relations which enable the organism to respond appropriately to changing environmental conditions. However, to explain representational content, an explanation of both reference and predication is required (because logical structure is required), and the teleofunctionally determined correspondence relations are, by themselves, insufficient to explain both components. However, I've suggested that causal etiology is the aspect of a representational vehicle that determines the thing to which it refers. Furthermore, isomorphism between systems of representations and systems of representeds determines the specific property predicated of the thing to which the representation refers. The correspondence relations that the state has the teleofunction of bearing to energy states at the periphery just are the mappings between relational systems that determine isomorphism and match up, one-to-one, states of the representational

me, the receptor?"). This objection is somewhat strange in that what *constitutes* veridical representation is precisely the question. Thus, in order to say that sensory systems are not veridical, one must first be committed to some theory of representational content. Her claims that thermoreceptive systems are not veridical, therefore, cannot be used as an objection to the very project of understanding veridicality itself. Akins, apparently, considers thermoreceptors and the neural machinery attached to them to be narcissistic and non-veridical because they do not have linear response profiles, but instead have very complicated response profiles depending on local context. This doesn't show that they are not veridical, just that they behave according to complicated nonlinear correlations to the environment, and can change in different contexts. These complicated response profiles nonetheless describe mapping functions between relational systems composed of neural activity and relational systems composed of energy states, and bearing these response profiles may very well be what these thermoreceptors and other neural machinery are *supposed to do*; that is, have the teleofunction of doing.

system (e.g., specific voltages) to states of the represented system (e.g., specific concentrations of chemoattractant). Henceforth, I'll refer to the theory as the *structural preservation theory* of representation.

4 The Neurophysiological Mechanisms of Vibrotactile Discrimination

In what follows, I describe a research program aimed at delineating the neural and cognitive mechanisms that underlie vibrotactile discrimination. I then use these results to illustrate the structural preservation theory of representation and furthermore to show how the theory helps in interpretation of the empirical results. The basic, classical task (LaMotte and Mountcastle 1975; Mountcastle et al. 1990) is as follows. A seated Macaque monkey has its left hand secured, palm up. A stimulator tip is lowered, indenting the skin of one of the monkey's fingertips; it is not vibrating at this point. The monkey then presses a key with its free right hand and holds the key down. The stimulator then produces a sinusoidal vibration, between 5 and 50 Hz, to the left hand fingertip (this is the *base stimulus*, or f_1 for first frequency), followed by a delay period (or *interstimulus interval*), followed again by a second vibration (the *comparison* or f_2), also between 5 and 50 Hz. At the offset of the comparison stimulus, the monkey releases the key with its right hand and signals its choice on which frequency was faster by pressing one of two push buttons located at eye level. The monkey is rewarded with a drop of juice for correct discrimination.

A schematic of the neural events that occur during this task is as follows. Rapidly adapting, superficially located mechanoreceptors in the finger known as *Meissner's corpuscles* transduce the mechanical energy into action potentials, which travel up the spinal cord, through the thalamus, into primary somatosensory cortex (S1), and thence to the secondary somatosensory cortex, or S2 (Gardner and Kandel 2000; Gardner et al. 2000; Vallbo 1995). The outgoing signal from S2 then gets widely distributed, to at least the prefrontal cortex (PFC), the ventral premotor cortex (VPC), and medial premotor cortex (MPC); PFC and VPC both appear to be serially connected to MPC. Then MPC transmits activity to the primary motor cortex (M1), whose activity ultimately results in the monkey's button-pressing behavior signaling its choice (Romo et al. 2004a). These cortical areas are typically associated with cognitive activities as follows. Primary and secondary sensory areas are involved in sensory processing. PFC is widely implicated in short-term or working memory processes, and MPC/VPC are considered to be premotor areas, which begin the transformation of signals from sensory and memory processes into motor plans. Primary motor areas are associated with the implementation of generalized motor plans, which then get refined into more specific muscle commands, taking into account various feedback mechanisms by the basal ganglia, cerebellum, and spinal cord.

The neural activity that occurs during the presentation of the stimulus is as follows. In the periphery, neural firing is phase-locked to the stimulus, where the neuron fires a spike or burst of spikes for each amplitude peak of the sinusoidal stimulus (Mountcastle et al. 1969, 1990; Salinas et al. 2000). Traveling into the cortex, there appear to be two subpopulations in S1.[6] In the first, subpopulation-1, neural activity is no longer phase-locked to the stimulus, but the temporal structure of neural firing correlates with the stimulus frequency, as follows. Periodicity is the property of exhibiting regular, repeating characteristics. Using a Fourier decomposition of the firing pattern, it is possible to deconstruct the function describing that pattern into its component sine and cosine functions, as well as determine their "power," or determine which frequency contributes most to the original function. In subpopulation-1 of S1, the power spectrum frequency at peak (*PSFP*), which is the frequency that contributes most to the firing pattern, matches the frequency of the tactile stimulus (Hernandez et al. 2000; Salinas et al. 2000). In subpopulation-2 of S1, the firing pattern becomes less periodic, and the PSFP is no longer matched to the frequency of the stimulus. However, the aperiodic firing pattern now correlates with stimulus frequency in terms of its rate, approximating a monotonic linear function of rate (Salinas et al. 2000).

In S2 and beyond, the rate correlation remains prominent, and the temporal, periodicity-based, or phase-locked code is no longer evident. An important difference emerges in S2. As in S1, there are subpopulations characterized by their differential responses to sensory stimuli; however, in S2 and in all of the more central areas of this circuit, the subpopulations are oppositely "tuned" (Salinas et al. 2000; Romo et al. 2004a). In S1, all neurons increase their firing with increases in stimulus frequency. In more central areas, approximately half increase firing rate as a monotonic increasing function of increasing stimulus frequency, whereas the other half decrease their rate as a monotonic decreasing function of increasing stimulus frequency. Thus, as stimulus frequency gets slower, the negatively tuned neurons increase their firing rate. Oppositely tuned subpopulations responsive to sensory stimuli are found in S2, PFC, VPC, and MPC (Romo et al. 2004a).

The above events occur during the presentation of the base and comparison stimuli. During the interstimulus interval (of 3–6 s, although this can be increased to 10–15 s without a significant difference in performance), no stimuli are presented. To successfully discriminate the first from the second tactile stimulus, and decide which

[6] The primary somatosensory cortex is composed of four areas: 1, 2, 3a, and 3b. Each area has a complete topographic map of the body's surface composed of the receptive fields of the respective neurons. Further, the specialization of peripheral fibers seems to continue in S1; neurons are classified in S1 as rapidly adapting, slowly adapting, or Pacinian, because their firing activities are similar to their respective primary afferents (Romo and Salinas 2001, 109). The areas associated with the rapidly adapting circuit here under consideration are areas 1 and 3b. *Within* those areas, there are subpopulations, one of which appears to encode stimulus information using a temporal, periodicity-based code (described in the text), and the other using an aperiodic firing rate code (also described in the text). The terms 'subpopulation-1' and 'subpopulation-2' should not be confused with areas 1, 2, 3a, and 3b. The subpopulations here under consideration are defined by their behavior in this task and are subpopulations of anatomical areas 1 and 3b.

has a greater frequency, the animal must maintain something like a mnemonic trace of the first stimulus. During this period, neurons in PFC correlate their firing rate with the frequency of the base stimulus, with approximately half showing a monotonic increasing relationship to frequency and the other half showing a monotonic decreasing relationship (Romo et al. 1999). Correlated neural responses during the delay period are also found in S2, VPC, and MPC, also with oppositely tuned subpopulations (Hernandez et al. 2002; Romo et al. 2004b; Salinas et al. 1998, 2000).

The comparison stimulus is then presented, whereby neural activity correlates as before in terms of phase-locking and periodicity in the periphery and early S1, and transformed into a rate code in S1 and then S2. Rate is also correlated with the stimulus in PFC, VPC, and MPC. Additionally, something like a comparison and decision process now occurs, whereby the system decides which of the two frequencies is greater. The relationship of firing rate R to the base and comparison frequencies is given by the regression equation (Hernandez et al. 2002; Romo et al. 2002, 2004a):

$$R = a_1 f_1 + a_2 f_2 + c,$$

where c is a constant, f_1 and f_2 are the frequencies of the base and comparison stimulus, respectively, and a_1 and a_2 are coefficients that determine the strength of the relationship between R and frequency. When either of the coefficients is zero, there is no detected correlation between rate and that coefficient's frequency. Importantly, when $a_1 = -a_2$, then firing rate is now correlated with neither f_1 nor f_2, but with the *difference*, $f_2 - f_1$.

During the comparison period, neurons in S1 only show correlation to f_2 throughout the stimulation period; hence, the neural activities act as sensory representations of the comparison frequency. In S2, some neurons begin the period correlated with f_2, then the population as a whole shifts towards correlation with the difference, $f_2 - f_1$ (i.e., $a_1 = -a_2$) (Romo et al. 2002). In VPC and MPC, there are several different populations. Some neurons begin the comparison period correlating with the base frequency; thus, they are something like mnemonic traces, whereas others begin the period correlating with the comparison frequency as if they were sensory representations. Toward the end of the comparison period, the majority of the responsive neurons in MPC and VPC correlate with the difference, $f_2 - f_1$ (Hernandez et al. 2002; Romo et al. 2004b). Additionally, firing rates correlated with $f_2 - f_1$ are found in PFC (Romo et al. 2004a).

As with neural activity that correlates with the base or comparison frequency, the neural responses correlated with $f_2 - f_1$ (in S2, VPC, MPC, and PFC) show opposite slopes, where approximately half fire more strongly when $f_2 - f_1$ is positive, and the other half fire more strongly when $f_2 - f_1$ is negative.

Finally, M1 plays a crucial role in the animal's behavior during this task. While M1 shows no significant response above baseline activity during the base stimulus, delay period, or early in the comparison period, it does show neural activity correlated with $f_2 - f_1$, similar to the activity found in earlier areas, with subpopulations differentially responsive to the case where $f_2 > f_1$ and where $f_1 > f_2$ (Romo et al. 2004a).

In a different task, monkeys must categorize rather than discriminate the same type of tactile stimuli, simply saying whether a stimulus belongs to arbitrary categories of *high* or *low* which were learned during training (Salinas and Romo 1998). In this instance, firing rates had a sigmoidal shape: For a neuron that "preferred" higher speeds, its firing rate was essentially the same for stimulus speeds of 22–30 Hz. For a neuron that "preferred" lower speeds, its rate was essentially the same for stimulus speeds of 12–20 Hz (see Salinas and Romo 1998, figures 3 and 4). Thus, as found earlier, there are two subpopulations, each of which is selective for either high or low speeds. The sigmoidal shape of the firing rate as a function of tactile speed suggests that these neurons correlate with arbitrary, learned categories ("high" or "low"). Whether or not that analysis should be applied to the tactile discrimination task is uncertain. However, M1 does appear to play a role in the decision procedure for at least the categorization task, and it does have differential activity selective for the different decisions the animal may make (i.e., base greater than comparison or vice versa). Whether that differential activity participates in the comparison and decision procedure, or simply receives a copy of a decision already made, is unclear.

5 Applying Structural Preservation Theory

It should be apparent from the above discussion that neurons in this circuit use a variety of mechanisms for encoding information about the stimuli. From the periphery and centrally inward, neurons use a simple one-to-one burst code, followed by a temporal code in which periodicity is the operative mechanism, followed by a variety of rate codes, some with opposite slopes, and some reflecting neither the base nor comparison frequency, but rather their difference. In motor cortex, a binary outcome (pressing the medial or lateral button) is reflected in the sigmoidal shape of the firing patterns. A theory of biological representation, if it is to be empirically useful, ought to be able to unify these various encoding mechanisms under an overarching conceptual framework that explains what biological representation is and how representational content is determined, from a general standpoint. I suggest that structural preservation theory does do this, mostly as a result of the versatility of the concept of isomorphism and, more broadly, structural preservation.

The first step is to establish *that* these neural mechanisms are representations; this aligns with what I've called the metaphysics of representation, or, what makes something a representation at all. I've argued that a state is a representation if it has the teleofunction of bearing certain correspondence relations such that its doing so is adaptive for the organism of which that state is a part. I'll only discuss this question with respect to burst rate in the periphery since the arguments are both simple and immediately applicable to the other neural areas and firing patterns.

The tactile sensitivity of the glabrous areas of primate skin makes possible various evolutionarily adaptive behaviors, such as grasping objects and tactile recognition, which in turn aid us in getting food into our mouths. We primates do all sorts of things with our hands, which contribute to behavior that is conducive to survival and procreation. Furthermore, the kinds and levels of energy needed to activate this circuit are very specific. Due to the microanatomy of Meissner's corpuscles, only vibrating mechanical energy in the 5–50 Hz range, at the superficially located level (around 500 μm beneath the surface), will generate trains of action potentials. Faster or deeper vibrations simply won't activate the Meissner's circuit, but will instead activate Pacinian corpuscles, and slower indentations in the form of constant pressure will activate the slowly adapting mechanoreceptors and their associated afferents (Gardner et al. 2000; Gardner and Kandel 2000). And these are each forms of tactile, mechanical energy. Electromagnetic, chemical, thermal, or acoustic mechanical energies won't activate this circuit at all. While we should always be wary of just-so stories about evolution, it is reasonable to presume that burst rate covaries with vibrotactile frequency because, in the course of evolutionary history, there was selection for peripheral nerves that emitted a burst at a rate equal to frequency of a sine wave of pressure on the fingertip, for the specific frequency and depth ranges mentioned above, at specific anatomic locations. Therefore, the teleofunction of the primary, secondary, and tertiary afferents associated with the rapidly adapting circuit is to covary with mechanical deformations at their respective receptive fields, according to the simple function $r_1 : A \rightarrow B$, where A consists of vibrotactile frequencies, B consists of burst rates, and $r_1(x) = x$. This function maps frequencies to rates, where x Hz vibrotactile frequency maps to x bursts/s. A similar argument applies to the other correspondence relations defined by periodicity and rate; therefore, they are each representational states of the organism. However, the explanation of representational *content*, allowing for accuracy and error, is given in terms of causal etiology and isomorphism.

I'll discuss four different kinds of sensory representations: the peripheral burst code, the periodic/temporal code in subpopulation-1 of S1, and both the positively and negatively sloped rate codes in S2 and beyond. We begin by defining some simple mathematical functions and relational systems. These functions are the empirically discovered correspondence relations between neural activity and ambient energy, which serve two purposes in the theory. First, these are the correspondence relations that the neural states have the teleofunction of bearing to external states; by bearing these correspondences that reflect the varying states of ambient energy, other neural processing mechanisms are able to use that correspondence to compute appropriate behavioral responses. These patterns of neural firing are representations in virtue of having the teleofunction of bearing these correspondence relations. Second, the mapping functions between relational systems define isomorphisms between those systems and match up states of neurons with energy states at the periphery, serving to determine predication. Further, as mentioned previously, for any two isomorphic relational systems, there always exists numerous if not infinitely many mapping functions between them that preserve structure equally well. However, the empirically discovered correspondences serve to rule out every other transformation on the mapping function, thus avoiding one of the key problems for isomorphism-based theories of representation.

Relational systems consist of sets with relations on them. Let $\mathfrak{A}$ = the stimulus relational system and $\mathfrak{B}$ = the physiological relational system, in each case that follows. Each relational system is an ordered pair consisting of a set (or domain) and a relation on that set. Hence, $\mathfrak{A} = \langle A, r \rangle$, with r being a relation on A, the domain of $\mathfrak{A}$. Isomorphism is defined by defining a bijective function[7] from the domain of one relational system to the domain of the other, such that the relational structure of one system is preserved in the other (though the relations themselves need not be the same).[8] The domain of the stimulus relational system, A, consists of vibrotactile frequencies and is ordered by $>_A$, the empirical higher-frequency-than relation. The domain of the first physiological relational system, B, consists of burst rates. We define a *burst* in terms of interspike intervals: A burst is "a group of spikes in which all intervals between consecutive spikes [is] less than τ msec" (Salinas et al. 2000, 5504). The shorter that τ gets, the closer burst rate will be to firing rate. For our purposes here, whatever τ maximizes the linear fit of the function from frequency to burst rate should be chosen. B is ordered by $>_B$, the empirical greater-burst-rate relation. The first mapping function was introduced above, with $r_1 = A \rightarrow B$:

$$r_1(x) = x.$$

The second physiological relational system will define neural activity in subpopulation-1 of S1 which, recall, does not correspond to peripheral frequency either in terms of burst rate or firing rate, but rather in its temporal structure. In this case, again let $\mathfrak{B}$ = the physiological relational system. To define $\mathfrak{B}$, we'll define the members of B in terms of PSFP, or power spectrum frequency at peak (Salinas et al. 2000). Briefly, recall that PSFP is calculated with a Fourier decomposition of the time course of neural activity, then the frequency bin with the peak power is found, and its median taken. This is the frequency that contributes most to the oscillatory activity of the particular neuron under consideration. Each member of B is a *frequency*, and so the natural ordering relation is the greater-frequency-than relation, $>_B$. Like r_1, r_2 is exceedingly simple, with r_2: $A \rightarrow B$:

$$r_2(x) = x.$$

Note that r_1 is distinct from r_2: The first is a function from frequencies to burst rates, while the second is a function from frequencies to PSFP. Furthermore, PSFP is not a measurement of "more or less" periodicity, in the way that firing rate is a measure of how many spikes fire per second. It is rather a measurement of which frequency component of the overall activity of the neuron contributes most to its oscillatory activity. The final two functions I'll define describe the relationship between firing rate in subpopulation-2 of S1 and frequency, and then the firing rate of neurons farther downstream with negative slopes, relative to frequency. In each

[7] A function is bijective if it is *injective* and *surjective*. A function is injective (or one-one) if each member of the range is mapped to by only one element of the domain. A function is surjective (or onto) if every member of the range is mapped to by some element of the domain.

[8] More specifically, $\mathfrak{A}$ and $\mathfrak{B}$ are isomorphic if there exists a bijective function f: $A \rightarrow B$ such that for every a and b in A,

$$aRb \text{ iff } f(a)Sf(\text{b}).$$

If f is surjective but not injective, then $\mathfrak{A}$ and $\mathfrak{B}$ are *homomorphic*. A variety of other kinds of structure-preserving mappings can also be defined, by selectively loosening certain criteria. See (Swoyer 1991) for some examples.

case, the domain of B now consists of firing rates, and it is ordered by $>_B$, the greater-firing-rate relation. Let $r_3: A \rightarrow B$:

$$r_3(s) = 22 + 0.7s,$$

where s is stimulus frequency and $r_3(s)$ is rate described as a function of frequency. As reported in Salinas et al. (2000, 5506), this equation describes the relation between firing rate in S1 and stimulus frequency. (The equation also includes a noise term, but since noise is by definition not a signal, I've deleted the final term. Nonetheless, noise is a significant issue to be addressed; on this, see fn. 13.) Neurons in this population fire at a baseline rate of 22 spikes/s and increase linearly with a slope of 0.7 as vibration frequency increases. Finally, there are populations of neurons in S2 and beyond, which are oppositely tuned, whereby increasing frequencies generate decreasing firing rates (Salinas et al. 2000; Hernandez et al. 2000). To my knowledge, the specific equations describing the relations between the negatively sloped subpopulations and vibration frequency have not been published, though they are noted to be monotonic linearly decreasing functions.[9] For concreteness then, I'll stipulate $r_4: A \rightarrow B$ as

$$r_4(s) = 65 - 0.5s.$$

Although stipulated, r_4 should be considered as the equation that describes the activity of neurons in a population (either in S2, PFC, VPC, or MPC) with a negative slope relative to stimulus frequency.

Each of these four equations is an empirically discovered correspondence relation (with the exception of r_4 which is stipulated; I'll omit that qualification henceforth) between neurons in specific populations and mechanical stimulation of the fingertip. These are the "specific correspondence relations" I've appealed to above in determining the teleofunctions of the neurons. Furthermore, the equations each define bijective functions that in turn define an isomorphism between the stimulus relational system $\mathfrak{A}$ and their respective physiological relational systems $\mathfrak{B}$.[10] The key idea here is that we find *systems* of representations, and *systems* of properties

[9] Furthermore, note that r_3 only describes the specific relationship discovered among neurons in subpopulation-1 of S1 with vibration frequency. Presumably, the populations of neurons in S2, PFC, VPC, and MPC, which also show positively sloped response profiles, admit of different specific relationships with stimulus frequency (i.e., different baselines and different slopes). They have not however been published (to my knowledge). Note that these different equations don't change the overall philosophical analysis of biological representation presented here; the theory easily accommodates differing correspondence relations between neural states and represented states, due to the versatility of the concept of structural preservation.

[10] Proving isomorphism is not trivial, and furthermore, measurement theory is concerned with one empirical and one numerical relational system, not two empirical relational systems as I've described here. But the technical details are outside the scope of this chapter, so I've made simplifying assumptions. Namely, I'll assume that $\mathfrak{A}$ and $\mathfrak{B}$ both have uncountable domains with countable order dense subsets, and their respective relations generate a total order on the domains. This suffices for isomorphism between two empirical relational systems $\mathfrak{A}$ and $\mathfrak{B}$ (Collins 2010, 406). Whether these assumptions are justified depends on whether making idealizing assumptions in general are justified.

represented, each organized in such a way that individual members from each domain map to members in the other, mapping specific firing patterns to specific vibration frequencies. I'll refer to these four functions as *representation functions*.

To determine representational content, recall that both causal etiology and structural preservation (e.g., isomorphism) are required. In each of the sensory representations throughout the vibrotactile discrimination circuit, the causal antecedent of the particular pattern of firing is the experimental stimulator. Thus, the thing to which each representation refers, determined by causal etiology, is the stimulator.[11] But causal etiology alone is not enough to determine predication, that is, to determine what property the representation predicates of the stimulator. For this, the representation functions for each respective neural population define which property is predicated of the stimulator and, crucially, determine which neural patterns would constitute accurate representation, and which would constitute error.

For example, assume that primary afferents in the rapidly adapting circuit are firing at a burst rate of 50 bursts/s and that this was caused by the stimulator. From r_1, we see that the representation function matches up frequencies to burst rate one-to-one; therefore, the representational content of this activity is something like *the stimulator is vibrating at 50 Hz*.[12] If the stimulator is indeed vibrating at 50 Hz, then the representation is accurate; if the stimulator is not vibrating at that speed, then the representation is inaccurate. But for neurons in subpopulation-2 of S1, where neurons have the teleofunction of corresponding to such external stimuli in terms of their firing rate rather than burst rate, and according to a different

[11] There are a variety of intermediate events between the stimulator's vibrating and a particular pattern of neural firing that it caused, say, in S2. For example, ion channels have opened and closed, neurotransmitters have been released, a variety of firing patterns have occurred in upstream areas in the spinal cord, brainstem, thalamus, internal capsule, S1, and so on. Determining which of these causal antecedents is the one to which the representation refers is known as the *causal chain problem*, which is a problem for any theory of representation that appeals to causation. While I won't attempt detailed discussion here, a reasonable solution (at least in this instance) is to appeal to teleofunction. The correlation of neural activity in S2 with upstream neural activity is not what confers survival advantage. Rather, by covarying with energy states at the periphery of the organism, in well-defined ways, distinct neural mechanisms can use that activity to perform transformations and computations which ultimately result in behavior that is appropriate to the environment. Hence, it is not arbitrary to claim that the neural activity refers to the stimulator and not some other link in the causal chain.

[12] Notice I write that the content is *something like* ... (rather than that the content *is* ...). It is unjustified to assume that the representational content of the lowest-level biological representations instantiated in the firings of individual neurons can be translated straightforwardly into a natural language. Rather, we should be satisfied with *describing* the content using natural languages, though should not expect a straightforward translation. Furthermore, note that it is equally justified to describe the content as "*that thing* is vibrating at..." as compared with "*the stimulator* is vibrating at...." The neural activity under question does not predicate the property of being a stimulator, only the property of vibrating at a certain frequency. Again, for the purpose of describing the content, rather than expressing or translating it, either rendering is acceptable because both expressions refer to the stimulator in this context.

representation function (r_3), if these neurons fire at a firing rate of 50 spikes/s, it does not imply that they have the same representational content. Rather, a neuron from subpopulation-2 of S1, whose teleofunction is to accord with external stimuli according to r_3, would, if firing at 50 spikes/s, have the representational content that *the stimulator is vibrating at 40 Hz* because r_3 maps the property of vibrating at 40 Hz to the firing rate of 50 spikes/s. If the stimulator is not vibrating at 40 Hz, then the representation is inaccurate. Similarly, a neuron that is part of an oppositely tuned subpopulation, say, in PFC, which has the teleofunction of corresponding to external stimuli according to r_4, would have a different representational content. Assuming again that it was firing at 50 spikes/s, this neural activity would have the content that *the stimulator is vibrating at 30 Hz* because this is the property that r_4 maps to 50 spikes/s firing rate. Similar comments apply to the temporal codes that use periodicity in S1.

In general, although the monkeys are quite good at the task (with about a 90% accuracy rate), they do occasionally make behavioral errors. When this occurs, there is a correlation between standardized measures of firing rate in S1 and S2 with behavioral error (Salinas et al. 2000). For example, if the monkey presses the lateral button, signaling that it believed that the comparison was *lower* when in fact it was higher than the base, the firing rates of its neurons in S1 and S2 are *less* than they would have been, had the animal made an accurate discrimination and *mutatis mutandis* for the opposite mistake. For example, assume that the comparison frequency is 40 Hz and that the base frequency was lower at 30 Hz. Since neurons in subpopulation-2 of S1 have the teleofunction of corresponding to superficial vibration pulses in their respective receptive fields according to r_3, in order to correctly represent the comparison stimulus of 40 Hz, the neurons should be firing at 50 spikes/s. Assume however that a neuron is firing at 40 spikes/s in this circumstance; in this case, its representational content is something like *the stimulator is vibrating at 25.7 Hz*, thus misrepresenting the frequency of the stimulator, which then leads, ultimately, to a behavioral error. In other words, sometimes a well-trained animal makes a mistake, signaling that it believes the comparison was lower on a trial in which the comparison was in fact higher. When this occurs, the neural firing patterns in early sensory areas (S1 and S2) fire at a rate that is lower than what it would have been, had the neurons accurately represented the stimulus frequency.

It thus appears that the behavioral error is a result, at least partially, of an early stimulus encoding error, where the sensory representations misrepresent the frequency of the stimulus. If only one or two neurons misrepresent that frequency, the animal's behavior as a whole will likely be unaffected. But as the number of neurons in error begins to mount, it becomes increasingly likely that the animal will behaviorally signal in error. Crucially, without accounting for the logical structure inherent in the representational content of neural activities, there is no way to make sense of the idea that the early sensory encoding mechanisms had *misrepresented* the stimulus, that is, that there was a stimulus encoding *error*. By accounting for both components of representational content, however, the structural preservation theory provides a theoretical framework that allows for such an interpretation.[13]

Structural preservation theory also applies to the sigmoid response profiles in motor cortex, which constitute generalized motor plans to press either the medial or lateral push buttons. These generalized plans become refined downstream by neural mechanisms in the basal ganglia, cerebellum, spinal cord, and motor neurons at the periphery. As with the sensory representations discussed above, we begin with the question of whether the neural activities in M1 are representations (at all), before addressing their content.

The behavioral output of pressing the medial versus lateral button in response to a comparison of two vibrating stimuli is learned, not evolved. Nonetheless, the animals do achieve high accuracy levels, and a reasonable teleological argument can be made on these grounds: The monkeys have learned that pressing the medial button when and only when the comparison stimulus is higher results in the acquisition of juice, and *mutatis mutandis* for the lateral button. Further, after learning, certain neural activities have come to be regularly correlated with the muscular motions associated with medial and lateral button-pressing. It is reasonable to conclude that the consumers of the neural activity in M1 (i.e., the neural mechanisms downstream of M1 in the basal ganglia, cerebellum, spinal cord, and motor neurons at the periphery) have the teleofunction of producing the state of affairs corresponding to the motor plan in M1. Or in other words, if the motor plan says something like *my right arm is pushing the medial button*, then the consumers of that motor plan have the teleofunction to make that true. This is analogous to my intention to pick up the coffee cup, which can be either satisfied or not. Thus, unlike sensory representations, whose teleofunction is to correspond to energy impinging on the periphery so that doing so is adaptive for the organism, the teleofunction of procedural representations or motor plans is to play a role in *bringing about* the states to which they correspond. In this case, the "direction of fit" is the reverse: Sensory representations "fit" the world; motor representations make the world "fit" them (cf. Searle 1992).

---

[13] As mentioned in the text above, the equation published in Salinas et al. (2000) includes a noise term, so should be written as: $r(s) = 22 + 0.7s + \sigma\epsilon$, where ϵ is noise with zero mean and unit variance and σ is the standard deviation of the mean firing rate. Since noise is by definition not a signal, I've deleted the final noise term. Nonetheless, noise in neural systems is a significant conceptual and practical issue to be addressed by a theory of representation; any plausible view must be able to account for it because there is no such thing as a noiseless signal in the brain. Many biochemical mechanisms such as ion channel opening, vesicle release, and ion diffusion are stochastic processes, so there will always be "random" electrical activity which is not a result of stimulus representation or neural computation. Although I don't have space for an in-depth discussion of this here, the theory on offer does have the resources to account for noise in neural systems. The general idea is to distinguish those alterations in the content-bearing properties of a vehicle of representation (e.g., firing rate) which are due to alterations at the source (e.g., vibrotactile frequency) from those alterations which are not due to alterations at the source; these latter alterations constitute noise. A firing rate that is within the range of noise, given its particular (empirically discoverable) noise range, representation function, and the value of its represented parameter, is a noisy-but-true signal, whereas one that is outside the noise range is a noisy-and-false signal. For more detail see Collins (2010, 359–363).

Recall that at the end of the comparison period, neurons in M1 correlate with neither the base nor comparison frequencies, but rather instead correlate with the difference, $f_2 - f_1$. Furthermore, there are again subpopulations with affinities for $f_2 > f_1$ and $f_1 > f_2$, respectively. Consider, for example, a positively sloped subpopulation (i.e., which "prefers" $f_2 > f_1$). As above, the specific equations defining the relationship between firing rate and $f_2 - f_1$ have not been published, to my knowledge, so I stipulate one for concreteness (and define a linear rather than sigmoid function for simplicity, but the conceptual points do not change). Notice that $a_1 = -a_2$, and that the constant is the point at which the function crosses the *y*-axis. Thus, if $f_2 = f_1$, the neuron will fire at the constant rate, and as f_2, the comparison stimulus, gets increasingly greater than the base, the firing rate increases as well.

$$g_1(f_1, f_2) = -2f_1 + 2f_2 + 44.$$

Notice that in this subpopulation, 44 spikes/s is the baseline rate, which increases or decreases depending on whether and by how much the base and comparison stimuli differ from each other. Unlike the sensory case however, these generalized motor plans only map to two outcomes: pressing the medial or lateral buttons. Thus, the mapping function from the set of firing rates to the set of behavioral outcomes very simply maps every firing rate from 0 to 44 spikes/s to something like *is pushing the lateral button*, and all rates above 44 spikes/s to something like *is pushing the medial button*. Note that this does not define an isomorphism between relational systems. It does however counter-preserve (but does not preserve)[14] the greater-firing-rate relation in the relational system composed of the two behavioral outcomes related very simply by the ordered pair, $\langle M, L \rangle$ (with *M* abbreviating "is pushing the medial button" and *L* abbreviating "is pushing the lateral button"). Thus, this mapping function fits within the broader construct of structural preservation and is an analogue of the technically more restrictive isomorphism.

Assume that a neuron in this subpopulation is firing at 55 spikes/s. Since g_1 maps this rate to the property *is pressing the medial button*, it follows that this neural activity predicates the property of pressing the medial button, of whatever it refers to. However, as before, reference is determined by causal history. Rather than referring to what caused them, however, procedural representations refer to what they caused. This reflects the reversed "direction of fit" of motor plans relative to sensory representations. Since the neural activity in M1 currently under consideration causes

[14] A function *preserves* a relation *R* only if $aRb \rightarrow f(a)Sf(b)$. A function *counter-preserves R* only if $f(a)Sf(b) \rightarrow aRb$, and thus, a function *respects R* only if it preserves and counter-preserves *R*; for isomorphism between relational systems, the mapping function needs to respect the relation *R*. As I mentioned earlier, there are good reasons to relax the strict requirements on isomorphism when using this tool to construct a theory of representation while keeping the basic idea of the preservation of internal relational structure across systems. The type of structural preservation appealed to in the text is a Δ/Ψ-morphism (Swoyer 1991), which preserves a subset of relations in one system while counter-preserving a subset of relations in the other (in this case, identity is preserved, while greater-firing-rate is counter-preserved; see Collins 2010, 329–330 for the details).

changes in the contraction levels of the various muscle groups of the animal's right arm, it follows that the representation refers to the animal's right arm. Hence, the representational content is something like, "my right arm is pressing the medial button." As with sensory representations, motor plans are semantically evaluable in the sense that they are satisfaction-evaluable; they can be satisfied or not. If the animal does in fact press the medial button, then the motor plan has been carried out; if not, then the motor plan or intention remains unsatisfied. This is the analogue of an inaccurate sensory representation. Note, as above, that different aspects of the representation determine different aspects of its content. Its bearing certain correspondence relations to behavioral outcomes, and having the teleofunction of producing the outcomes to which they correspond, makes them representations. The different firing rates are part of an ordered system, which correspond to a set of behavioral outcomes which also form a (very simple) ordered system, and the rates match up to the behavioral outcomes to which they correspond, determining an analogue of predication. Finally, causal etiology determines that the property of pressing the medial button is to be realized by the right arm.

The analysis of motor representations in monkey M1 is given at a far more abstract level than, say, the five pairs of motor neurons in the chemotaxis circuit of *C. elegans* discussed previously. In the latter case, the voltages of the motor neurons bear a continuous and specific relationship of proportionality to the degree of extension of muscles in the neck, which determine the neck's turning angle (and hence the direction in which the worm moves). This is due to the relative complexity of the different nervous systems (*C. elegans* has only 302 neurons). However, as the monkey's neural signals travel down the motor circuit and get closer to the periphery, the analysis of the content of motor representations will get more specific, analogous to the specificity of the sensory representations in early sensory processing areas. I consider this result – that structural preservation theory would analyze the neural activity in M1 in terms of abstract, generalized motor plans – to speak in favor of the theory. As I mentioned earlier, structural preservation is a versatile conceptual tool, and anything can be a member of a relational system, including relatively abstractly described behavioral outcomes.

6 Conclusion

The concept of representation, or at least *aboutness*, is the foundation upon which all other concepts of mental states and processes are built. To understand the place of mind in nature, we must understand what representation is and how living biological systems realize it. In this chapter, I have presented a sketch of a theory of biological representation and have illustrated it by appealing to the neurophysiological mechanisms involved in a sensory discrimination task. There are a variety of open questions that must be dealt with, including noise in neural systems and the causal chain problem. My main purpose for this chapter however was to outline and illustrate a *theoretical framework* that I think might be useful for making progress on a theory of representation in biological systems. Whether that framework can support the detailed conceptual analysis required of a philosophically viable theory remains to be seen.

References

Akins, K. (1996). Of sensory systems and the "aboutness" of mental states. *The Journal of Philosophy, 93*(7), 337–372.

Bargmann, C. I., & Horvitz, H. R. (1991). Chemosensory neurons with overlapping functions direct chemotaxis to multiple chemicals in *C. elegans*. *Neuron, 7*(5), 729–742.

Bechtel, W. (2001). Representation: From neural systems to cognitive systems. In W. Bechtel, P. Mandik, J. Mundale, & R. S. Stufflebeam (Eds.), *Philosophy and the neurosciences: A reader*. Malden, MA: Blackwell Publishers.

Collins, M. (2010). *The nature and implementation of representation in biological systems*. PhD dissertation, Department of Philosophy, CUNY Graduate Center, New York.

Devitt, M., & Sterelny, K. (1999). *Language and reality: An introduction to the philosophy of language* (2nd ed.). Cambridge, MA: MIT Press.

Dretske, F. I. (1981). *Knowledge and the flow of information* (1st MIT Press ed.). Cambridge, MA: MIT Press.

Dretske, F. I. (1988). *Explaining behavior: Reasons in a world of causes*. Cambridge, MA: MIT Press.

Ferree, T. C., & Lockery, S. R. (1999). Computational rules for chemotaxis in the nematode *C. elegans*. *Journal of Computational Neuroscience, 6*(3), 263–277.

Fodor, J. A. (1975). *The language of thought*. New York: Crowell.

Fodor, J. A. (1987). *Psychosemantics: The problem of meaning in the philosophy of mind*. Cambridge, MA: MIT Press.

Fodor, J. A. (1990). *A theory of content and other essays*. Cambridge, MA: MIT Press.

Fodor, J. A. (2008). *LOT 2: The language of thought revisited*. Oxford/New York: Clarendon Press/Oxford University Press.

Gardner, E. P., & Kandel, E. R. (2000). Touch. In E. R. Kandel, J. H. Schwartz, & T. M. Jessell (Eds.), *Principles of neural science*. New York: McGraw-Hill.

Gardner, E. P., Martin, J. H., & Jessell, T. M. (2000). The bodily senses. In E. R. Kandel, J. H. Schwartz, & T. M. Jessell (Eds.), *Principles of neural science*. New York: McGraw-Hill.

Hernandez, A., Zainos, A., & Romo, R. (2000). Neuronal correlates of sensory discrimination in the somatosensory cortex. *Proceedings of the National Academy of Sciences USA, 97*(11), 6191–6196.

Hernandez, A., Zainos, A., & Romo, R. (2002). Temporal evolution of a decision-making process in the medial premotor cortex. *Neuron, 33*(6), 959–972.

LaMotte, R. H., & Mountcastle, V. B. (1975). The capacities of humans and monkeys to discriminate between vibratory stimuli of different frequency and amplitude: A correlation between neural events and psychological measurements. *Journal of Neurophysiology, 38*, 539–559.

Mandik, P., Collins, M., & Vereschagin, A. (2007). Evolving artificial minds and brains. In A. C. Schalley & D. Khlentzos (Eds.), *Mental states, Vol. 1: Nature, function, evolution*. Philadelphia: John Benjamins Publishing Company.

Millikan, R. G. (1984). *Language, thought, and other biological categories: New foundations for realism*. Cambridge, MA: MIT Press.

Millikan, R. G. (1989). Biosemantics. *The Journal of Philosophy, 86*(6), 281–297.

Millikan, R. G. (2004). *Varieties of meaning, the Jean Nicod lectures*. Cambridge, MA: MIT Press.

Mountcastle, V. B., Talbot, W. H., Sakata, H., & Hyvarinen, J. (1969). Cortical neuronal mechanisms in flutter-vibration studied in unanesthetized monkeys: Neuronal periodicity and frequency discrimination. *Journal of Neurophysiology, 32*, 452–484.

Mountcastle, V. B., Steinmetz, M. A., & Romo, R. (1990). Frequency discrimination in the sense of flutter: Psychophysical measurements correlated with postcentral events in behaving monkeys. *The Journal of Neuroscience, 10*, 3032–3044.

Nicolelis, M., & Ribeiro, S. (2006). Seeking the neural code. *Scientific American, 295*(6), 70–77.

Romo, R., & Salinas, E. (2001). Touch and go: Decision-making mechanisms in somatosensation. *Annual Review of Neuroscience, 24*, 107–137.

Romo, R., Brody, C. D., Hernandez, A., & Lemus, L. (1999). Neuronal correlates of parametric working memory in the prefrontal cortex. *Nature, 399*, 470–473.

Romo, R., Hernandez, A., Zainos, A., Lemus, L., & Brody, C. D. (2002). Neuronal correlates of decision-making in secondary somatosensory cortex. *Nature Neuroscience, 5*(11), 1217–1225.

Romo, R., DeLafuente, V., & Hernandez, A. (2004a). Somatosensory discrimination: Neural coding and decision-making mechanisms. In M. Gazzaniga (Ed.), *The cognitive neurosciences*. Cambridge, MA: A Bradford Book. MIT Press.

Romo, R., Hernandez, A., & Zainos, A. (2004b). Neuronal correlates of a perceptual decision in ventral premotor cortex. *Neuron, 41*(1), 165–173.

Salinas, E., & Romo, R. (1998). Conversion of sensory signals into motor commands in primary motor cortex. *The Journal of Neuroscience, 18*(1), 499–511.

Salinas, E., Hernandez, A., Zainos, A., Lemus, L., & Romo, R. (1998). Cortical recording of sensory stimuli during somatosensory discrimination. *Society for Neuroscience Abstracts, 24*, 1126.

Salinas, E., Hernandez, A., Zainos, A., & Romo, R. (2000). Periodicity and firing rate as candidate neural codes for the frequency of vibrotactile stimuli. *The Journal of Neuroscience, 20*(14), 5503–5515.

Searle, J. R. (1992). *The rediscovery of the mind*. Cambridge, MA: MIT Press.

Swan, L. S., & Goldberg, L. J. (2010). How is meaning grounded in the organism? *Biosemiotics, 3*(2), 131–146.

Swoyer, C. (1991). Structural representation and surrogative reasoning. *Synthese, 87*(3), 449–508.

Vallbo, A. B. (1995). Single-afferent neurons and somatic sensation in humans. In M. Gazzaniga (Ed.), *The cognitive neurosciences*. Cambridge, MA: A Bradford Book. MIT Press.

Beyond Embodiment: From Internal Representation of Action to Symbolic Processes

Isabel Barahona da Fonseca, Jose Barahona da Fonseca, and Vitor Pereira

Abstract In sensorimotor integration, representation involves an anticipatory model of the action to be performed. This model integrates efferent signals (motor commands), its reafferent consequences (sensory consequences of an organism's own motor action), and other afferences (sensory signals) originated by stimuli independent of the action performed. Representation, a form of internal modeling, is invoked to explain the fact that behavior oriented to the achievement of future goals is relatively independent from the immediate environment. Internal modeling explains how a cognitive system achieves its goals despite variations in the environment with insufficient and noisy sensory–perceptual data. In a self that acts intentionally on the environment, knowledge is dependent upon the necessity to guide actions directed toward an aim. The self-inner model, a representation of internal and external environments (including reafferent and afferent messages) and also of the behavior plans and desirable future states (aims) and efferent intentions (motor planning and motor command messages), is intrinsically linked to a thinking capacity, which is supposed to emerge from the binding of multiple influences. Thinking emerges when higher behavior strategies are considered possible and capable of leading to aims or the fulfillment of intentions. In this model, symbolization processes are projective and anticipatory and, in this way, beyond present referents. Symbolization occurs linked to action planning, command, and regulation in mental simulation. Meaning is related to an inner sense of a self that acts over the environment.

I.B. da Fonseca (✉)
Department of Psychophysiology, Faculty of Psychology, Alameda da Universidade, 1613-049 Lisbon, Portugal

Faculty of Psychology, University of Lisbon, Lisbon, Portugal
e-mail: isabelbf@fpce.ul.pt

J.B. da Fonseca
Faculty of Sciences and Technology, New University of Lisbon, Lisbon, Portugal

V. Pereira
Faculty of Psychology, University of Lisbon, Lisbon, Portugal

L. Swan (ed.), *Origins of Mind*, Biosemiotics 8, DOI 10.1007/978-94-007-5419-5_9,

Keywords Cognitive model • Anticipation • Embodiment • Symbolic processes • Neurophysiological functions

1 Introduction

The relation between physiological processes and psychic events remains an unsolved problem. It is proposed in this chapter that meaning and symbolic processes are created in an internal space of representation involving the binding between internal and external sensory information, motor command and regulation models, planning of behavior, and anticipation of future aims. Meaning and symbolic processes occur in a projective way linked to efferent processes associated with the planning and command of behavior directed to the external environment and anticipation of future desirable states. From a neurophysiological point of view, these processes occur in a widely distributed network involving cortical and subcortical regions in systems that are massively parallel and interactive (in a neurophysiological sense). The efferent component linked to planning, executive functions, and anticipation depends on prefrontal, cingulate, and parietal networks and also on networks involving basal ganglia.

Another interpretation of the hypothesis that mind is dependent upon an interactive, internal space is the notion that what is represented is a type of information that is not exclusively sensory or motor but involves the interaction between messages of different origins: perceptual, motor programs, and intentions. In other words, a representation emerges in an interactive context in which sensory and motor events are submitted to a compatible frame of coordinates that allows for the creation of internal organism–environment models that are endowed with intention and meaning. For an organism that interacts with the environment, it is crucial to have an internal model at the neuronal level that represents the internal and external environment, and also future desirable states. This model is embodied, as far as it guides the purposeful action of the organism, and independent from physical immediate constraints—although the properties of the environment are represented in a way that is somehow homomorphic with physical constraints, at least in the functional way that allows for adaptive and successful behavior.

The mind builds an internal functional space that represents the characteristics of the external and internal environments as they occur in perception. For an organism to be able to successfully interact with its external environment, perception functions by identifying invariants that are further translated in the nervous system (NS) into well-executed motor actions that delivered back into the external world (Llinás 2001) or in cognitions that are not expressed in motor actions.

For an organism that behaves and moves, taking into account the constraints from the external environment, the distinct properties of its internal space and the properties of the external world should have a continuity in which the coordinates of the external world are translated (transduced) in the internal functional space, preserving homomorphic continuity (Llinás 1987, 2001).

2 The Self as an Agent and the Formation of an Internal Model that Allows Symbolic Processes

When looking for the hypothetical origins of the symbols utilized in the processes involved in the planning, execution, and regulation of an organism's movement, meaning is created by a sense of agency experienced by the organism. This self integrates multiple influences, sensory and motor, in an anticipatory model of the action planning, an internal simulation, which includes also motor plans and desirable future states that direct decision-making and behavioral planning.

The concept of self (a proto-self or core self) corresponds to the binding of diverse sensorimotor transformations into a single, internal representational model crucial for symbolic processes.

Such anticipation is a fundamental function of the NS: to the organism, especially as regards its adaptation to the environment, what is interesting is what is going to happen in the future, not what has already happened. Past experiences have been memorized and integrated in the internal model and are automatically considered in the mental simulation of an action. This classical hypothesis has been formulated in modern science by Llinás (1987, 2001) who has proposed that thinking capacity emerges from movement internalization. In other words, thinking emerges when higher behavior strategies are considered in terms of potentially leading to the fulfillment of intentions. Movement is related not only to body parts but also to objects from the external world, perceptions, and complex ideas.

For Llinás (2001), if we were able to study action internalization, perhaps we would be able to understand something about our nature—the way we think, learn, and represent ourselves in a self-composite and complex manner.

3 The Brain as Simulator

The fundamental function of the NS is action planning and regulation. Action regulation achieved in low-level loops integrates a feedback with an anticipatory component (feed-forward): feedback loops that involve a sensorimotor process of error detection and correction are regulated by feed-forward mechanisms. So feedback and feed-forward loops act on a wide group of synergies that regulate motor primitives. Internal models can be understood as neural mechanisms in the motor systems that reproduce a subset of input/output characteristics, or their inverse. Feed-forward internal models predict sensory consequences from efferent signals (also called corollary discharge) of motor commands issued but not yet executed. Inverse internal models can calculate necessary feed-forward motor commands from a desired final state.

Anticipation is adaptive: it saves time and effort in the execution of a motor program. In anticipation of desired final states, motor programming and execution are independent from a coordinate system, a kind of internal premotor invariance. An example is found in the apparent proficiency, for example, when signing your name using elbow and shoulder joints when writing on a boar, or using any other articulations, such as fingers and hand when signing in a paper, or even foot and leg. The point is that in all these cases, involving such diverse articulations, the results of motor executions, the signing, despite different scales and precision, are similar (Llinás 1987). It has been proposed that this constitutes a manifestation of a kind of motor invariance, in which actions are represented in an abstract form associated to the final intended result.

Parallel to this control function, neuronal loops of high level in the motor hierarchy, and also in the phylogenetic scale, begin to be progressively more complex and to function in an anticipatory and projective way. In this projective anticipatory process, brain signals are used to generate action plans (or internal models) in internal loops without direct relation to a present stimulus. The neuronal operations are noncontinuous and occur in neuronal maps whose parameters are the topographical and functional relations between neurons. This mode of prediction about future states, a mental simulation, does a kind of preselection of action strategies and, in general, guides decision-making. In this respect, the brain functions as a simulator projecting future states and strategies.

These processes are foundational for cognitive representation; they integrate and bind multiple signals such as action planning and command. The integration of the meaning of these multiple messages is referred to as a self in the environment, which becomes the center of the phenomenic experience.

4 The Self as an Agent: The Contribution of Efferent Copy (Efferenze Copie, Von Holst)

Knowledge is integrated in the internal model, creating conditions that are necessary for symbolic processes to occur within a preconceptual sense of an agent: a proto-self that is nonconscious, but without which, more sophisticated self-experiences cannot occur. Such functioning that creates an inner cognitive model is linked to the sense of agency—the experience that the subject has of being himself the cause and generation of action (Gallagher 2000, 2012).

Above, we discussed the contribution of the binding of multiple neuronal signals in the creation of the conditions for symbolic processes. These influences are inner sensory experiences related to body (somatosensory signals) and external stimuli (some dependent and others independent from the subject's action), activation of memories of past experiences, action plans, efferent commands, the representation of future desirable states, and projective and anticipatory models of action.

The sense of agency that the organism feels when engaged in voluntary action is created by the correspondence between three kinds of neuronal signals: (1) somatosensory signals resulting directly from movement, (2) visual and auditory signals that may result indirectly from the organism's movements, and (3) the corollary discharge—the copy of the efferent motor command that generates the movement.

In sensorimotor loops, what is distinctive about the processes that specify the self as an agent, distinguishing between self and nonself, is that the sensory signals with an external origin independent of the organism's own actions are noncontingent and uncorrelated with efferent action command signals; that is, a match between efferent and reafferent signals creates self-specifying meanings.

Dependent upon receptors and neuroanatomic pathways, reafference is distinguished from afferent signals in the process of comparing or matching these signals with efferent commands. The reafference is self-specifying because it is intrinsically related to a self-initiated action and it will originate reafferent signals that match the corollary discharge or efferent command. It is this correspondence between the efferent command signals and their reafferent consequences that signals that the information is self-specific and distinct from nonself-sensory afferent signals that are uncorrelated and noncontingent with efferent command.

5 The Experiential Self, Interoceptive Loops, and Internal Cognitive Models

Another kind of self-specifying processes can be found in the regulation of the organism's internal environment, in which loops of efferent–reafferent signals regulate the internal conditions for survival. In this case, efferent and afferent signals involve different structures from those related to voluntary action: brain stem nucleus and midbrain structures, somatoautonomic adjusting with low-level autonomic reflexes and high-level loops involving the lymbic structures, the hypothalamus, the insula, and the anterior cingulus. It is a homeostatic interoceptive system integrated in the vertical neuroaxis that specifies the state parameters of an experiential phenomenic self.

As opposed to the sensorimotor integration, which defines the relation between the organism and the external world, the homeostatic regulation specifies the organism's relation with its own environment and gives rise to subjective interoceptive feelings. The experience of feeling emerges from the binding of neuronal activities in a highly distributed system. Hypothetically, this subjective experience is related to a coherent matching between cognitive–affective states of higher-order, undifferentiated sensory signals processed by subcortical pathways involving the thalamus and limbic structures, and loops that regulate the organism's internal environment in structures such as the hypothalamus, insula, anterior cingulus, and other brain stem and midbrain structures, as well as low-level autonomic and somatic reflexes.

One method to study internal cognitive models of movement is based on the predictive effects of the sensory consequences of the subject's own actions. These effects consist in sensory suppression or attenuation of the reafferent signals and are produced in loops in which intentional commands modulate sensory feedback (Tsakiris and Hagaard 2005). The sensory suppression consists in the phenomenon of attenuating the reafferent sensory consequences of a self-generated movement. It has been thought that the reduction of sensory feedback of subject's own actions results from the voluntary nature of the movement. Numerous studies demonstrate the attenuation of the perceptual consequences of self-generated actions (Blakemore et al. 1999).

It is hypothesized that the perceptual consequences of self-generated actions are attenuated because internal models of the motor system use the efference copy (corollary discharge) to predict the consequences of the subject's own actions. This information is integrated in an internal "forward model" (Wolpert 1997) which is created and compares the predicted sensory outcome of the subject's own actions with the actual somatosensory reafferent feedback and other afferent messages that co-occur. The hypothesis of "efference copy" or copy of the motor command (Sperry (1950); Von Holst and Mittelstaedt 1950) was initially proposed to answer to Helmholtz's question: "How is it that, when we move our eyes, the world remains stable, despite the fact that the retinal image has moved?"

Von Holst and Mittelstaedt (1950) suggested that motor actions are accompanied by an efference copy of the action, which sends a "corollary discharge" to the sensory cortex signaling that impending signals are self-initiated or self-generated. The efference copy/corollary discharge mechanism works to suppress or reduce the perception of events that result from a self-generated action. Thus, it may allow an automatic distinction between internally and externally generated percepts. In the visual system, this system may serve to stabilize the visual image during eye movements, maintaining visuospatial constancy.

It is hypothesized that in sensory attenuation of the consequences of self-initiated actions, the process consists in analyzing a copy of an efferent motor command, an "efferenze copie" of a planned action, which is sent through a "feed-forward" mechanism to the appropriate sensory cortex, preparing it for the arrival of the feedback sensation—the efference copy works to suppress (or to reduce) perception when it results from a self-generated action.

These processes allow the organism to recognize that it has produced an action, and this information is used to modulate sensory consequences of movement. It is hypothesized that the prediction of sensory reafference and its integration in an internal model, relating the efferent, the afferent, and the behavioral intention, is expressed in a sensory suppression of inputs resulting from self-initiated actions.

6 Development of Self-Awareness

The consideration of these processes allows a hypothesis of defining the contribution of innate factors to the experience of self and finding indirect evidence about the way that meaning is influenced by innate factors linked to the structure and functions of the NS. By linking meaning and symbol formation to internal models of self-created in action planning and execution, it can be said that meaning and symbol formation originate from a sense of self as an agent.

There exists some evidence that the process of distinguishing the self-generated sensory reafferents from externally generated afferents, which indicate a sense of a proto-self and of agency, seems to begin early in life.

Meltzoff and Moore (1977) describe imitative behavior in infants within 42 min of being born; for example, babies imitate a tongue protrusion gesture performed by an adult. Meltzoff and Moore (1997, 1999) claim that perceivers, including infants, establish "supramodal representations" of bodily parts and their interrelations (in the case a tongue protruded between teeth) and, thus, that they have a type of proto-self or body schema that allows them to reproduce behaviors they observe. Other imitation behaviors, such as vocal imitation, observed in infants from 12 weeks of age are based essentially in intramodal comparisons (Kuhl and Melzoff 1996; Kuhl and Moore 1977).

In what concerns the distinction between sensory consequences of self-action and sensory consequences of stimuli independent of self-action, Rochat and Hespos (1997) observed that the rooting response of newborns (i.e., head orientation with mouth opening in the direction of a tactile stimulation on one of the cheeks) is significantly more frequent and predictable when the tactile stimulation comes from outside (single touch stimulation) than when results from spontaneous self-stimulation from the baby's own hand touching the cheek. This evidence of a differentiated rooting response in newborns suggests that they are capable of discriminating, at a very basic perceptual level, what corresponds to the sensory consequences of their own body movements from what corresponds to the external stimulation.

Developmental studies suggest that explicit self-awareness in infants comes much later. Between the 14th and 18th month, infants become embarrassed when they see in a mirror that there is a red spot on their face (Bertenthal and Fischer 1987). In this case, when children manifest shame or embarrassment, they take a meta-evaluative stance toward the embodied self. By the end of the second year, children begin to show self-consciousness—a meta-step in development that correlates with significant brain maturation, particularly in regions of the prefrontal cortex (Rochat 2010).

Nevertheless, manifestations of a self can be found in much earlier ages. There exists evidence that 4-month-olds start playing in front of mirrors (Tasakiris and Hagaard 2005) and are able to discriminate between their own and other's mirror images. Discrimination between self and others is interpreted as a proto form of self-awareness (Rochat and Striano 2002). The examples of imitative behavior or of distinct reactions to self and to external stimulation suggest that there exists a pre-reflexive form of experience of self, a proto-self, innate, present very early in ontogeny from birth, that allows a rudimentary distinction from self and nonself.

The existence of a proto-self in infants can further be conceived as a manifestation of an innate tendency to establish ties with a caregiver that will ensure safety, security, and protection. Meltzoff says that "we are born social"—that is, there exists an attachment to a caring figure that ensures proximity between the infant and the attachment figure.

In what concerns meaning and proto-symbolic processes linked to action planning and anticipation integrated in the sense of self, these considerations point to some aspects that are innate and depend on the structure of the NS, which can be thought as structures of knowledge and meaning that are further elaborated in higher level semantic processes acquired during development in the interaction with the environment and also in linguistic processes.

7 What Are the Unique Characteristics of Self-Representation?

The first and most primordial representation of the self is a body representation. The experience of the body has some characteristics that distinguish it from all other experiences, and it is the maximum invariant of the phenomenal and behavioral space.

The physiological sensory origin of this perceptual experience of body can be attributed to multiple sensory messages: pressure on and stretching of skin and deep tissues, friction and vibration on the skin, information about the body from neuromuscular and articulatory receptors, vestibular and balance information from the inner hear, the disposition and body volume from stretch receptors, nutrition and other homeostatic states from internal receptors, neuromuscular fatigue, and cerebral systems sensible to blood composition.

This systematization of somatic, interoceptive, and exteroceptive sensory systems shows that the body self doesn't rely on a single modality and neither is the information provided from a single modality. What distinguish the self-representations from all other phenomenal representations is the unique representational structure in the brain that receives a permanent sensory input (Kinsbourne 1995). What makes the body representation unique among all the percepts and phenomenal experiences is that the body representation is the maximum invariant—the center of the phenomenal space.

For all phenomena that can occur in consciousness, the body afferences are continuous and co-occurring permanently, some with a very slow or even nonexistent rate of sensory adaptation (such as proprioception, joint receptors, nociception). Although the relations in space and the movement can vary widely, the body remains a perceptual object that constantly generates afferent stimuli. Only the subject has first-person access to this ongoing sensory flux, which contributes to the subjective phenomenal experience of the self in a way that differs from the experience that results from an external object, which can be immediately socially shared (Zahavi 2002).

8 Body Representation: The Integration Between Peripheral Sensory Stimulation and Central Neuronal Mapping in Somatic and Motor Cortex

Peripheral sensory factors as well as central factors seem to play a role in the subjective feeling of embodiment. The consideration of some pathological conditions, such as "phantom phenomena," points to the contribution of central factors in body representation.

Having an experience of a part of the body that no longer exists, such as what occurs in phantom limb phenomena and phantom pain, has been attributed to a peripheral stimulation and also to a central factor. The body's inner representation at the neuronal level of the missing limb can be activated by intrinsic nervous activity or by activity that results from stimulation in other parts of the body. Whatever its origin, neuronal activity in the body's inner model (in cortical somatic maps) will be projected to the periphery that doesn't exist. The explanation of phantom limb phenomena depends on the activation of a central body neuronal model (Halligan 2002; Ramachandran and Hirstein 1998). This body model is innate but modified during development and later in adulthood by social interactions and behavioral interactions with the environment.

Other clinical observations of phantom limbs symptoms in 20% of children born without one limb suggests that they develop a complex body model that includes the parts of the body that never existed (Ramachandran and Hirstein 1998). This phantom experience is attributed to a central origin and also suggests the existence of an innate body model or body schema.

9 Heterogeneity of the Experiences of Self

The self is the author, actor, and executor of its own actions; it acts and perceives from its own perspective. In this chapter, we have considered a sense of self that is related to the concept of body schema. Nevertheless, even within the sense of self as an agent, it is possible to distinguish different subjective experiences.

Although the sense of agency has been considered short-lived and phenomenologically recessive, the thin phenomenology of action has been analyzed. Pacherie (2005, 2008) identifies three cascading "stages" of action specification: F intentions (intentions directed to the future), P intentions (intentions directed to the present), and M intentions (motor intentions). For Pacherie, the sense of agency is complex and contains a variety of aspects: an experience of intentional causation, the sense of initiation, and the sense of control.

The F or future intentions are formed before the actions and represent the whole plan of actions. Their content is detached from the specific situation and therefore is conceptual and descriptive. The F intentions are means–end coherent, that is, consistent with the agent's beliefs and intentions.

The P intentions serve to implement action plans defined in F intentions. They anchor the action plan both in time and in the situation of action. They involve a transformation of the descriptive contents of the action plan into perceptual–movement contents constrained by the present spatial characteristics of the agent, the target of action, and the surrounding environment. The final stage is action specification which involves the transformation of the perceptual action contents of P intentions into sensorimotor representations (M intentions) through a precise specification of the spatial and temporal characteristics of the constituent elements of the selected motor program (Pacherie 2007).

From a sense of agency, it is considered that the F—intentions that are relatively abstract and conceptual—may be spontaneously formulated and occur prior to the action. The P intention, which is more specific to the situation, occurs with higher temporal proximity to the action; involves a dynamic monitoring of the action; and can implement F. The P intentions, which have an initiating function as they trigger the intended action and a sustaining function until completion of action, guide the function and monitor its effects. It can be supposed that each of these stages specifies a distinct agentive self-experience.

The neurophysiology of motor planning and regulation is well known within multiple neuronal systems. It seems possible to establish a parallelism between Pacherie's fine phenomenology of agency and the CNS's (central nervous system) hierarchical regulation of motor functions, and it should be noted that many functions of behavior planning, command, and execution operate at an unconscious level.

In the CNS, motor planning begins with a general outline of behavior and is translated into concrete motor responses through processing in the motor pathways. The regulation is hierarchical with levels of regulation (interdependent, parallel, with feed-forward and feedback neuronal loops): a superior level with functions in the definition of objectives or aims in behavior involving associative areas of the cortex and premotor cortex and interactions with basal ganglia; the next level associated with primary motor cortex of precentral gyrus and the cerebellum and with the function of specification of a motor program in which the kinetics and dynamics of movement is planned and commands issued; and an execution level involving brain stem nuclei and circuits of spinal cord, interneurons, and motor neurons that regulate a variety of automated movements that control posture and locomotion (Kandel 2000).

10 Beyond Embodiment: Internal Representation of the Model of Action

In sensorimotor integration, representation is tentatively defined as a form of intentional internal modeling. This internal modeling is invoked to explain the fact that behavior is oriented to the achievement of future goals and is relatively independent from immediate environmental stimuli or specific sensorimotor representations. Internal modeling occurs in a projective and anticipatory way, and what is represented are anticipated states or intentional goals in an abstract form.

One of the most fundamental properties of cognition is, as Kenneth Craik put it, its power to predict events (Craik 1943).

In representations, there are three essential processes: (1) translation of external processes and internal data into words, numbers, or other symbols; (2) emergence of other symbols by a process of reasoning, deduction, and inferences, that is, the process of prediction; and (3) retranslation of these symbols into external processes—or at least a correlation between these symbols and external events (as in realizing that a prediction is fulfilled), the result of which is translated into the world.

The process of reasoning produces a final result similar to that which might have been reached by causing the actual physical process to occur. The thought processes have homomorphic properties with external events and so can be used to predict these external events (on the condition, there is a time delay between the two).

Thus according to Craik, the essence of thought is to provide a model of the external world. The mental prediction (or anticipatory model) in internal modeling is flexible and versatile—sensing, modeling, planning, and acting.

To invoke Liz Swan and Louis Goldberg (2010a) about symbols such as words, icons, or signals, symbols are elements that map signifiers to that which they signify. These mappings can be either arbitrary or transparent. Words are signifiers arbitrarily related to a significant; an icon is a transparent signifier that is linked by resemblance to the things they refer to, and signals are transparent signifiers that have a physical or mechanical connection to other objects.

The model they propose is one wherein symbol formation has a sensory–perceptive origin, that is, sensory receptors detect the presence of and respond to stimuli, which are processed and coded by perceptual symbol formation. The symbols induce effector processes (Swan and Goldberg 2010a, b).

In this chapter, we have tried to link symbol formation to a sense of self that invokes meaning and symbolic processes that occur in a nonconscious proto-self that constitutes a first-order representation. The second-order representation includes the relation between the self and the object. The third-order representation involves meta-representation of autoreflexive processes.

The model we propose takes an efferent–anticipatory point of view in which symbolic meanings are created in an interactive internal space, referred to as the agent or self, that is, an internal model that binds perceptual present, past memories, and also future desirable states. It is proposed that symbols are projective, anticipatory, or beyond immediate instantiations. They are abstract and intentional, and in this sense, symbols are beyond embodiment.

11 Conclusion

We have proposed an individual-centered perspective for symbolization and meaning processes. The embodied ground of meaning, linked to the formation of an internal model, allows phenomenic agentive first-person experience and shapes judgments. Meaning arises in this internal model, which integrates external and internal influences and recruits neural systems involved in perception, movement, and emotion. The embodied model of symbolization processes points to mental simulation, anticipatorily driven by a complex interplay between sensory and motor components, in which intentions and future aims or desirable states are represented. These successively more complex and higher-order behavioral strategies are recursively generated independent of their embodiment.

References

Bertenthal, B., & Fischer, K. (1987). Development of self recognition in the infant. *Developmental Psychology, 14*, 44–50.

Blakemore, S.-J., Frith, C. D., & Wolpert, D. (1999). Spatio-temporal prediction modulates the perception of self produced stimuli. *Journal of Cognitive Neuroscience, 11*, 551–559.

Craik, K. (1943). *The nature of explanation*. Cambridge: Cambridge University Press.

Gallagher, S. (2000). Philosophical conceptions of the self: Implications for cognitive science. *Trends in Cognitive Science, 4*(1), 14–21.

Gallagher, S. (2012). Multiple aspects in the sense of agency. *New Ideas in Psychology, 30*(1), 15–31.

Halligan, P. W. (2002). Phantom limbs: The body in the mind. *Cognitive Neuropsychiatry, 7*(3), 252–268.

Kandel, E. R. (2000). From nerve cells to cognition: The internal cellular representation required for perception and action. In E. R. Kandel, J. H. Schawrtz, & T. M. Jessell (Eds.), *Principles of neural science* (pp. 381–402). New York: McGraw Hill.

Kinsbourne, M. (1995). Awareness of one's own body: An attentional theory of its nature, development and brain basis. In J. L. Bermúdez, A. Marcel, & N. Eilan (Eds.), *The body and the self* (pp. 205–223). Cambridge, MA: MIT Press.

Kuhl, P., & Moore, M. (1977). Infant vocalizations in response to speech: Vocal imitation and developmental change. *The Journal of the Acoustical Society of America, 100*, 2425–2438.

Kuhl, P. H., & Melzoff, A. N. (1996). Infant vocalizations in response to speech: Vocal imitation and developmental change. *Journal of the Acoustic Society, 100*, 2425–2438.

Llinás, R. R. (1987). 'Mindness' as a functional state of the brain. In C. Blakemore & S. Greenfield (Eds.), *Mindwaves. Thoughts on intelligence, identity and consciousness*. New York: Basil Blackwell.

Llinás, R. R. (2001). *I of the vortex. From neurons to self*. Cambridge, MA: MIT Press.

Meltzoff, A., & Moore, M. K. (1997). Explaining facial imitation: A theoretical model. *Early Development and Parenting*, 6, 179–192.

Meltzoff, A., & Moore, K. (1999). Persons and representation: why infant imitation is important for theories of human development. In J. Nadel & B. Butterworth (Eds.), *Imitation in infancy* (pp. 9–35). Cambridge: Cambridge University Press.

Pacherie, E. (2005). Perceiving intentions. In J. Saagua (Ed.), *A explicação da intrerpretação humana* (pp. 401–414). Lisbon: Edições Colibri.

Pacherie, E. (2007). The sense of control and the sense of agency. *Psyche, 13*(1), 1–30.

Pacherie, E. (2008). The phenomenology of action: A conceptual framework. *Cognition, 107*, 179–217.

Ramachandran, V. S., & Hirstein, W. (1998). The perception of phantom limbs. *Brain, 121*, 1603–1630.

Rochat, P. (2010). The innate sense of the body develops to become a public affair by 2–3 years. *Neuropsychologia, 48*, 738–745.

Rochat, P., & Hespos, S. J. (1997). Differential rooting response by neonates: Evidence for an early sense of self. *Early development and Parenting, 6*, 105–112.

Rochat, P., & Striano, T. (2002). Who's in the mirror? Self-other discrimination in specular images by four- and nine-month- old infants. *Child Development, 73*, 35–46.

Sperry, R. W. (1950). Neural basis of the spontaneous optokinetic response produced by visual inversion. *Journal of Comparative and Physiological Psychology*.

Swan, L. S., & Goldberg, L. J. (2010a). Biosymbols: Symbols in life and mind. *Biosemiotics, 3*(1), 17–31.

Swan, L. S., & Goldberg, L. J. (2010b). How is meaning grounded in the organism? *Biosemiotics, 3*(2), 131–146.

Tsakiris, M., & Haggard, P. (2005). Experimenting with the acting self. *Cognitive Neuropsychology, 22*(3/4), 387–407.

Von Holst, E., & Mittelstaedt, H. (1950). Das Reaffernzprinzip echselwirkungen zwichen zentrainervensystem und peripherie. *Naturwissenschalten, 37*, 464–476.

Wolpert, D. M. (1997). Computational approaches to motor control. *Trends in Cognitive Sciences, 1*, 209–216.

Zahavi, D. (2002). First-person thoughts and embodied self-awareness. *Phenomenology and the Cognitive Sciences, 1*, 7–26.

Part III
Consciousness

Imitation, Skill Learning, and Conceptual Thought: An Embodied, Developmental Approach

Ellen Fridland

Abstract It is the goal of this chapter to offer a strategy for moving from imitation to conceptual thought. First, I accept that imitation plays a vital role in accounting for the facility with which human beings acquire abilities, but I argue that successful task performance is not identical to intelligent action. To move beyond first-order behavioral success, I suggest that the orientation that humans have toward the means of intentional actions, that is, the orientation required for imitation, also drives us to perfect our skills in a way that produces fertile ground for florid thought.

In Section "What Is So Special About Human Imitation?", I propose that the difference between animal and human copying lies in what I call the "means-centric orientation." In Section "Imitation Is Great, but It Ain't Everything", I explore three characteristic features of intelligence and claim that the first-order behavioral success that results from imitation is not characterized by these features. In the final section of this chapter, I argue that the means-centric orientation, when inverted onto itself, motivates skill refinement and, as such, allows us to reach the intermediate level of cognitive development. It is at this level, through the individuation and recombination of action elements, that we see a basic syntax of action arise and, with it, the characteristic features of intelligence emerge.

1 Introduction

In the search for that special something that might account for the difference between human cognition and the cognition of nonhuman animals, imitation has received a lot of attention. This is especially true in developmental and social psychology

E. Fridland (✉)
Berlin School of Mind and Brain, Humboldt University of Berlin,
Luisenstr. 56, Berlin 10099
e-mail: ellenfridland@gmail.com; ellen.fridland@philosophie.hu-berlin.de

L. Swan (ed.), *Origins of Mind*, Biosemiotics 8, DOI 10.1007/978-94-007-5419-5_10,

circles where imitation, an arguably unique human capacity, has been deemed crucial to the development of social cognition and higher-order executive function (Tomasello et al. 2005; Tomasello and Rokoczy 2003; Meltzoff 2005). It is thought that imitation fosters in humans the capacity to form tight social bonds, to share in joint attention, joint action, linguistic communication, shared intentionality, an understanding of other minds, and finally, an understanding of ourselves. These interpersonal connections are meant to pave the way to full-fledged, florid, higher-order, human-style thinking. The problem remains, however, that it is not at all obvious how imitation alone is going to guide us into these lofty cognitive realms.

In this chapter, my goal is to offer a theoretical strategy for moving from imitation to conceptual thought. After accepting that imitation plays a vital role in accounting for the facility with which human beings acquire abilities, I argue that successful task performance is not identical to intelligent action. To move beyond first-order behavioral success, I suggest that the motivation driving imitation, when applied intrapersonally, acts as a parsimonious and powerful force. Specifically, I argue that the orientation that humans have toward the means of intentional actions, that is, the orientation that drives imitation, also propels us to perfect our skills in a way that produces fertile ground for florid thought. I develop this account by presenting a theory that grounds the flexibility, manipulability, and transferability of mature human cognition in embodied skill.

In Sect. 2, I propose that the difference between animal and human copying lies in what I call the "means-centric orientation." In Sect. 3, I explore three characteristic features of intelligence and claim that the first-order behavioral success that results from imitation is not characterized by these features. In the final section of this chapter, I argue that the means-centric orientation, when directed at one's own actions, motivates skill refinement and, as such, allows us to reach the intermediate level of cognitive development. It is at this level, through the individuation and recombination of action elements, that we first see a basic syntax of action arise and, with it, the characteristic features of intelligence emerge.

2 What Is So Special About Human Imitation?

Everyone involved in the imitation debate agrees that human imitation is special. By this, I do not mean to suggest that there is a lack of disagreement about whether imitation is an exclusively human affair.[1] My point is, rather, that even those who deny that imitation is proprietary to humans admit that human imitation is importantly distinct from the imitation of nonhuman animals.[2] Notably, nonhuman

[1] For instance, Tomasello (1996, 1999; Call and Tomasello 1998) claims that imitation is proprietary to humans, while others (Byrne 2002; Horner and Whiten 2005) claim that imitation can be observed in the behavior of nonhuman primates.

[2] For an instance of such a position, see Byrne and Russon's (1998) distinction between action and program-level imitation.

primates, our closest evolutionary relatives, neither imitate as often as human children nor do they reproduce the particular detailed style with which an observed action is instantiated (Byrne 2002; Byrne and Russon 1998; Call et al. 2004; Tomasello 2009). Additionally, the role of imitation in cultural learning and transmission has no comparable function anywhere outside of human society (Tomasello 2005; Boesch and Tomasello 1998; Tomasello and Rokaczy 2003). As such, even if some nonhuman animals are found capable of imitation, we will still need an account of human imitation that explains its prominence and uniqueness as a learning strategy for children.

2.1 Reworking the Definition of Imitation

In this section, my goal is to argue that the means or instrumental strategy of goal-directed actions plays an essential role in forming the intention motivating imitation. In this sense, I'd like to amend the preferred definition of imitation by highlighting the significance for the imitator of the instrumental strategy with which an observed and reproduced action is instantiated. In particular, I suggest that the efficient cause of imitation, that is, the reason why an individual imitates, is fundamentally connected to the imitator's irreducible interest in or concern for the means of an observed intentional action. I call this general perspective "the means-centric orientation."

The means-centric orientation is best understood as the not-merely-instrumental interest in or preference for the means of an intentional action. Specifically, my claim about the means-centric orientation amounts to the following: when a subject S imitates some action A, which is aimed at accomplishing a goal G, it is both the means M that are used to accomplish G and G itself that hold inherent value for S. For example, if an agent models for a child how to open an umbrella, both the end of opening the umbrella and the means that the model uses to open the umbrella become objects of intrinsic concern for the child who imitates.

Importantly, the means-centric orientation turns the means of goal-directed actions into a locus of significance. It makes the means of an observed and imitated action important and interesting in their own right; it makes the details of an observed behavior contain value that is not necessarily reducible to its practical payoff or purpose. This is not to say that the "not-merely-instrumental" concern for means is necessarily reducible to the means themselves, but it is to say that the value of means overflows their capacity to facilitate goal satisfaction.[3] Notably, focusing on this aspect of imitation also allows me to present a clear strategy for relating imitation to higher-order cognition in later sections of this chapter.

[3] I use "not-merely-instrumental" value and not simply "inherent" value in order to leave open the possibility that means are a locus of value or significance as a result of their role in offering opportunities for social connection and intersubjective rewards. In this sense, the concern for means would be not-merely-instrumental for the goal at hand, but still offers other kinds of important payoffs.

To be clear, I understand my emphasis on the means-centric orientation as compatible with conventional definitions of imitation. In fact, if we take Michael Tomasello's definition of imitation, the means-centric orientation should be seen as a refinement and not a replacement of it. Boesch and Tomasello write that the "the archetype of imitative learning… [is the] reproduction of both behavior and its intended result" (1998, p. 599). This definition of imitation requires that the imitator exhibits sensitivity both to the goals of the observed demonstration and also to the particular behavioral strategy that the model uses in order to achieve her goals.[4]

To better understand the nature of imitation, and why my proposed amendment is necessary, it may be helpful to contrast it, as Tomasello famously does, with emulation.[5] Boesch and Tomasello define emulation as "the process whereby an individual observes and learns some dynamic affordances of the inanimate world as a result of the behavior of other animals and then uses what it has learned to devise its own behavioral strategies" (1998, p. 598). For Tomasello, the primary distinction between imitation and emulation is that imitation requires the imitator to recognize and reproduce the intentional goal state of the demonstrator, while emulation only requires reproducing the observed behavior in order to manipulate the world. What Tomasello overlooks, however, by focusing on the shared psychology of imitator and demonstrator is the fact that an imitator must show concern not only for the mental states of the demonstrator but also for the actual actions that the demonstrator performs.[6] That is, the imitator can not only be interested in the intentional constitution of the demonstrator but must also be interested in the task or action that the demonstrator models. To reflect this point, on my account, imitation learning differs from emulation learning in two ways: (1) in sharing a goal with the demonstrator, and (2) in expressing a noninstrumental preference for reproducing the behavioral strategy that the demonstrator models.

We should note that while for Tomasello the particular details of an observed behavior must be reproduced in order for some action to count as imitation, he does not require that the imitator have a special interest in or intention for reproducing the behavior.[7] In contrast, on my account, it is not simply that the imitator happens

[4] Importantly, studies on rational imitation show that it is not just movements, but actions that are recognized as intentional, which are imitated by children. See Meltzoff (1995); Carpenter et al. 1998; Bellagamba and Tomasello 1999; Gergely and Csibra(2005); Schwier et al.(2006).

[5] Tomasello (2009) has now admitted that, in rare cases, nonhuman primates do in fact imitate. However, he still holds that in most circumstances, the copying behavior of nonhuman primates is emulation and not imitation.

[6] In fact, ideally, the interest in the action should form the path by which the imitator can learn about intentional states. She should not already know about the demonstrator's mental states if imitation is meant to be a strategy by which she is going to learn about them. See Meltzoff (2005) for a defense of this position.

[7] To be fair, in 2009, Tomasello has written that a concern with action itself may be crucial for differentiating between animal and human copying. This admission, however, is not reflected in a new definition of imitation. As such, my proposal constitutes a significant change in what is taken to be necessary for imitation.

to reproduce the same behavioral sequence that the model demonstrates as a result of sharing a goal with the demonstrator, but that the imitator's reason for producing the behavior makes the reproduction of the observed behavior part of the goal of her action—it becomes part of the intentional state driving imitation. In short, the means-centric orientation drives imitation by making sure that the imitator has the reproduction of the means of an observed action incorporated into her objective for acting.

As such, this preoccupation with the means of action poises humans for imitation by overriding the more pragmatic concerns of action, such as implementing whichever strategy will most efficiently lead to the satisfaction of one's desires. The saliency of the means of action keeps humans focused on and attentive to the instrumental strategy of an observed action rather than on the world or the goal at which the action is aimed. And this keeps us hooked specifically on imitation in a way that simply sharing goals with a demonstrator cannot.[8] It keeps us reproducing the detailed, particular strategies that we see others perform because it is the means by which we achieve our goals, and not only the goals, that are interesting and meaningful for us.

2.2 Empirical Evidence of the "Not-Merely-Instrumental" Preference for Means

Happily, empirical research on imitation supports my claim that humans have a not-merely-instrumental preference for the means of intentional action. A great many studies have clearly demonstrated that humans imitate regardless of whether imitation produces the most efficient route for achieving an end. I will present just one of these studies here.[9]

In a particularly elegant study, Victoria Horner and Andrew Whiten (2005) presented chimpanzees and 2-year-old human children with a demonstration of a complex sequence of actions aimed at opening a box containing a food reward in two conditions: one opaque and one transparent. In the opaque condition, the causal structure of the interaction between the experimenter and the box was hidden from the subjects, and so, when the demonstration included a causally irrelevant behavior, the subjects were unable to see it as such. Alternatively, in the transparent condition, the subjects were able to see how the experimenter's actions were causally related to the opening of the box. Horner and Whiten found that chimpanzees reproduced the observed behavioral sequence, including the useless movement, in the opaque condition but not in the transparent condition. That is, once the

[8] After all, the sharing of goals with another person may lead to numerous kinds of behaviors that are neither identical to nor connected with imitation.

[9] In addition to this study, especially notable is the work of Gergely and Csibra (2005).

chimpanzees determined that the movement was causally irrelevant for opening the box, they no longer incorporated that movement into their behavioral repertoire.

In contrast, children continued to reproduce the causally irrelevant action in both the opaque and the transparent condition. That is, even after identifying a movement as causally irrelevant, children continued to reproduce it when opening the box. Importantly, both chimpanzees and humans, in separate experiments, were shown to have the capacity to appreciate the relevance of causal information for achieving some end. These findings then clearly demonstrate that children will imitate even when imitation is not the most efficient way for them to achieve their goals. Further, this is not at all an isolated result. Children regularly display their impractical orientation toward imitation. This is especially evident in children's imitation and over-imitation of the detailed style with which an action is performed, a feature that is often completely irrelevant for task success (Byrne 2002, Lyons et al. 2007; McGuigan et al. 2007; Whiten et al. 2009).

The take-home point is this: for children, but not for nonhuman primates, the reproduction of the means of an observed action has a value that is not simply reducible to its value as a means to an end. Whatever else is true about the ultimate explanation of this orientation, we must admit that humans imitate as a result of a not-merely-instrumental preference for reproducing an observed behavior. This must be the case because if the value of reproducing an observed action were only instrumental, then when some means did not serve as the most efficient path to a goal, it would be abandoned. Since this does not always happen,[10] we must conclude that human beings have some interest in reproducing means, which is divorceable from the role of those means as a strategy for achieving some end. And it is precisely this nonstandard preoccupation with means, I claim, that gives us insight into what is special about the copying behavior of children.

2.3 A Few More Considerations

I hope to have shown that a preoccupation with the means of goal-directed actions is central to explaining the motivational structure that drives imitation. My claim is that by not acknowledging that means themselves enjoy a certain kind of impractical celebrity as part of the intentional content driving imitation, we overlook a crucial aspect of imitative behavior.

Lastly, we should note that it is thoroughly surprising when compared to the rest of the animal world that the human concern for action is often not reducible to the goal at which the action is aimed. This imprudence, this impracticality, I claim, is what makes human imitation special. Notably, this orientation can also explain the curiously impractical nature of many human activities. After all, it is only humans

[10] Of course, there will be times when humans are concerned with the goals of an action more so than with the means of that action. The main point, however, hangs on the fact that humans are *not always* so concerned with action, while nonhuman primates are.

that spend vast amounts of time and energy pursuing hobbies and skills that have no obvious evolutionary payoff. Think of playing video games, crocheting, creating miniatures, or solving a Rubik's cube puzzle with one's feet.[11] Only humans spend countless hours practicing and perfecting abilities and skills that are, on almost any practical measure, useless. On my account, the reason for this odd human characteristic is easy to explain. After all, a not-merely-instrumental preference for the means of intentional behavior accounts for why so many different activities could themselves become sources of interest, curiosity, and pursuit.

3 Imitation Is Great, but It Ain't Everything

In this section, my goal is to elucidate that the development of many features characteristic of human-level cognition cannot be accounted for with imitation or shared intentionality alone. My goal is not to downplay the importance of imitation in human cognitive development, but merely to highlight the additional work that needs to be done if we are going to be able to establish anything resembling a full account of human cognition.

First, it is vital to recognize that imitation is a great way to account for the transmission of highly complex and idiosyncratic practical and cultural knowledge. By imitating, humans acquire a huge number of skills that target the very specific needs of our geographical and historical situations. In fact, there seems to be no better way to transmit the infinite variety of methods required to master technology, ritual, and culture than to provide an innate "do as I do" mechanism (Meltzoff 2005). The problem, however, is that this mechanism alone cannot breed higher-order cognition. That is, imitation can account for task success and even cooperative, shared action, but it isn't obvious how either of these is meant to produce *our* kind of cognition.

3.1 Imitation: Task Success and Understanding

One of the most obvious examples of imitation's insufficiency for explaining the emergence of human understanding and intelligence comes from the fact that children are capable of imitating long before they are capable of understanding how their imitated actions are related to the world. The fact is that children can successfully act on objects in their environment by using an imitative strategy without thereby understanding much about the nature of the objects on which they are acting. For instance, Want and Harris (2001) show that at age two, children "blindly imitate," while by the age of three, they imitate in an "insightful" fashion. Want and Harris

[11] Yes, people actually do this and hold competitions!

establish this conclusion by demonstrating that 3-year-olds benefit from observing a mistaken or incorrect action while 2-year-olds do not. Thus, they reasonably conclude that only 3-year-olds imitate in a way that reveals an understanding of the causal relations between their actions and the environment.

Importantly, if successful imitation exists in the absence of task-specific knowledge, then we must conclude that, developmentally, imitation alone is not sufficient for understanding. This does not mean that imitation doesn't offer us a parsimonious strategy to gain such knowledge, but it does mean that imitation must be coupled with additional mechanisms, if it is to do any cognitive work. That is, imitation must work in conjunction with other cognitive learning processes if it is to account for our knowledge of objects, the environment, the self, others, and the causal and conceptual connections between these.

The mechanisms of imitation, if they are to provide us with the powerful tools that many theorists think they can, must be cashed out in such a way as to make clear how the appropriate connections, associations, and causal structures are formed as a result of their implementation. If imitation is to get us to knowledge, then imitation must work together with processes that can gather and connect the right kind of information with the right kinds of expectations.

We should notice, however, that these kinds of connections, associations, and expectations alone do not even begin to approach what is unique about human cognition. After all, the requirement for basic learning mechanisms will most certainly be held in common with nonhuman animals. Even emulation learning, after all, requires the subject to develop an understanding of environmental features and their causal affordances. Whatever accounts for that, coupled with imitation, should suffice for a basic explanation of "insightful imitation," or imitative learning that yields an understanding of the causal structure of the environment. Further, by focusing on the requirement that imitation is rational (Meltzoff 1995; Carpenter et al. 1998; Bellagamba and Tomasello 1999; Gergely and Csibra 2005; Schwier et al. 2006), we can even accept that imitation lays the groundwork for a basic understanding of other minds. But even if this then allows for shared attention, cooperation, and joint action, it still isn't clear how these are sufficient to explain the fantastic heights that we reach in abstract, conceptual thought?

That is, what should we say about our human cognitive capacities that go well beyond learning about the causal structure of the world or the recognition of actions as intentional? How might imitation be involved in the flexibility, manipulability, and transferability of human thought, our fine-grained recombinatorial abilities, our capacity for meta-representation, or the development of a sense of agency? Is it at all possible that this lofty grab bag of cognitive virtues has any connection to imitation? Before offering some guidance on how such a connection might be established, I'd like to take a moment to clarify how the above-listed capacities are distinguishing characteristics of human thought and also to elucidate why imitation, even if it can foster cooperative action and shared intentionality, cannot give us an explanation of them.

3.2 *Intelligence and the Three Sisters: Flexibility, Manipulability, Transferability*

Flexibility, manipulability, and transferability are related concepts that highlight important features of intelligence. In this section, I attempt to give an overview of the contributions that each makes to the notion of intelligence and also, where necessary, to point out the conceptual connections between them.

3.2.1 Flexibility

As we begin to consider some of the key features of human intelligence, flexibility quickly comes to mind. It seems that a behavior, no matter how sophisticated, which is rote, rigid, or inflexible, could not possibly qualify as intelligent. In fact, definitionally, intelligence is often contrasted with fixed, automatic, or stimulus–response behaviors. As José Bermúdez writes, "a distinguishing mark of the cognitive is that it is variant, and not stimulus–response" (2003, p. 8). He contrasts this with cognitively integrated "behavior that is flexible and plastic and tends to be the result of complex interactions between internal states learning and adaptation contributing and determining present responses" (Bermúdez 2003, p. 9). It follows that a lack of flexibility undermines the possibility of a behavior qualifying as genuinely intelligent. But what constitutes the special relationship between flexibility and intelligence? Is all flexible behavior intelligent? Could unintelligent behavior be flexible? After only a moment's consideration, I think that we will all agree that the answer to the first question is "no" and to the second, "yes."

After all, a random behavior or event, though it might be flexible to the point of being unpredictable, carries no guarantee of intelligence. Shouting the lyrics to a Dylan song in the middle of the library might not be something that is fixed in your instinctual behavioral repertoire, but that doesn't make it smart. The fact is that intelligence presupposes a degree of freedom, but it also requires a healthy dose of constraint. This is because intelligence is about doing the right thing at the right time and not just about doing anything whatsoever.[12] So, intelligent behavior must be simultaneously flexible and grounded. Intelligent behavior must be variable within the confines of the environment, a creature's goals, and the possibilities for instrumental action afforded thereby.[13] If this is correct, then we see that flexibility

[12] Dennett makes a similar point when he says that "The criterion for intelligent storage is then the appropriateness of the resultant behavior to the system's needs given the stimulus conditions of the initial input and the environment in which the behavior occurs" (1969, p. 50).

[13] There are obvious parallels to the point that I am making here and Hume's classic compatibilist critique of liberty (1961, section VIII). That is, as Hume points out, being free, uncaused, or random cannot ground responsibility since one cannot be responsible for a random or uncaused event.

isn't sufficient for intelligence, but merely necessary for the kind of behavioral changes about which we care. Namely, it is a prerequisite for appropriateness, learning, improvement, adaptation, and success. And we take these processes to be indicative of intelligent systems.

As such, we should conclude that flexibility is not by itself a mark of intelligence, but rather a sort of pointer to it. Flexibility's value is derived from the role that it plays in affording the possibility for a certain kind of behavior, namely, for affording the possibility of appropriate behavior in response to changing environmental conditions.

3.2.2 Manipulability

In addition to flexibility, manipulability is often cited as a characteristic of intelligent behavior. Manipulability requires a certain kind of flexibility, since that which is to be manipulated cannot be fixed; however, manipulability demands something more as well. Manipulability highlights the fact that when we speak of intelligence, we want behavior that is not only flexibly related to the world but flexible as a result of its being under the control of an agent. As such, the flexibility required for appropriate environmental responses, learning, and improvement should not just result from various parallel processes, but it should be hierarchical; it should be top-down. Intelligent behavior is behavior that an agent can access. It is behavior that an agent plans, organizes, reorganizes, guides, and controls.

Jesse Prinz (2004)[14] goes so far as to *define* cognition in terms of this kind of control, and Richard Byrne and Anne Russon write

> [W]e would be reluctant to describe as intelligent any sequence of behavior whose mental organization is a single unit or action connected to a goal representation, a long sequence of linear associative connections or a rigid hierarchical structure. Thus whether a behavioral structure is modifiable by the individual becomes crucial in diagnosing it as "intelligent." (1998, p. 671)

One crucial implication that follows from the requirement that intelligent behavior be manipulable is that intelligence becomes a personal-level phenomenon. That is, though it is possible that subpersonal systems respond flexibly to various environmental and internal circumstances, they are ruled out as intelligent because they are not under the control of an agent. The requirement that intelligent systems be manipulable entails that intelligence is a phenomenon that occurs on the level of

---

The connection between the agent and the action must be fundamental if agents are going to be responsible for their actions. Likewise, being flexible is not enough for being intelligent, but behaviors must be connected to their environments in the right way if these behaviors are to qualify as intelligent.

[14] Prinz writes that "[c]ognitive states and processes are those that exploit representations that are under the control of an organism rather than under the control of the environment" (2004, p. 45).

persons and not subsystems precisely because the kind of control demanded here is only available to whole agents. As such, we see that intelligence requires central integration that is impossible at lower levels of cognitive processing.

As a brief aside, I'd like to point out that at this stage, we are not required to decide whether or not the cognitive capacities that I am discussing here are necessary features of intelligence. This question is not immediately relevant because even if we decide that manipulability is not a necessary condition for some event to qualify as intelligent, we must still admit that paradigmatically intelligent behaviors often possess this feature. So, at the end of the day, even if we decide that our definition of intelligence makes room for intelligent acts that are *not* manipulable by the agent, we will still have to provide an account of those particularly intelligent acts that *are* thus manipulable. As such, an account of manipulability will be part of our theory of intelligence whether or not manipulability is deemed to be a necessary condition of intelligent action.

3.2.3 Transferability

In addition to flexibility and manipulability, transferability or generality is also frequently invoked as a distinguishing characteristic of intelligent behavior. We can think of transferability as the requirement that intelligent behaviors possess the potential for wide application. If instrumental learning occurs in one domain but cannot be transported to another, then we should wonder if such changes are really intelligent. For example, if I can add jelly beans but not match sticks or sheep, then maybe I'm not really adding.

As with manipulability, we should notice that even if transferability does not turn out to be a necessary feature of intelligent events, paradigmatically intelligent behaviors possess this feature. That is, paradigmatically intelligent behaviors are largely context independent. Take propositional thought as an example: I can believe, desire, or fear that it is raining. I could do this yesterday, today, and tomorrow. I can do it in Boston, in Hawaii, or in Berlin—in the morning or at night. I can compare rain with snow. I can remember the summer rain of my childhood, and I can predict how rain will affect my weekend plans. Crucially, the emphasis on transferability points to the fact that we want intelligence to play a general role in our cognitive economy. We insist that knowledge and skills are accessible to an agent in a large number of circumstances. It follows from this that the information upon which intelligent behaviors depend will be stored in a form that is abstract enough to be applied at different times and places. It follows that such information cannot be bound to particular stimuli.

We should also notice that transferability is intimately related to both flexibility and personal-level processing. Transferable behaviors must be flexible if they are to break free from a particular domain in order to be utilized in others. In fact, we can think of transferability as a kind of diachronic or horizontal flexibility. But also, transferability must be person or agent level because to be transferred to various independent domains, information or skills must be centrally accessible. This point is especially clear if we think of the mind as composed largely of various modular, informationally encapsulated systems. In such a mind, transferring information between independent domains requires a central process that will be responsible for the appropriate extraction and application of information. We are confronted with the fact that information that is *in* a system, but not available *to* a system (Karmiloff-Smith 1992, p. xiv; Clark and Karmiloff-Smith 1993), that is, information that is subpersonal but not agent accessible, is not information that can be used by intelligent processes.

3.3 Imitation and the Three Sisters

Imitation functions as an important mechanism accounting for how children acquire abilities and skills, but we should be careful to notice that success at a task by no means entails the presence of flexibility, manipulability, or transferability. That is, developing the capacity to *a* does not mean that one can *a* flexibly, that one can manipulate the way in which one *a*s, or that one can transfer the knowledge required to *a* into another independent domain. As such, if imitation can guarantee task success but not flexible, manipulable, or transferable behaviors, then we must conclude that imitation alone cannot account for intelligence.

This fact about imitation becomes especially salient, if we turn to Annette Karmiloff-Smith's model of representational redescription (RR) (Karmiloff-Smith 1986, 1990, 1992). According to this model, human cognitive development progresses in three basic stages. Movement through these developmental stages "involves multiple levels of redescription, leading to increasing accessibility and flexibility" (Clark and Karmiloff-Smith 1993, p. 496). That is, as representational states are redescribed at higher levels, they begin to express more and more features characteristic of higher-order intelligence.

For our purposes, it is especially important to take note of the nature of representation at the first level of redescription. The first level of representational redescription, the I-level or implicit level, is "procedural and must be run in its totality. It cannot be accessed or operated on" (Clark and Karmiloff-Smith 1993, p. 495–496). I-level procedures are context dependent, inflexible, informationally encapsulated, and not accessible to consciousness. They are procedures that are rigid, sequentially constrained, difficult to interrupt, individuate, change, and control (Karmiloff-Smith 1990). However, and this is vital for our purposes, I-level procedures support practical success. That is, behavioral mastery is achieved at the I-level, and in fact, "behavioral mastery is a prerequisite for subsequent representational change" (Karmiloff-Smith 1990, p. 60).

This means that at the I-level, a child is capable of successfully performing a task, but the child cannot reorganize, reorder, shuffle, manipulate, or access the procedures responsible for successful task performance. The performance hits its mark, but it is not flexible, manipulable, or transferable. As Karmiloff-Smith writes about linguistic development:

> Despite the limitation of the implicit representations symptomatic of phase 1, it is essential to recall that by the end of the first phase for a particular linguistic form, children have achieved communicative adequacy in their use of the particular linguistic form. (1986, p. 106)

As such, the presence of flexibility, manipulability, and transferability in human thought does not immediately follow from practical success. This has severe implications for imitation because it suggests that imitation, as a basic mechanism, can only account for a child's acquisition of first-order representations but not for later representational change. After all, we have no reason to posit that imitation, by facilitating the acquisition of task-specific capacities, provides children with anything beyond first-order, implicit, procedural states. The central point is that imitation can account for task success, but task success does not entail intelligence. So, though the kinds of practical behaviors acquired through imitation are impressive in breadth and complexity, they turn out to be fairly low-level cognitive achievements in terms of the spectrum of their intellectual characteristics. As such, we must conclude that though imitation can account for ability acquisition, it cannot account for the higher-order cognitive features that are part and parcel of intelligent behavior.

Of course, at this stage, it wouldn't hurt to ask what we need to add to behavioral success in order to get to intelligence. One proposal that seems plausible is that what is needed for intelligence is the capacity to "develop explicit representations which allow a system to become more manipulable and flexible" (Clark and Karmiloff-Smith 1993, p. 503). That is, "explicit representations provide a system with a kind of flexibility and generality not possible in any first order network" (Clark and Karmiloff-Smith 1993, p. 492). It isn't entirely clear why explicit representations get us this sort of payoff, but one possibility is that explicit representations, since they are represented outside of the subsystems in which they are run, can be entertained off-line in various independent settings. As such, with explicit representation, we get a dissociation from the immediate stimulus environment, which offers us the possibility of entertaining representations whether or not they are immediately relevant. It seems that with explicit representation, we become what Dan Dennett (1996) has termed "Popperian animals." That is, we become the kind of animals that can do trial and error in our heads; an animal that can let its hypothesis die in its stead. As Ruth Millikan writes, "The Popperian animal is capable of thinking hypothetically, of considering possibilities without yet fully believing or intending them. The Popperian animal discovers means by which to fulfill its purposes by trial and error with inner representations" (Millikan 2006, p. 188).

But we should notice that representation into explicit form is not a straightforward consequence of the behavioral mastery that is acquired through imitation. After all, there is nothing in the specifications of imitation that seems even remotely poised to guarantee that the results of imitative learning are represented explicitly. Therefore, it becomes impossible to hold, without further refinement, that the mechanisms of imitation will be able to account for the development of the explicit representations that underwrite the flexibility, manipulability, and transferability of human thoughts and behaviors.

4 Imitation and Skill Refinement: Making Our Way up the Cognitive Ladder

As we have seen, imitation can provide an account of the facility with which children pick up various practical and cultural competencies. We see that the imitative faculty is crucial in accounting for the easy transmission of highly nuanced human knowledge and skill, and in creating the circumstances for shared intentionality and cooperative action. Imitation goes a long way in explaining how children become proficient in relating to both objects and other people in an impressive variety of ways in an incredibly short period of time. Despite the impressiveness of this kind of learning, however, we must be careful not to overstate the work that imitation can do in our theory of cognition. Specifically, we must be careful not to confuse the social and behavioral mastery that imitation affords with the higher-order, full-fledged, florid, flexible, manipulable, transferable, recombinable, agent-directed intelligence present in fully mature, conceptual thought.

Though imitation alone cannot ground a theory of human cognition, in this section, I will elucidate how the means-centric orientation, which I have argued is central to imitation, can be employed in order to explain movement up the cognitive ladder. I propose that the means-centric orientation, which drives imitation in an intersubjective context, when inverted onto one's own actions, can provide us with a way to move from the first-order stage of implicit, procedural, practical success to the intermediate level of cognitive development. In particular, I claim that shifting the means-centric orientation from the intersubjective realm into an intrasubjective arena endows children with the capacity to move beyond ability acquisition and into a stage of skill refinement. And it is through skill refinement, as I explain below, that the first signs of intelligence begin to appear.

The sort of transition from the interpersonal to the intrapersonal that I am suggesting should not be altogether startling to those familiar with classic childhood development literature. In fact, this is a fairly straightforward application of Lev Vygotsky's conjecture that "[e]very function in the child's cultural development appears twice: first, on the social level, and later, on the individual level; first, between people (interpsychological) and inside the child (intrapsychological)" (1978, p. 57). Even if this claim turns out to be false as a general principle, we can see that it is quite apt in this particular context. By embracing the shift from the *interpersonal* means-centric orientation to the *intrapersonal* means-centric orientation, we find ourselves in a position to explain how it is that a child first begins to control, guide, attend to, and refine her own actions. By embracing this transition, we are in a position to explain how a child's own abilities and behaviors become a "problem space"[15] for her. And once we have done this, as I will argue below, we are in a position to explain the birth of the agentive features characteristic of cognition.

We can conceptualize the above transition in the following manner: the intersubjective means-centric orientation present in imitation highlights children's concern with reproducing the particular detailed manner or style of an observed intentional behavior. When imitating, we see that children are concerned with the strategies of an observed action, not merely insofar as they are instrumental for reaching some end but as objects of interest and concern themselves. Now, if we reapply this means-centric orientation intrapersonally, what results is a concern for and attention to the particular detailed manner or style in which *one executes one's own* actions and abilities. As such, a child's own abilities become a source of attention and curiosity. So, just as imitation makes the particular detailed means of an observed action salient, valuable, and interesting, the intrapersonal means-centric orientation makes the detailed means of *one's own* actions salient, valuable, and interesting. Crucially, at this stage, the previously transparent, instrumental means by which various ends were achieved are now poised to become ends in themselves. And this transition from means as ends in the world to means as ends in oneself, I claim, holds special explanatory power.

This is because when a child's own actions become ends in themselves, the particular way in which she performs a task becomes something for her to attend to, manipulate, and control. With this shift, she becomes able to invert her attention onto herself in order to take her own actions as objects to be transformed, improved, and perfected. As such, the means-centric orientation grounds a child's motivation to rearrange, reorganize, replace, refine, guide, and control the means by which she performs certain tasks. And this transition, I claim, provides us with a foundation upon which to explain the transition from first-order behavioral mastery to the limited flexibility, manipulability, and transferability that arises at the intermediate stage of cognitive development. It is precisely this transition, I claim, that paves the way for substantial cognitive change.

We should notice that as a result of the inversion of the means-centric orientation, children become engaged in what I call skill refinement. After all, this is exactly what skill refinement requires—that agents express a concern for their own actions and attempt to improve not only the probability that they'll reach some end but also the particular manner or style employed to reach that end. As such, we see that the means-centric orientation, applied to oneself, provides an explanation of why humans have a special interest in developing their own abilities. The inversion of the means-centric orientation onto one's own actions allows us to account for the peculiar human habit of expending huge amounts of energy on the practice and perfection of abilities long after they have reached the point of proficiency. But it also offers us a naturalistic, embodied explanation of the ontogeny of intelligence.

[15] This is Karmiloff-Smith's term (1990, p. 139).

4.1 Skill Refinement and the Intermediate Stage of Cognitive Development

At the intermediate stage of cognitive development, through recurrent cycles of redescription, representational states begin to take on novel properties. Karmiloff-Smith describes the intermediate stage of the RR model as composed of two transitions (Ei and Eii). At the Eii stage, a child first has conscious access to her own implicit procedures, and she begins to "gain some control over the organization of her internal representations" (Karmiloff-Smith 1990, p. 107). It is here, in a primitive and limited way, that flexibility, manipulability, and transferability characteristic of intelligent processes first make their appearance.[16]

Though I rely heavily on Karmiloff-Smith's model of representational redescription in order to support my own claims about skill refinement and cognitive development, my model differs from hers in an important way. Whereas Karmiloff-Smith claims that children at the intermediate stage of cognitive development are primarily concerned with their own internal representations, I claim that the object of concern for children at this stage of cognitive development is their own abilities and actions. On my account, it is not her internal representation that a child attends to and tries to control but the way, manner, or style in which she performs intentional actions.

As such, pace Karmiloff-Smith, I claim that at this middle stage of cognitive development, "a child turns her focus onto refining her abilities and not onto refining the representation of those abilities" (Fridland forthcoming). On my way of understanding this intermediate stage, the major shift from the implicit level to the intermediate stage of cognitive development is best described as a shift in concern from actions that are directed at the world to the way or manner in which one performs those actions. It is not, as Karmiloff-Smith suggests, a shift from actions directed at the world to one's internal representations of those actions.[17] On my account, the child at the intermediate level of redescription is involved in skill refinement.

We should also note that the choice between identifying a mental state as having a representation of an action or ability as its intentional object and a mental state having an action or ability itself as its intentional object is not simply a semantic one. This is because when we are concerned with intentional states, we are concerned with states that have both intensionality (with an s) and extensionality. That is, we are concerned with states that, in Fregean terms, are subject to a sense-reference distinction (Frege 1892). As such, we cannot simply conclude that since an action or ability is actually a kind of representation, then that in attending to that action or ability, the child is attending to it *as a representation.* And it is the question

---

[16] See Karmiloff-Smith (1986, 1990) for evidence of the systematic limitations on flexibility and transferability present at the intermediate level of redescription.

[17] See Fridland (forthcoming) for an argument diagnosing why Karmiloff-Smith makes this mistake.

of what the child is attending to, from the child's perspective, which is of central concern for us here. As such, this distinction that I make above is a crucial one for this theory.

Returning to my account, the intermediate stage of cognitive development is marked with a transition from a concern with means as ends located in the world to a concern with one's own means as ends. As a result of this transition, we can first see fixed, first-order, implicit, procedural action sequences break apart and become individuated and reidentifiable action elements that are capable of showing up in a variety of contexts. The procedural behaviors that once went unnoticed but served as perfectly good ways to achieve certain ends now become sources of attention and concern themselves. When these fixed, instrumental behaviors become ends in their own right, through a kind of practical trial and error, they are refined into individuated elements out of which a basic syntax of action can be composed.

4.2 The Labor of Skill Refinement Spawns the Three Sisters

In the following section, I provide an explanation of how it is that limited flexibility, manipulability, and transferability emerge out of skill refinement. I try to show how skill refinement is a process that grounds the compositionality, combination, and recombination of action elements, making room for the characteristic features of cognition that I have discussed above.

4.2.1 Trial and Error

At the intermediate stage of cognitive development, the child's objective becomes the improvement or refinement of the way in which she instantiates her abilities. These attempts to refine the way or manner in which she performs certain tasks require that the child interferes with the fixed action sequences that have up until that point been used for reaching her ends. In order to improve, the child must change the way in which she performs her actions. As such, skill refinement requires intervention for the sake of variation. Through the process of skill refinement, the child quite literally breaks up her procedural knowledge and introduces the seeds of flexibility into her actions as a result.

Implementing the kind of interference required for skill refinement is best construed, I claim, as a process of practical trial and error. The child, at this stage, begins experimenting with the way in which she instantiates her abilities. In order to figure out how to improve upon the way in which she performs some action, the child must play with different ways of producing the action. In order to do things better, she must figure out how to do things differently.

As we reflect on embodied expertise and skill refinement, we see that before acquiring the kind of control that is required for high-level skills, children must sacrifice basic proficiency. We see evidence of the primitive decomposition process that results from trial and error in the mistakes that children make in domains in which they have previously achieved behavioral mastery. Specifically, there is evidence that after attaining procedural success, children begin to exhibit marked errors (Karmiloff-Smith 1986). These sorts of mistakes offer clear evidence that an interference and reorganization of the implicit procedures responsible for first-order task success is taking place:

> This kind of trade-off between success and flexibility is easy to understand. To improve the way in which one performs some task requires shuffling, shifting, adjusting, and altering the way in which the task is instantiated. The once fixed but successful sequence is tweaked through trial and error and, as a result, the child makes various errors when instantiating it. (Fridland forthcoming)

In this way, we see that trial and error introduces flexibility into an action sequence, but at first, it does so at the cost of efficacy. In order to gain control over her own abilities, that is, in order to gain the capacity to flexibly manipulate her actions, a child must interfere with her automatic, fixed, implicit behaviors. She must apply effort and attention in order to perfect her actions, but this means overriding and thus sacrificing her reliable, first-order, procedural behaviors.

We should notice that because the child interferes with her actions through a process of effortful trial and error, we see the most basic shoots of manipulability arise in this context. That is, refining one's own abilities is a process that begins and ensues because of the child; it is the child that instigates, engages, and controls the process of ability refinement. And it is precisely this kind of effort and control that constitutes the property of being manipulable or under the control of the agent. So, in order to reorganize the means by which she achieves certain goals, the child must manipulate her actions. It is through a coarse kind of top-down control applied to her first-order behaviors that fixed actions sequences begin to break apart and acquire a degree of flexibility.

Importantly, in order for a child to treat her abilities as objects to be changed and manipulated, she must be able to take them as objects of interest. As such, we see that without the basic conditions that the means-centric orientation provides, the refinement of abilities would be impossible. This isn't to say that the means-centric orientation is the only driving force behind skill refinement. The social setting of the child can certainly be a motivation as well. The child may want to improve a certain ability because she sees her older brother doing it, her classmates, or a celebrity on TV. Still, it is the capacity to produce an inverted perspective onto one's own actions that will underpin the child's ability to practice and perfect the ways in which she performs particular tasks.

The takeaway point here is that as a result of the trial and error process required for skill refinement, a child manipulates her behavioral repertoire and introduces a degree of flexibility into her action patterns. As a result of this limited, crude kind of flexibility and manipulability, through recurrent cycles of repetition, a child creates the conditions for more and more fine-grained flexibility, manipulability, and transferability.

4.2.2 Individuation and Recombination

The process of practical trial and error breaks up fixed action patterns and allows behavioral procedures to relax in various limited ways. This kind of intervention allows for, at first, coarse-grained action elements to emerge out of whole behavioral sequences. That is, out of fixed, rigid, uninterruptable procedures, individuated action elements emerge. For example, a procedure goes from being one whole sequence to being composed of two parts: a beginning and an end. These parts, freed in this small way from their former procedural rigidity, take on the capacity to combine and recombine in limited ways.

As action elements attain a degree of freedom and independence, they also acquire the capacity to become the intentional objects of further trial and error, attention, effort, and control. As the boundaries of individuated action elements become more pronounced, the parts can then be manipulated further, which injects more flexibility and further individuation into the behavioral sequence. As such, the process of skill refinement produces more fine-grained elements that can be further combined and recombined in various contexts. Individuation and recombination break behavioral sequences into fine-grained action elements, which, through practical trial and error, can become subject to further individuation and recombination. Thus, individuation spawns freedom for recombination, which spawns further individuation, which spawns further recombinatorial freedom, and so on.

Happily, through the process of skill refinement, we notice the development of a basic syntax of action, which requires the features of flexibility, manipulability, and transferability. Like the concept "RAIN" must be able to show up in different thoughts, in different positions, and propositions, we see that skill refinement allows action elements to do the same. We see that skill refinement produces action elements that can play various roles in the constitution of various actions. So, for example, the kick before a cartwheel can show up as the kick before a handstand, in between a front walkover and an ariel, or at the end of full turn. The kick can take different positions in different actions, once it becomes an identifiable and reidentifiable element. Another way to put this point is that the individuated elements out of which skills are composed become transferable from one task to another. They become capable of playing a general role in the domain of skilled action.

From this discussion, we should conclude that skill refinement plays a central role in producing the distinguishing characteristics of intelligence. This is because skill refinement is responsible for the individuation of first-order behavioral sequences into combinable and recombinable parts. Importantly, we should notice that (1) the more fine-grained the individuated elements constituting a skill become, the more flexible, responsive, and adaptable the skill is, and (2) the more fine-grained the action elements constituting a skill become, the easier they are to manipulate and control. Finally, (3) as the sequences responsible for ability instantiation break down into more and more fine-grained, identifiable action elements, the easier it is for these elements to break free from any one particular sequence to be transferred to other tasks and behaviors. It should be clear, then, that at this intermediate stage of skill refinement, we enter into a realm where the features of intelligence can truly be said to apply to the behaviors of children. Through skill refinement, we are able to give a naturalized, embodied, developmental account of the flexibility, manipulability, and transferability of cognitive states and processes.

5 Conclusion

In this chapter, I attempt to connect imitation to the development of higher-order cognition by isolating and identifying the means-centric orientation as the motivation for imitation. Once this motivation is identified, I show how it can be used to account for skill refinement. I also hope to have convinced the reader that skill refinement offers us a naturalized strategy for accounting for some characteristic features of intelligent states and behaviors.

In the second section of this chapter, I argue that in order to develop an adequate account of human imitation, we must take seriously the means-centric orientation. The means-centric orientation, I claim, makes the means of intentional actions salient and interesting for not-merely-instrumental reasons. This orientation gives us an explanation of the human preoccupation with imitative learning in a way that an account that makes reference to social, cooperative reinforcement alone cannot.

In the third section of this chapter, I investigate three characteristic features of intelligence: flexibility, manipulability, and transferability. By relying on Karmiloff-Smith's theory of representational redescription, I argue that imitation alone, though impressive as a strategy by which to gain behavioral mastery, cannot provide us with an account of these three central features of intelligence.

In the final section of this chapter, I propose that by inverting the means-centric orientation onto oneself, one can move from the first level of procedural task success to the intermediate stage of cognitive development. I argue that this intermediate stage is one of skill refinement, where a child's goal is to practice and perfect the way or manner in which she instantiates her abilities. Through this process, the first signs of intelligence emerge. This is because as children work on their abilities, they begin to break apart their fixed action patterns into identifiable and reidentifiable action elements, which can then be combined and recombined in various ways and contexts. This process, I claim, is the process through which flexibility, manipulability, and transferability develop.

I hope that this brief overview has elucidated how skill refinement, underpinned by an inverted means-centric orientation, accounts for the emergence of flexibility, manipulability, and transferability by producing a basic syntax of action. Though more work needs to be done in order to get us to completely abstract, conceptual thought, I take it that this naturalized story of skill refinement and intelligence puts us on a productive path.

Works Cited

Bellagamba, F., & Tomasello, M. (1999). Re-enacting intended acts: Comparing 12- and 18-month-olds. *Infant Behavior & Development, 22*(2), 277–282.

Bermúdez, J. (2003). *Thinking without words*. Oxford: Oxford University Press.

Boesch, C., & Tomasello, M. (1998). Chimpanzee and human cultures. *Current Anthropology, 39*(5), 591–614.

Byrne, R. W. (2002). Emulation in apes: Verdict 'not proven'. Emulation in apes: Verdict "Not Proven". *Developmental Science, 5*, 21–22.

Byrne, R., & Russon, A. (1998). Learning by imitation: A hierarchical approach. *The Behavioral and Brain Sciences, 21*(5), 667–721.

Call, J., & Tomasello, M. (1998). Distinguishing intentional from accidental actions in orangutans (*Pongo pygmaeus*), chimpanzees (*Pan troglodytes*), and human children (*Homo sapiens*). *Journal of Comparative Psychology, 112*, 192–206.

Call, J., Hare, B. H., Carpenter, M., & Tomasello, M. (2004). 'Unwilling' versus 'Unable': Chimpanzees' understanding of human intentional action? *Developmental Science, 7*, 488–498.

Carpenter, M., Akhtar, N., & Tomasello, M. (1998). Fourteen-through-18-month-old infants differentially imitate intentional and accidental actions. *Infant Behavior & Development, 21*(2), 315–330.

Clark, A., & Karmiloff-Smith, A. (1993). What's special about the development of the human mind/brain? *Mind & Language, 8*(4), 569–581.

Dennett, D. (1969). *Content and consciousness*. London: Routledge & Kegan Paul.

Dennett, D. (1996). *Kinds of minds*. New York: Basic Books.

Frege, G. (1892) Über Sinn und Bedeutung. *Zeitschrift für Philosophie und philosophische Kritik*, 100, 25–50. Translated as Black, M. (1980). On sense and reference. In P. Geach & M. Black (Eds., Trans.) *Translations from the philosophical writings of Gottlob Frege* (3rd ed.). Oxford: Blackwell.

Fridland, E. (forthcoming). Skill learning and conceptual thought: Making our way through the wilderness. In B. Bashour & H. Muller (Eds.), *Contemporary Philosophical Naturalism and Its Implications*. Routledge.

Gergely, G., & Csibra, G. (2005). The social construction of the cultural mind: Imitative learning as a mechanism of human pedagogy. *Interaction Studies, 6*(3), 463–481.

Horner, V., & Whiten, A. (2005). Causal knowledge and imitation/emulation switching in chimpanzees (*Pan troglodytes*) and children (*Homo sapiens*). *Animal Cognition, 8*, 164–181.

Hume, D. (1961). An enquiry concerning human understanding. In *The empiricists* (pp. 307–430). New York: Doubleday.

Karmiloff-Smith, A. (1986). From meta-processes to conscious access: Evidence from children's metalinguistic and repair data. *Cognition, 23*, 95–147.

Karmiloff-Smith, A. (1990). Constraints on representational changes: Evidence from children's drawing. *Cognition, 34*, 57–83.

Karmiloff-Smith, A. (1992). *Beyond modularity: A developmental perspective on cognitive science*. Cambridge, MA: MIT Press.

Lyons, D., Young, A., & Keil, F. (2007). The hidden structure of overimitation. *Proceedings of the National Academy of Sciences, 104*(50), 19751–19756.

McGuigan, N., Whiten, A., Flynn, E., & Horner, V. (2007). Imitation of causally opaque versus causally transparent tool use by 3 & 5-Year-old children. *Cognitive Development, 22*, 353–364.

Meltzoff, A. N. (1995). Understanding the intentions of others: Re-enactment of intended acts by 18-month-old children. *Developmental Psychology, 31*(5), 838–850.

Meltzoff, A. N. (2005). Imitation and other minds: The "Like Me" hypothesis. In S. Hurley & N. Charter (Eds.), *Perspectives on imitation: From neuroscience to social science* (Vol. 2, pp. 55–77). Cambridge, MA: MIT Press.

Millikan, R. G. (2006). Styles of rationality. In S. Hurley & M. Nudds (Eds.), *Rational animals?* (pp. 117–126). Oxford: Oxford University Press.
Prinz, J. (2004). *Gut reactions: A perceptual theory of emotion*. Oxford: Oxford University Press.
Schwier, C., van Maanen, C., Carpenter, M., & Tomasello, M. (2006). Rational imitation in 12-month-old infants. *Infancy, 10*(3), 303–311.
Tomasello, M. (1996). Do apes ape? In C. M. Heyes & B. G. Galef Jr. (Eds.), *Social learning in animals: The roots of culture* (pp. 319–346). San Diego: Academic.
Tomasello, M. (1999). Emulation learning and cultural learning. *The Behavioral and Brain Sciences, 21*(5), 703–704.
Tomasello, M. (2009). The question of chimpanzee culture, plus postscript, 2009. In K. Laland & B. Galef (Eds.), *The question of animal culture* (pp. 198–221). Cambridge, MA: Harvard University Press.
Tomasello, M., & Rokaczy, H. (2003). What makes human cognition unique? From individual to shared to collective intentionality. *Mind and Language, 18*(2), 121–147.
Tomasello, M., Carpenter, M., Call, J., Behne, T., & Moll, H. (2005). Understanding and sharing intentions: The origins of cultural cognition. *Behavioral and Brain Sciences, 28*(5), 675–735.
Vygotsky, L. S. (1978). *Mind and society: The development of higher mental processes*. Cambridge, MA: Harvard University Press.
Want, S., & Harris, P. (2001). Learning from other people's mistakes: Causal understanding in learning to use a tool. *Child Development, 72*(2), 431–443.
Whiten, A., McGuigan, N., Marshall-Pescini, S., & Hopper, L. (2009). Emulation, imitation, over-imitation and the scope of culture for child and chimpanzee. *Philosophical transactions of the Royal Society of London. Series B, Biological Sciences, 364*, 2417–2428.

Evolving Consciousness: The Very Idea!

James H. Fetzer

Abstract Discovering an adequate explanation for the evolution of consciousness has been described as "the hard problem" about consciousness that we would like to understand. The difficulty becomes compounded by the introduction of such notions as the unconscious or the preconscious as its counterparts, at least for species of the complexity of human beings. An evaluation of the prospects for unconscious factors as exerting causal influence upon human behavior, however, depends upon understanding both the nature of evolution and the nature of consciousness. This paper sketches a theoretical framework for understanding both phenomena in general with regard to their various forms and suggests the evolutionary function of consciousness in genetic and in cultural contexts. It becomes increasingly apparent that, given a suitable conceptual framework of minds as semiotic systems, the evolution of consciousness may not be such a "hard problem", after all.

Philosophers spend most of their time dealing with vague and imprecise notions, attempting to make them less vague and more precise (Fetzer 1984). When we are dealing with notions like "the unconscious mind," where we have only a vague notion of consciousness and an imprecise notion of the mind, it may be appropriate to propose a few suggestions in an effort to sort things out a bit better, especially when the role of evolution in producing mentality and consciousness appears to be poorly understood. This study attempts to shed light on these problems by exploring how consciousness of different kinds might contribute to evolution in relation to its causal mechanisms.

"Why did consciousness evolve?" has been called *the hard problem* and some have even denied that consciousness itself can qualify as an adaptation (Harnad 2002). So "What are the adaptive benefits of consciousness?" and "How does consciousness

This is a slightly revised and expanded version of Fetzer (2002a).

J.H. Fetzer Ph.D. (✉)
McKnight Professor Emeritus, Department of Philosophy,
University of Minnesota Duluth, MN 55812, Duluth
e-mail: jfetzer@d.umn.edu

L. Swan (ed.), *Origins of Mind*, Biosemiotics 8, DOI 10.1007/978-94-007-5419-5_11,

enhance the prospects for survival and reproduction for species that possess it?" are therefore crucial questions. But their answers necessarily depend upon the nature of consciousness itself. In his *Kinds of Minds* (1996), for example, Daniel Dennett suggests that consciousness is sensitivity plus some additional factor "x," yet he also thinks there might be no such "x." But if consciousness is merely the capacity for sensation and sensation is no more than a propensity to undergo change, then consciousness might even be separable from mentality, with no discernable motive for its evolution.

If consciousness were instead the sensory awareness of the sensible qualities of things, such as their colors, shapes, and sizes, by comparison, it might make a difference and even imply the presence of mind. In *The Evolution of Culture in Animals* (1980), for example, John Bonner describes E. coli bacteria as moving toward 12 chemotactic substances and away from 8 others. Assuming the ones it moves toward are nutrient or beneficial, while the ones it moves away from are harmful or deleterious, it is not difficult to imagine how evolution could have produced this result at this stage for those bacteria. Perhaps "the hard problem" might turn out not to be such a hard problem, after all.

1 The "Black Box" Model

We tend to operate on the basis of a rather simple model—a "black box" model—for organisms. We have a stimulus S that brings about a response R by an organism O with a fixed probability or propensity p (Fetzer 1981, 1993a). The propensity p for response R by an organism O, when subjected to stimulus S, can be formalized as: (Fig. 1) where "==>" is the subjunctive *were/would* conditional, "=p=>" is a causal conditional for *would (with propensity p) bring about,* and the universal strength causal conditional "=u=>' stands for *would bring about,* where the same effect always occurs under those conditions. Alternatively—and probably more intuitively—by simply exchanging the positions of the organism and the stimulus S (Fig. 2), which means that organism O (or any organism of that specific kind) has a propensity p to display response R when subjected to stimulus S, where different species and different organisms O', O",... within the same species may be subject to different ranges of stimuli S and of response R with different propensities p, where the properties that make a difference require explicit specification.

This model does not offer any analysis of processes internal to O, which makes it a "black box" model. A more refined analysis would take into account the possible existence of links that relate an initial INTERNAL response R1 to the occurrence of one or more possible additional INTERNAL responses Ri, where these responses may lead to EXTERNAL responses Rj of motion or sound by the organism formalized as (Fig. 3) displays.

Fig. 1 The black box

Stimulus S ==> [Organism O =p=> Response R]

Fig. 2 The black box (reversed)

Organism O ==> [Stimulus S =p=> Response R]

(EXTERNAL) O ==>
(INTERNAL) [(S =p1=> R1) & (R1 =p2=> R2) &
(R2 =p3=> R3) & . . .] =pj=>
(EXTERNAL) Rj

Fig. 3 A more refined model

MOTIVES (m1, m2, . . . , mn) &		
BELIEFS (b1, b2, . . . , bn) &		
ETHICS (e1, e2, . . . , en) &	=p=>	RESPONSES (r1, r2, . . . , rn)
ABILITIES (a1, a2, . . . , an) &		including motion and sounds
CAPABILITIES (c1, c2, . . . , cn)		

Fig. 4 Human behavior as a probabilistic effect

Thus, for an ordinary organism of kind O, under suitable circumstances, an external stimulus S, which might be a sight or a sound, causes a pattern of neural activation R1, which in turn may (probabilistically) bring about a pattern of neural activation R2, which in turn may (probabilistically) bring about other patterns of neural activation, which may eventually lead to (public) external responses Rj, such as motion or sounds. The simpler the organism, the simpler these internal links (Fetzer 1990, 1996, 2005).

This approach invites the introduction of at least three measures of complexity that could distinguish between species or even conspecifics as members of the same species, based upon various properties of such links as possible internal causal chains, namely, (a) the complexity of these internal chains, especially with regard to (1) number of possible links and (2) their deterministic (same cause/same effect) of probabilistic character (same cause/one or another possible effect within a fixed set); (b) the temporal interval between the initial stimulus S and the ultimate behavioral response R, if any; and (c) the complexity of those possible responses that organisms display themselves.

2 Human Behavior

A simple example in the case of human behavior might be making a date, such as to attend a conference. We may do so months in advance, but our behavior responses to our commitments are only displayed when the time draws near. This reflects the consideration that human behavior arises as a result of a complex causal interaction between multiple factors of the kinds, motives, beliefs, ethics, abilities, and capabilities, where behavior may be a probabilistic manifestation of their interaction (Fig. 4).

Some of those factors may not even be accessible to conscious memory, however, and the effects of unique events during our lives may not even be adequately understood, which makes non-trivial anticipatory predictions and simulations—ones that are not simply retrospective representations, which even video-tapes provide, or even scripted sequences of actions, which depend on satisfying the premises of the script—of human behavior virtually impossible, where knowledge engineers cannot possibly possess the kind of information that would be necessary to produce them (Fetzer 2011).

While one mental state may bring about another mental state through a series of transitions between links of the kind described above, the totality of factors that interact to (probabilistically) bring about our behavior consists of specific values of variables of each of these kinds, where one complete set of values for the variables motives, beliefs, ethics, abilities, and capabilities constitutes *a context.* The concept of a context turns out to be fundamental to meaning and mind (Fetzer 1991, 1996, 2005).

The difference between deterministic and indeterministic behavior can then be spelled out as follows. Relative to a context, when the same behavior would occur in every case, without exception, then that behavior is *deterministic*. When one or another behavior within a fixed class would occur in every case without exception, with a constant probability, then that behavior is *indeterministic*. Consequently, even persons in the same context C can manifest different behavior so long as it is among the possible outcomes that occur with fixed propensity within that context.

With regard to motives, for example, if you like Heavenly Hash twice as much as you do Peppermint BonBon, where they are your clear preferences in ice cream, then we would expect that you would choose Heavenly Hash about twice as often as Peppermint BonBon when you enter Baskin Robbins. You would not know which you would pick on any single visit, but over time, you would pick one about twice as often as you pick the other. Frequencies are produced by propensities across trials, which can explain them and for which they function as evidence (Fetzer 1981, 1993a, 2002a).

3 Meaning and Behavior

What holds for motives also holds for beliefs, ethics, and the other variables that affect our behavior. With regard to beliefs, for example, I happen to live at 800 Violet Lane, Oregon, WI 53575. If someone were to believe instead that I lived at 828, that would have multiple manifestations in their behavior, such as the directions they might give to get to my house, what they would write on a letter they wanted to mail to me, where UPS and FED/EX deliveries to me would be made, and the like.

This approach supports a dispositional theory of meaning, according to which the meaning of a belief, Bi, is the difference that Bi makes over alternatives Bj, Bk…, relative to every context consisting of specific values of motives, of other beliefs, and so forth, where, when there is no difference in the totality of behavior that would be displayed given Bj as opposed to Bi across every context, then the meaning of Bj is the same as the meaning of Bi (Fetzer 1991, 1996, 2005). And it turns out that meaning itself is amenable to degrees.

Those who know that my home is the corner house on the block on the north-east side, for example, might be able to find it without great effort because of their other beliefs about how to get around in Oregon, but for other purposes, the street number would be required. Some but not all of the same behavior would result from those overlapping beliefs. Two half-dollars, four quarters, ten dimes, and so forth all have the same purchasing power, but in some contexts carrying a bill rather than bulky change might matter.

This account of meaning, which connects stimuli S with responses R by means of internal dispositions of an organism O, comports with a theory of concepts and even of mind. If we think of concepts as constellations of habits of thought and habits of action, then when an experience is subsumed by means of a concept, the expectable outcome is whatever behavioral effects would (probably) be produced in a context. Some concepts, no doubt, will be innate, while others may—for higher species—be acquired (Fetzer 1991, 1996, 2005).

Another species that exemplifies these notions is that of vervet monkeys, which make at least three different kinds of alarm calls. In his *Introduction to Ethology* (1985), P. J. B. Slater reports that one such call warns of a land-borne predator in the vicinity and, when the monkeys hear this call, they climb up into the trees to evade it. Another warns for an air-borne predator in the vicinity and, when they hear it, they crawl down under the brushes for protection. The third is for things on the ground, where they climb down and poke around so they can see just what is going on.

Our behavior, especially voluntary, turns out to be a partial manifestation of meaning to us, where the meaning of meaning to us turns out to be the multiple potentialities for behavior in the presence of something S and where I want to identify that S more precisely as a stimulus of a certain special kinds, which makes a crucial different to our behavior. The suggestion I am going to make is that an approach, which has not yet received a lot of attention as yet, but that was advanced by Charles S. Peirce—whom I consider to be the only great American philosopher—can help to clarify and illuminate the nature of mind.

4 The Nature of Signs I

According to Peirce, *a sign* is a something that stands for something else in some respect or other for somebody. A simple example is a red light at an intersection. For qualified drivers who know the rules of the road, that light stands for applying the breaks and coming to a complete halt, only proceeding when the light changes and it is safe to do so. Under ordinary circumstances—in a "standard context," let us say—that is precisely the behavioral manifestation that we expect to occur (Fetzer 1988, 1991).

This would be an example of an appropriate behavior response for someone who understood the rules of the road and is not incapacitated from exercising that ability, as might be the case, for example, if they were blindfolded. And of course there can be other signs with the same meaning, such as, in this case, a stop sign or an officer with his palm extended, which have essentially the same meaning (of applying the breaks and coming to a complete halt, but only proceeding when the officer tells you to do so). Peirce called the complex of dispositions of a user to respond to a sign its "interpretant."

Perice suggests there are three different ways in which signs can be "grounded" or related to those things for which they stand. The first is on the basis of *resemblance relations*, where the sign looks like (tastes like, smells like, feels like, or sounds like) that for which it stands. Examples include statues, photographs, and paintings, when realistically construed. (This Picasso achieved a niche in the history of art when he violated the canons of representation of the nude female.) Perice called these "icons."

My driver's license exemplifies an important point about icons. As you might or might not see (when I hold it up), my license photo looks a lot like me—maybe on not such a great day—but if you turn it on its side, it no longer resembles me, because I am just not that thin. What this implies is that even the use of the most basic kind of sign, an icon, presupposes *a point of view*. Anything incapable of having a point of view, therefore, is incapable of using signs or of possessing a mind, a point to which I shall return.

The second mode of grounding that Peirce introduced is that of *causal connections*, where a cause stands for its effects, effects stand for their causes, and so forth. Thus, smoke stands for fire, fire for smoke, ashes for fire, and so no, while red spots and an elevated temperature stand for the measles—which means that there may be special classes of individuals who are practiced in reading signs of certain kinds, such as scientists and physicians, but also those whose parallel claims may be suspect, such as palm readers and crystal-ball gazers. Peirce called these signs "indices" (as the plural of "index").

5 The Nature of Signs II

The third mode of grounding Peirce introduced involves mere *habitual associations* between signs and that for which they stand, where the most familiar examples are the words that occur in ordinary language, such as "chair" and "horse" in ordinary English. These words certainly do not look like or resemble nor are they causes or effects of that for which they stand. Unlike icons and indices, which might be thought of as "natural signs" because they are there in nature, whether we notice them or not, these signs are ones we have to make up or create. These "artificial signs" are known as "symbols."

In order for a specific something to stand for something else in some respect or other for somebody on a specific occasion, that somebody must have the ability to use signs of that kind, s/he must not be incapacitated from the exercise of that ability, and that sign must stand in an appropriate causal relationship to that sign user. If a red light were invisible to a driver because of a driving rain (a dense fog, overgrown shrubbery, or whatever), it could not exert its influence on that sign user on that occasion any more than if s/he had been temporarily blinded by a flash of lightning or an oncoming car (Fetzer 1990, 1996, 2005).

Even more interesting, perhaps, is the realization that the specific something for which something stands in some respect or other need not even exist. We can have signs for persons who do not exist, such as Mary Poppins and Santa Claus, or for species of things, such as unicorns and vampires, that do not exist, without

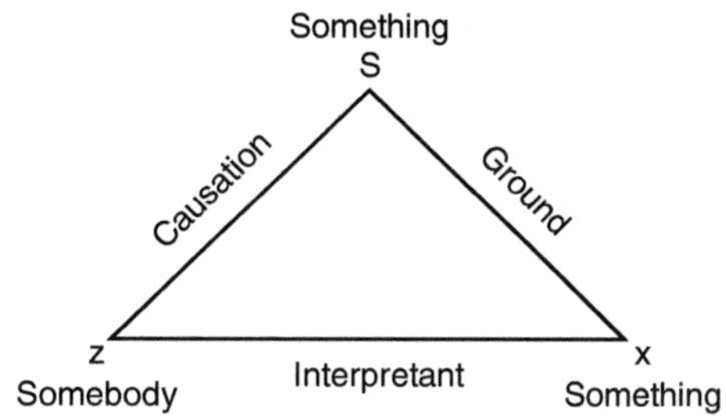

Fig. 5 The triadic sign relationship

incapacitating those signs from standing for things of those kinds. We can even make movies about alien visitations and American werewolves in London, which means that the use of signs has enormous scope and range with respect to those things for which they stand. They do not even have to exist!

6 The Nature of Minds

The sign relationship, therefore, is three-placed (or "triadic"), where a something, S, stands for something else, x (in some respect or other) for somebody, z. The meaning of a sign for somebody is therefore the totality of causal influences it would exert for that somebody across every possible context, Ci (Fig. 5). When we pause to consider more precisely the kind of thing for which something can stand for something else, however, it becomes extremely attractive to entertain the hypothesis that *the ability to use signs* might be exactly what distinguished minds.

Let us focus on the sign user z rather than the sign S and avoid taking for granted that the kinds of things for which something can stand for something else have to be human by abandoning the term "somebody" and use the more neutral term "something." Then anything, no matter whether it happens to be human being, (other) animal, or inanimate machine, for which something (a sign) can stand for something else in some respect or other *possesses a mind.* And let us refer to systems of this kind that are capable of using signs as *semiotic systems* (Fetzer 1988, 1989, 1990).

7 A Semiotic Systems

"Interpretant" thus stands for a system's semiotic dispositions as the totality of ways it might respond (probabilistically) to the presence of a sign within different contexts. Its behavior in context Ci can therefore differ from its behavior in Cj in the presence of the same sign (Fetzer 1991). And a semiotic system z can be diagrammed as shown by (Fig. 6).

The grounding relations between signs and that for which they stand (by virtue of relations of resemblance, of cause-and-effect, or of habitual association, as we have discovered) are therefore crucial to the nature of semiotic systems. Unless that

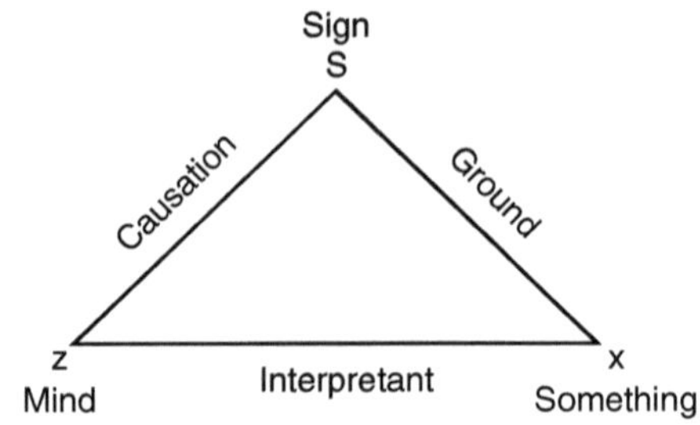

Fig. 6 A semiotic system

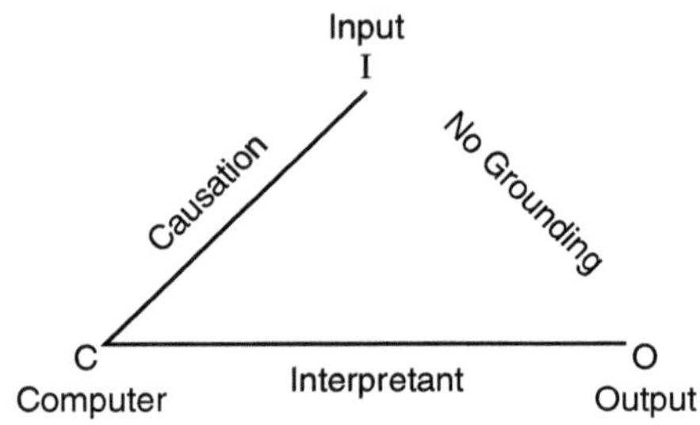

Fig. 7 An input-output system

causal connection between the presence of something—which could be an icon or an index or a symbol—and the (potential or actual) behavior of a system obtains *because it functions as an icon, an index, or a symbol for that system* (by virtue of its grounding relation of resemblance or of causation or of habitual association), it cannot be a semiotic connection (Fetzer 1990, p. 278).

Semiotic systems for which things function as signs afford a basis for separating systems that have minds from others that do not, such as digital machines, which lack the grounding relationship relating signs to those things for which they stand. This difference can also be diagrammed to display this crucial difference as follows (Fig. 7).

Thus, although they are designed to process marks on the basis of their shapes, sizes, and relative locations, those marks mean nothing to those digital machines, say, as inventories or as dollars and cents. They should therefore be characterized not as *semiotic systems* but as *input/output systems* instead, where the inputs that exert causal influence upon them are properly understood to function merely as stimuli rather than as signs. They can be called "symbol systems," provided that does not imply that they use symbols in Peirce's sense (Fetzer 1988, 1990, 1996, 2002b).

8 Communication and Convention

Another important distinction that can be drawn is that communication between semiotic systems is promoted when those systems use signs in similar ways. When a sign-using community reinforces common sign-using behavior by means of some

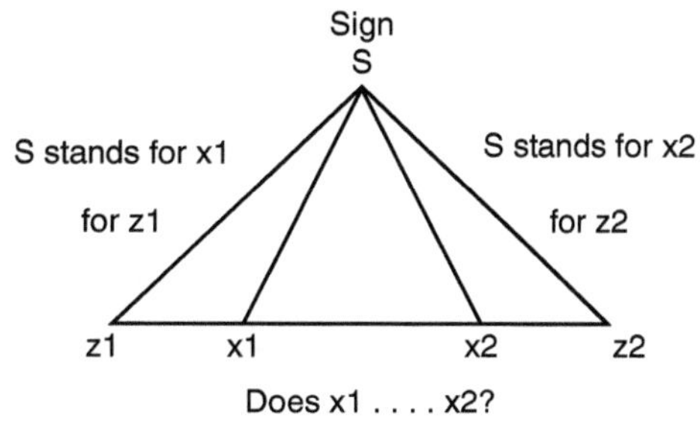

Fig. 8 Communication situations

system of institutions, such as schools, those customs, traditions, and practices take on the status of conventions, which promote the objectives of communication and cooperation, thereby facilitating the pursuit of community goals (Fetzer 1989, 1991, 2005) (Fig. 8).

When one semiotic system uses signs to communicate with another semiotic system, then those signs assume the character of *signals.* There thus appears to be a hierarchy between mere stimuli, signs, and signals, because every signal is a sign and every sign is a stimulus, but not vice versa. Causes that can produce changes in inanimate objects, for example, are stimuli but not signs, just as things that stand or other things are signs for those systems, even if they are not signals. While all three—stimuli, signs, and signals—are possible causes that can affect the behavior of different systems, only signs and signals entail the presence of minds.

9 Consciousness and Cognition

Even more important, however, the theory of minds as semiotic systems also provides illuminating conceptions of consciousness and of cognition, where both turn out to be adequately defined only relative to signs of specific kinds. Thus, a system z is *conscious (with respect to signs of the specific kind S)* when (a) z has the ability to use signs of kind S and (b) z is not incapacitated from using signs of that kind within its present context C. *Cognition (with respect to a sign S of a specific kind)* thus occurs as the effect of a causal interaction between a system z and a sign S when (a) z is conscious with regard to signs of kind S and (b) a sign of kind S occurs in suitable causal proximity to z, which brings about as the outcome of a suitable opportunity (Fetzer 1989, 1990, 1996) (Fig. 9).

The conception of minds as semiotic systems (sign-using systems) thus not only brings with it the definition of mentality as semiotic ability but useful concepts of consciousness and of cognition. Informally expressed, consciousness (with respect to signs of kind S) combines the *ability* to use signs of that kind with the *capability* of exercising that ability, while cognition combines *consciousness* with respect to signs of that kind with the *opportunity* for causal interaction with a sign of that

Fig. 9 Consciousness and cognition (informal)

Consciousness (with respect to signs of kind S) = df ability + capability (within a context)

Cognition (of a specific sign of kind S) = df an effect of consciousness + opportunity

	Mentality		
	Type I	**Type II**	**Type III**
Definition	iconic	indexical	symbolic
Criterion	type/token recognition	classical PAVLOVIAN conditioning	SKINNERIAN operant conditioning

Fig. 10 Basic modes of mentality

kind. That definition can be combined with a general criterion of mentality, which is *the capacity to make a mistake,* since anything that can make a mis-take has the ability to take something to stand for something, which is the right result (Fetzer 1988, 1990).

The outcome of this approach is the introduction of a theory of mentality that is applicable to human beings, to (other) animals, and to inanimate machines, if such a thing is possible. It yields a theory of types of minds of increasing strength, from iconic to indexical to symbolic, where symbolic presupposes indexical and indexical iconic, but not vice versa (Fig. 10).

These types and criteria of their presence are shown here, where an evidential indicator of the presence of iconic mentality is the capacity for *type/token recognition* of instances as instances of specific kinds; of indexical is *classical Pavlovian conditioning* as the generalization of a cause inducing an effect; and of symbolic mentality *Skinnerian operant conditioning*, where one thing comes to stand for another based merely upon habitual association (Fetzer 1988, 1990).

10 Higher Modes of Mentality

This approach invites the evolutionary hypothesis that various biological species are predisposed toward mentality of specific types, which would be expected to be distributed as a reflection of their evolutionary standing, the lowest organisms with the lowest levels of mentality, the higher with the higher. Indeed, there appear to be at least two higher modes of mentality that are characteristic of human beings, which are the capacity to fashion arguments as *transformational mentality* and the ability to use signs as *metamentality*, especially for the purpose of criticism, where sign-users can subject signs to changes intended to improve them (Fig. 11).

Fig. 11 Higher modes of mentality

	Higher Mentality	
	Type IV	Type V
Definition	transformational	metamentality
Criterion	logical reasoning	criticism

Among the virtues of the conception of minds as semiotic systems is that it allows for the existence of modes of mentality that are less sophisticated than those involved in the use of language, which appears to be a relatively late development in evolution (Donald 1991; Fetzer 1993b, c). The extraordinary attention to which it has been subjected by Noam Chomsky's work on grammar as a species-specific innate syntax and Jerry Fodor's work on meaning as a species-specific innate semantics has reached its latest incarnation in work such as that of Stephen Pinker (1997), who hold that the human mind is a computer for survival and reproduction, and of Euan MacPhail (1998), who maintains the key to the evolution of consciousness is the evolution of language.

However, if the evolution of language were the key to the evolution of consciousness, then, insofar as language is a phenomenon emerging rather late in evolution, it would be rather difficult to imagine how consciousness could have evolved at all. Preoccupation with language truncates consideration of multiple modes of meaning and of nonhuman kinds of minds. Not only are iconic and indexical mentality more primitive than symbolic, but preoccupation with linguistic transformations and syntactical structures manages to focus on higher modes of mentality to the neglect of lower, while even placing the syntactic cart before the semantic horse. As Thomas Schoenemann (1999) has argued—and as I agree—that syntax evolved as an emergent response to semantic complexity affords a better explanation for the phenomena than its innate alternatives.

11 Conceptions of Consciousness

The idea that the mind is a computer that evolved through natural selection, of course, takes for granted that, at some appropriate level of description, both minds and machines operate on the basis of the same or similar principles, which already appears to be false given the difference in grounding relations. But modeling minds after machines also confounds languages as products of the evolution of culture with species as products of the evolution of genes. The relative adequacy of alternative theories (of consciousness, mentality, and language) may be assessed by the extent to which they are able to explain the full range of related phenomena (of consciousness, mentality, and language), where, I would submit, the semiotic conception encounters no serious rivals.

(C-1)	Sensitiviy
	stimuli with casual influence but does not imply mentality:thermostats,thermometers, litmus paper as a kind of mindless consciousness
(C-2)	Semiotic ability
	sensitivity regarding stimuli that stand for something in some respect for something: hence, (C-2) implies (C-1) and the presence of mind
(C-3)	Self-awareness
	semiotic ability that includes signs that stand for the sign user itself for the sign user; so (C-3) implies (C-2) with self-referential ability
(C-4)	Self-awareness with articulation
	semiotic ability that includes signs that stand for the user itself with the ability to articulate that self-awareness; so (C-4) implies (C-3) with articulative ability
(C-5)	Self-awareness with capacity for communication
	semiotic ability that includes signs standing for oneself and other conspecifies, which promotes cooperation, so (C-5) implies (C-4) with signals

Fig. 12 Five modes of consciousness

As an illustration, consider the multiple modes of consciousness that can be differentiated within the scope of this approach. Those that do not make reliance upon signs indispensable to mentality lack the semiotic dimension distinctive of mentality. The Dennett hypothesis that consciousness may be nothing more than sentience qualifies thermostats, litmus paper, and thermometers as "conscious," yet is not sufficient to endow them with mentality (Fetzer 1997). They are thus examples of sensitivity as the susceptibility to stimuli that does not imply mentality as a version of "consciousness without minds." Let's call this (C-1).

A stronger mode of consciousness would combine sensitivity with semiotic ability, which implies the presence of mind. Call this (C-2). A third mode of consciousness would combine semiotic ability with self-awareness, involving the use of signs to stand for the sign-user itself. Call this (C-3). Yet a fourth mode of consciousness would combine self-awareness with the capacity for articulation, which we shall call (C-4). A fifth mode of consciousness would combine self-awareness with the capacity for articulation and the ability to communicate with others using signs as signals. Let us call this final mode (C-5) (Fig. 12).

This schema does not represent the only possible kinds of consciousness but rather serves as a template for considering the prospective roles of consciousness in evolution. In this case, for example, each mode of consciousness implies each of the lower modes, where (C-5) implies (C-4), (C-4) implies (C-3), and so forth. If there are cases of communication involving signals, which presumably would be at the level of (C-5), such as vervet monkey alarm calls, where their use of signals may or may not be accompanied by self-referential ability at the level of (C-3), then this account would have exceptions that would display the desirability of deviant typological schemes.

12 Evolution and Consciousness

Evolution understood as a biological process should be characterized in terms of three principles, namely, that more members are born live into each species than survive to reproduce, that crucial properties of offspring are inherited from their parents, and that several forms of competition between the members of a special contribute to determining which of them succeeds in reproducing. The mechanisms that tend to produce genetic variation include genetic mutation, sexual reproduction, genetic drift, and genetic engineering, while the mechanisms that tend to determine which members of existing populations tend to survive and reproduce are natural selection, sexual selection, group selection, and artificial selection (Fetzer 2002c, 2005, 2007).

The question with which we began, you may recall, "Why did consciousness evolve?" is amenable to alternative formulations, which include "What are the adaptive benefits of consciousness?" but also "How does consciousness enhance the prospects for survival and reproduction of species that possess it?" Having clarified the nature of consciousness sufficiently to make these questions meaningful (or at least interesting) enough to pursue them, the objective becomes to consider each of these causal mechanisms in turn to ascertain whether consciousness in any of these five modes would provide adaptive benefits in order to answer "the hard question."

The following table reflects the big picture, in general, as the intersection of the eight different evolutionary mechanisms with those modes of consciousness that might enhance them or benefit from them. The first four are modes that promote variability in the gene pool. Consciousness beyond sensitivity would appear to make no difference to the occurrence of genetic mutation, which of course presupposes consciousness (C-1). Similar considerations obtain for sexual reproduction and genetic drift, understood as causal processes apart from the mechanisms that determine who mates with whom and under what conditions (Fig. 13).

Genetic engineering, by contrast, requires highly sophisticated mental abilities that would appear to benefit from reasoning skills and critical thinking up to the level of (C-5). The emergence of consciousness at levels far beyond (C-1) would provide adaptive benefits. In the case of natural selection, all these modes would be beneficial in competition with conspecifics for food and other resources. Success in sexual selection, moreover, would benefit from self-referential abilities and the capacity for articulation, not to mention the ability to transmit signals. Artificial selection and group selection could not operate without communication.

Fig. 13 Adaptive roles of modes of consciousness

Mechanism	Consciousness
(1) Genetic mutation	(C-1)
(2) Sexual reproduction	(C-1)
(3) Genetic drift	(C-1)
(4) Genetic engineering	(C-5)
(5) Natural selection	(C-1) to (C-5)
(6) Sexual selection	(C-2) to (C-5)
(7) Group Selection	(C-5)
(8) Artificial Selection	(C-5)

If these considerations are well-founded, then they suggest that the potential adaptive benefits of consciousness are both obvious and profound. In response to the question, therefore, different modes of consciousness appear to enhance the prospects for survival and reproduction by species that possess them. Intriguingly, the motives for consciousness to evolve differ in relation to different evolutionary mechanisms. It should come as no surprise that natural selection and sexual selection should both benefit from consciousness up to the highest kinds, where genetic engineering and artificial and group selection could not function without consciousness around (C-5).

13 Minds are Not Machines

What this exercise has secured it a plausibility proof that evolution can produce consciousness among its varied manifestations, since organisms with these kinds of abilities would secure advantages in competition with nature and with conspecifics across a wide range of evolutionary mechanisms. This means that there would be adaptive benefits for possessing consciousness of these various kinds that would enhance the prospects for survival and reproduction among those possessing them. It should also be observed, however, that this analysis could be improved upon by, for example, systematically integrating considerations for different kinds of minds.

There should not be much room for doubt, for example, that higher modes of consciousness tend to presume higher types of mentality, where transformational mentality and metamentality can greatly extend the abilities of organisms in dealing with conspecifics and their environments. All of this may even seem to reinforce the claim that the human mind is a computer for survival and reproduction. That claim, however, trades upon an ambiguity. There is some general sense in which the human mind is a processor for survival and reproduction, but this is a trivial claim. The sense in which the human mind is a computer, alas, implies that they operate on the basis of the same or similar principles, which is false.

We have already seen that digital machines lack a grounding relation that typifies not just human minds but every mind. (Compare Fig. 7 with Fig. 6.) So that is one important difference, which we might call "the static difference." Another is that these machines function on the basis of algorithms implemented by using programs, which execute operations in specific sequences of steps. They have definite starting points and definite stopping points, where their application is perfectly general and they always yield a correct solution to a problem in a finite number of steps (Fetzer 1994, 2002b, 2007). When you reflect upon it, these are important differences between computing and thinking.

How many kinds of thinking have these properties? Certainly neither perception nor memory nor dreams or daydreams come close. None of them ordinarily qualifies as "solving problems." None of them has a definite starting point and another definite stopping point. None of them can be counted upon to yield correct solutions in finite steps. We might call this "the dynamic difference," which means that they are systems of distinctly different kinds. Human beings surely are systems for survival and reproduction, but that does not turn them into computers. Pinker (1997) is wrong, because minds are not machines (1990, 1996, 2002b, 2005).

14 Genetic vs. Cultural Evolution

In an earlier book, Pinker (1994), he embraced the hypothesis of a uniquely human "language instinct," while acknowledging that this species-specific conception does not appear to be compatible with a modern Darwinian conception of evolution "in which complex biological systems arise from the gradual accumulation over generations of random genetic mutations that enhance reproductive success" (Pinker 1994, p. 333), which seems to finesses the theory of punctuated equilibrium in passing. His solution is to explain that the history of evolution produces a bushy structure, not an ordered sequence, where his account is not endangered by its incapacity, for example, to show that monkeys have language. But surely it would be more reasonable to suppose that our evolutionary relatives, including monkeys, chimpanzees, and gorillas, have some counterpart ability to use different yet comparable methods for communication. A broader semiotic framework would relate the use of signs to the subsumption of experience by means of concepts.

An adequate understanding of the evolution of language and mentality, moreover, heavily depends upon a firm grasp of the differences between genetic and cultural evolution. By adopting the common distinction between "genes" as units of genetic evolution and "memes" as units of cultural evolution, John Bonner (1980) already identified three important differences, where (1) genes can exist independently of memes, but not conversely (there are no disembodied thoughts); (2) genes are transmitted but once per organism, while memes can be acquired and changed across time; and (3) that the rate of change for genes is constrained by gestation, whereas the rate of change for memes approximates the speed of information transmission. Thus (Fig. 14),

Fig. 14 Genetic vs. cultural evolution (Bonner)

Genetic Evolution	Cultural Evolution
(1) Genes can exist independently of memes	(1') Memes cannot exist independently of genes
(2) One time transmission of information (conception)	(2') Multiple opportunities for information transmission
(3) Changes very slow (bound by rate of reproduction)	(3') Changes very fast (bound by speed of light)

Fig. 15 Genetic vs. cultural evolution (Fetzer)

(4) affect permanent properties	(4') affect merely transient properties
(5) mechanisms of genetic change are Darwinian, including: genetic mutation natural selection sexual reproduction ... artificial selection genetic engineering	(5') mechanisms of memetic change are Lamarekian, including: classical conditioning operant conditioning imitating others ... logical reasoning rational criticism

Other differences distinguish them as well, however, which in some contexts may be even more important. Thus, for example, the genetically heritable properties of organisms are ones that any organism with those genes could not be without (given fixed environmental factors) as *permanent properties*, while the memetic properties of organisms are often *transient and acquired.* The causal mechanisms underlying cultural evolution are rooted in the semiotic abilities of the species (Fetzer 1981, 2002a, 2005).

Ultimately, distinctions must be drawn between species for which their mental abilities are innate, inborn, and species-specific, and those for which their mental abilities can be enhanced through conditioning, learning, and even critical thinking. Low-level species, such as bacteria, may satisfy a conception of evolution where complex biological systems arise by the gradual accumulation over generations of random genetic mutations that enhance reproductive success. But other species far transcend the limitations that those constraints would impose. The only permanent properties related to language that humans have to possess are predispositions for the acquisition of concepts as habits of thought and habits of action, including the use of icons, indices, and symbols. There is no need for a "language instinct" as an innate disposition to use language (Fetzer 1991; 2005; Schoenemann 1999; Dupre 1999) (Fig. 15).

15 Concluding Reflections

In a broader sense, thinkers like Steven Pinker, Jerry Fodor, Noam Chomsky, and Euan MacPhail, who are preoccupied with language, have missed the boat by taking syntax to be more basic than semantics. When it comes to evolution, they have some general appreciation for the origin of species but little understanding of key differences between genetic and cultural evolution. They have developed their theories largely independent of the question, "But where did language come from?", as though it could arrive on the scene full-blown as a "language of thought" rich enough to sustain every sentence in every language—past, present, or future—that did not have to be a product of evolution!

The considerations adduced here, however, provide a fertile point of departure for other studies that carry this approach into new domains. While the theory of minds as semiotic systems clarifies and illuminates the very idea of consciousness as an evolutionary phenomenon, the elaboration of that approach for unconscious and preconscious phenomena requires further exploration (Fetzer 2011). At the very least, it makes clear that mental phenomena are semiotic phenomena involving the use of signs. When organisms are exposed to stimuli for which they lack corresponding concepts, for example, they are unable to subsume them and they remain merely "preconscious." When they are subsumed by concepts for which those organisms have no signals, then they are restricted to private use and might be said to be "unconscious."

This raises the possibility that the notions of "preconscious" and of "unconscious" may ultimately be envisioned *as relative to kinds of consciousness*. The study of Freud should contribute considerably within this context, since no one ever had a firmer grip of the intricacies of the human mind with regard to its conscious, unconscious, and preconscious dimensions (Smith 1999). Although the semiotic conception elaborated here supports appealing accounts of consciousness and of cognition, which have obvious evolutionary ramifications for the origin of species, its implications for the preconscious and the unconscious invite future investigation.

The theory of minds as semiotic systems presents an attractive alternative to models of the mind inspired by computers and language. Their respective merits should be assessed on the basis of the criteria of comparative adequacy for scientific theories, including (a) the clarity and precision of the language in which they are couched; (b) their respective scopes of application for explaining and predicting the phenomena to which they apply; (c) their respective degrees of empirical confirmation on the basis of suitable observations, measurement, and experiments; and (d) the simplicity, economy, or elegance with which their scopes of application happen to be attained (Fetzer 1981, 1993a). By this standard, the semiotic approach, which applies to human beings, (other) animals, and even machines, if such a thing is possible, provides a far superior framework for understanding consciousness and cognition including its ability to place "the hard problem" in proper evolutionary perspective.

References

Bonner, J. (1980). *The evolution of culture in animals*. Princeton: Princeton University Press.
Dennett, D. (1996). *Kinds of minds*. New York: Basic Books.
Donald, M. (1991). *Origins of the modern mind*. Cambridge, MA: Cambridge University Press.
Dupre, J. (1999). Pinker's how the mind works. *Philosophy of Science, 66*, 489–493.
Fetzer, J. H. (1981). *Scientific knowledge*. Dordrecht: D. Reidel Publishing.
Fetzer, J. H. (1984). Philosophical reasoning. In J. Fetzer (Ed.), *Principles of philosophical reasoning* (pp. 3–21). Totowa: Rowman & Littlefield.
Fetzer, J. H. (1988). Signs and minds: An introduction to the theory of semiotic systems. In J. Fetzer (Ed.), *Aspects of artificial intelligence* (pp. 133–161). Dordrecht: Kluwer.
Fetzer, J. H. (1989). Language and mentality: Computational, representational, and dispositional conceptions. *Behaviorism, 17*(1), 21–39.
Fetzer, J. H. (1990). *Artificial intelligence: Its scope and limits*. Dordrecht: Kluwer.
Fetzer, J. H. (1991). Primitive concepts. In J. H. Fetzer et al. (Eds.), *Definitions and definability*. Dordrecht: Kluwer.
Fetzer, J. H. (1993a). *Philosophy of science*. New York: Paragon.
Fetzer, J. H. (1993b). Donald's origins of the modem mind. *Philosophical Psychology, 6*(3), 339–341.
Fetzer, J. H. (1993c). Evolution needs a modern theory of the mind. *The Behavioral and Brain Sciences, 16*(4), 759–760.
Fetzer, J. H. (1994). Mental algorithms: Are minds computational systems? *Pragmatics and Cognition, 2*(1), 1–29.
Fetzer, J. H. (1996). *Philosophy and cognitive science* (2nd ed.). St. Paul: Paragon.
Fetzer, J. H. (1997). Dennett's kinds of minds. *Philosophical Psychology, 10*(1), 113–115.
Fetzer, J. H. (2002a). Evolving consciousness: The very idea! *Evolution and Cognition, 8*(2), 230–240.
Fetzer, J. H. (2002b). Propensities and frequencies: Inference to the best explanation. *Synthese, 132*(1–2), 27–61.
Fetzer, J. H. (2002c). *Computers and cognition: Why minds are not machines*. Dordrecht: Kluwer.
Fetzer, J. H. (2002d). Introduction. In J. H. Fetzer (Ed.), *Consciousness evolving* (pp. xiii–xix). Amsterdam: John Benjamins Publishing.
Fetzer, J. H. (2005). *The evolution of intelligence: Are humans the only animals with minds?* Chicago: Open Court.
Fetzer, J. H. (2007). *Render unto Darwin: Philosophical aspects of the Christian Right's crusade against science*. Chicago: Open Court.
Fetzer, J. H. (2011, January–March). Minds and machines: Limits to simulations of thought and action. *International Journal of Signs and Semiotic Systems, 1*(1), 39–48.
Harnad, S. (2002). Turing indistinguishability and the blind watchmaker. In J. H. Fetzer (Ed.), *Consciousness evolving* (pp. 3–18). Amsterdam: John Benjamins Publishing.
MacPhail, E. M. (1998). *The evolution of consciousness*. New York: Oxford University Press.
Pinker, S. (1994). *How the mind works*. New York: W. W. Norton.
Pinker, S. (1997). *The language instinct*. New York: William Morrow.
Schoenemann, P. T. (1999). Syntax as an emergent characteristic of the evolution of semantic complexity. *Minds and Machines, 9*(3), 309–334.
Slater, P. B. (1985). *An introduction to ethology*. Cambridge: Cambridge University Press.
Smith, D. L. (1999). *Freud's Philosophy of the unconscious*. Dordrect: Springer.

Mind or Mechanism: Which Came First?

Teed Rockwell

Abstract This chapter questions the reductionist assumption that bits of lifeless matter must have grouped themselves into complex patterns that eventually became living conscious beings. There is no decisive reason to question Peirce's suggestion that mind came first and that mechanical causality emerges when regions of a fundamentally conscious universe settle into deterministic habits. If we define consciousness in a way that ignores clearly accidental properties such as looking and behaving like us, some form of panpsychism is not only possible but plausible. Ignoring this possibility could cause us to subconsciously exclude legitimate avenues of research.

1 Peirce vs. Dawkins

> In the beginning was simplicity. It is difficult enough explaining how even a simple universe began. I take it as agreed that it would be even harder to explain the sudden springing up, fully armed, of complex order – life, or a being capable of creating life. Darwin's theory of evolution by natural selection is satisfying because it shows us a way in which simplicity could change into complexity, unordered atoms could group themselves into even more complex patterns until they ended up manufacturing people (Dawkins 1976, p. 12)

This volume is united around the attempt to solve the puzzle "how and why did organic mindedness come to exist in the natural world?" We might call this the question of *biogony*, as an analog to the question of *cosmogony*, which deals with how the universe as a whole came into being. I think it likely that many of the other authors will assume that questions about biogony can't even be asked without agreeing with the above Dawkins quotation. The question seems to presuppose that long

T. Rockwell (✉)
Department of Philosophy, Sonoma State University, USA
e-mail: teedrockwell@gmail.com

L. Swan (ed.), *Origins of Mind*, Biosemiotics 8, DOI 10.1007/978-94-007-5419-5_12,

ago there was a simple world of disconnected material bits, and that life emerged when these bits of the inorganic world were assembled in the appropriate way. What I am going to try to do in this chapter is to question whether this is the only possible biogony that philosophers and scientists should consider.

In the quotation above, Dawkins assumes above that the only way that complexity can emerge is if "unordered atoms could group themselves into even more complex patterns." However, once we accept this assumption, the "sudden springing up, fully armed, of complex order" is not just unlikely. It is impossible and self-contradictory, because it violates the metaphysical assumptions that make science possible (or so Dawkins believes). Things appear to "spring up suddenly" only when they are not fully understood. To understand a process is to analyze these "sudden" processes into discrete comprehensible steps. This is even more obvious in this quotation:

> …the hierarchical reductionist believes that cars are explained in terms of smaller units, which are explained in terms of smaller units, which are ultimately explained in terms of the smallest of fundamental particles. Reductionism, in this sense, is just another name for an honest desire to understand how things work. (Dawkins 1986, p. 13)

In other words, if you are not trying to explain complex things by breaking them down into simpler parts, you are not trying to explain them at all, because that is the only honest way to understand how things work.

We need not, however, make the reductionist inference about reality from this fact about scientific method. One can, after all, use the concept of a perfectly straight line as a regulative ideal without believing that there are any perfectly straight lines in the natural world. One can similarly use the techniques of analysis in scientific research without believing that there are ultimate fundamental parts that can be discovered using those techniques. More importantly for our question, the fact that reality is divisible into causally significant parts does not necessarily imply that it was assembled from those parts. Charles Peirce's metaphysics provided an alternative to reductionism by embracing two principles that he called *Synechism* and *Tychism.*

Synechism accepts "the necessity of hypotheses involving true continuity" (*Dictionary of Philosophy and Psychology*, vol. 2, Peirce 1931/1958a, pp. 6.169, 1902).[1] The important difference between synechism and reductionism is that the latter claims that there is only one way to divide up the universe that shows its fundamental causal relationships. For synechism, the ultimate reality is not an aggregate of bits, but a genuinely continuous process that can be divided up in a variety of scientifically useful ways, no one of which is the ultimate physical reality. Tychism is "the doctrine that absolute chance is a factor of the universe" (Peirce 1931/1958a, p. 6.201). For Peirce, this chance did not produce chaos, but rather "a spontaneity which is to some degree regular" (Peirce 1958b, p. 178). Regularity in

---

[1] In this passage, Peirce was referring only to the psychological continuity of ideas in experience. However, in Peirce 1892, he specifically endorsed applying these ideas to the external world as well (footnote p. 480).

the universe does not come from mechanical cause-and-effect connections but rather from this spontaneous force's tendency to settle into "habits" (Peirce 1958b, p. 177).

Note that these two principles of Peirce's metaphysics reverse the ontological priority expressed in the Dawkins quote above. (1) For Peirce, complex systems are not assembled from particles. Instead, both macro-objects and microparticles are moments in the flow of a fundamentally continuous reality. Reductionism claims that there are fundamental particles that possess all the causal power. The "true continuity" of synechism implies that the tiniest particles in the universe do not possess any more or less causal power than the medium-sized objects. Causal power resides in the process, not the particles, so whatever form the process takes can have its own causal power. Among other cash value differences, this makes free will possible, (although not necessary) because it implies that our beliefs and desires could control our neurons, rather than the other way around. (See Rockwell 2008). (2) For Dawkins-style reductionism, unpredictability is a function of our ignorance, because in reality all occurrences are governed by inevitable deterministic laws. For Peirce, spontaneity is a function of forces that are only to some degree regular. Although Peirce often used both "chance" and "spontaneity" to describe these forces, it's important to remember that Peirce's tychism is significantly different from the mechanical laws that govern coin flips. For Peirce, spontaneity is not chaos or randomness, but rather freedom of the same sort possessed by conscious agents:

> Tychism must give birth to an evolutionary cosmology, in which all the regularities of nature and of mind are regarded as products of growth, and to a Schelling-fashioned idealism which holds matter to be mere specialized and partially deadened mind. ("The Law of Mind," in Peirce 1940, p. 339)

Peirce is thus saying that living matter is ontologically prior to mechanism, because the latter emerges when spontaneous growing matter settles into deterministic patterns. For the reductionist, spontaneity is nothing but very complicated mechanism. For Peirce, mechanisms are spontaneity that has become simplified and rigid. This position could be described with a variety of terms sharing a single prefix: pantheism, panentheism, and panpsychism. All three of these terms imply that there are macro-patterns in the universe that are conscious in some sense. All three of these positions reject Dawkins' "Blind Watchmaker" theology, which holds that the only conscious entities in the universe are the medium-sized creatures with the largest brains.[2] Dawkins defines pantheism more broadly as a "metaphoric or poetic

---

[2] Many people, especially Dawkins himself, do not think of Dawkins' Blind Watchmaker theory as a theology. At one point, Dawkins even says that theology does not have a subject matter at all. This is an important mistake. It creates the illusion that Dawkins' position is only denying, rather than asserting, a fact about the world, which in turn gives the false impression that his theology is more parsimonious than its competitors. This assumption ignores the fact that the boundaries of an intellectual discipline are determined not by the answers it gives, but by the questions it asks. That is why Ptolemaic and Copernican astronomy are both forms of astronomy despite the vast differences in the answers they give to their shared questions. For the same reason, the Blind Watchmaker theology and Calvinist theology are both theologies because they ask the same questions and give radically different answers to them.

synonym for the laws of the universe" which enables him to say "Pantheism is sexed up atheism" (Dawkins 2006, p. 40). Perhaps some pantheists define the term this way, but I am concerned with the version which rejects this claim of Dawkins':

> Natural Selection…has no purpose in mind. It has no mind and no mind's eye. It does not plan for the future. It has no vision, no foresight, no sight at all. If it can be said to play a role of watchmaker in nature, it is the blind watchmaker. (Dawkins 1986, p. 5)

The form of pantheism/panentheism/panpsychism I will be defending is relatively cautious (for a theology, at any rate). I am claiming only that there is no reason to deny the existence of at least one other macro-pattern which is legitimately describable as conscious and purposive – something bigger than us, which does make plans and strives for some kind of future. As panpsychism makes the fewest claims about the nature of those patterns (i.e., no commitment to their omniscience, omnipresence, perfection, etc.), and I want to defend the smallest possible amount of territory, that is the term I will be using to label my Peirce-inspired position.

2 Panpsychism Defined and Defended

Panpsychism is easily confused with a few closely related straw persons. Among these is *vitalism*, which was the belief that physical science could not account for the behavior of living things, and that therefore biology needed principles that could not be reduced to physics. The panpsychism I am defending is the exact opposite of vitalism because it says that the subject matters of the physical and biological sciences are both fundamentally governed by the same principle. Vitalism was a form of dualism, which like most dualisms is supported by arguments based on gaps in our knowledge. Like most such arguments, it was defeated when those gaps were filled by scientific progress. My panpsychism, like materialism, is a form of monism and is thus untouched by any of these criticisms of vitalism.

Panpsychism is also sometimes misunderstood as the belief that each individual item in the universe is conscious, as if every tree, rock, and toaster were aware of something. I attacked this straw version of panpsychism on page 103 of Rockwell 2005. The more sophisticated forms of panpsychism acknowledge that we live in a world in which there are living things and nonliving things. This kind of panpsychism is consistent with reductionism in that it agrees we cannot designate certain spatiotemporal regions as purely conscious. Every conscious part of the universe can be divided up into parts which are mechanical, through and through. It is not the chemicals in our cells that are conscious, but rather some sort of pattern that supervenes on those chemicals. Neurotransmitters can partly embody thoughts and emotions when they chemically interact with nerve cells in an appropriately structured nervous system, but the neurotransmitters themselves do not think or feel anything. Panpsychism differs from reductionism only in claiming that even those parts of the universe that are not studied by contemporary biologists could very well be part of a larger conscious system, just as the chemicals in our bodies are part of larger conscious system. Charles Hartshorne puts it this way:

> A sand pile is loose-jointed so far as the pile taken as a whole is concerned. Its parts serve no imaginable unitary purpose enjoyed by the pile. But it does not follow that they serve no unitary purpose. There is no unity *of* action of the sandpile, but there is a unity of action *in* the sandpile,[3] a unity pervading the grains of sand but referring to a larger whole than the pile. (Hartshorne 1962, pp. 204–205, Italics in original)

Josiah Royce similarly asserts that "I do not suppose that any individual thing, say this house or yonder table is a conscious being, but only that it is part of a conscious process" (Royce 1901, p. 233). Panpsychists often express this distinction by saying that the sand pile or table is only an *aggregate*, and not a *system*. It is certainly possible that many things we see as mere aggregates are actually parts of overarching conscious systems. These systems may not be visible for us, but as Royce pointed out, there is no reason to think they should be. If they exist, they would probably take place at a time and spatial scale that was invisible to us. "I suppose that this process {of Consciousness} goes on with very vast slowness in inorganic nature, as for instance in the nebulae, but with great speed in you and me. But meanwhile, I do not suppose that slowness means a lower type of consciousness" (Royce 1901, p. 227). This supposition may be compatible with what science tells us, but do we have any positive reason for accepting it? Royce responds to this objection with an aggressive volley in the game of burden tennis:

> And I insist, meanwhile that no empirical warrant can be found for affirming the existence of dead material substance anywhere. What we find, in inorganic nature, are processes whose time rate is slower or faster than those which our consciousness is adapted to read or appreciate. (Royce 1901, p. 240)

I'm inclined to think that Royce has a valid point here. We know that we ourselves are both conscious and analyzable into mechanical parts. We know that there are other things (rocks, sand, etc.) which are also analyzable into mechanical parts. But where is the justification that these items possess this additional weird property we call mechanical unconsciousness? The structure of our language prejudices us into thinking that we are denying, rather than asserting, the existence of something when we describe a particular system as unconscious. But that should not blind us to the fact that claims of consciousness and unconsciousness are equally speculative. If conscious beings are analyzable into mechanical parts, the fact that rocks are similarly analyzable tells us nothing about their state of mind, or lack thereof. We have positive evidence that unconscious parts can be parts of conscious systems anytime we do a chemical analysis of a blood sample, etc. However, we have no evidence that there are any unconscious items that participate only in unconscious systems. I can't even imagine what such evidence would look like.

[3] I don't like the phrase "*in* the sandpile," because it conflicts with Hartshorne's more accurate reference to the "larger whole" which is the true determining factor of consciousness.

The Blind Watchmaker argument has the structurally fallacious form known as affirming the consequent.

> If organic life were created by a mindless mechanism, this process would be reducible to a system of causally interacting components.
>
> The process that created organic life is reducible to a system of causally interacting components.
>
> Therefore, organic life was created by a mindless mechanism.

This argument also contains the fundamental assumption of Deism, i.e., that the mechanical comprehensibility of the universe proves that God is not present in it. The only difference between Deism and Atheism is that Deism uses that assumption to kick God upstairs, instead of eliminating her altogether. Deism and Atheism are both compatible with our vast ignorance about theological matters, but this argument does not provide either position with any legitimate support. If we attempt to make the argument valid by reversing the order of the first two propositions in the first premise, then the first premise is no longer true. Here is the reversed version, minus the quantifiers that would apply it to the origins of life.

> If a process is reducible to a system of causally interacting components, it must be mindless.

We cannot assume that a system is unconscious just because it can be analyzed into unconscious parts, because we ourselves are conscious systems that are analyzable into unconscious parts.

3 Contemporary Panpsychism

Royce and Hartshorne are not the only defenders of panpsychism. Skrbina 2005 reveals that most philosophers who rejected panpsychism considered it seriously before doing so. During the nineteenth century, it was arguably the dominant position. Panpsychism has also undergone a resurgence in recent years (Rosenberg 2005; Seager 2004; Strawson 2006; Skrbina 2009). Although my conclusion is similar to many of these twenty-first-century panpsychists, I will not be relying on their central argument, which goes something like this:

(1) Science tells us that the world consists fundamentally of tiny distinct particles.
(2) Subjective experience cannot emerge from such particles therefore.
(3) Materialism cannot account for what Chalmers and Levine call "the explanatory gap," i.e., between the chemical structure of chocolate and the taste of chocolate therefore.
(4) We must conclude that subatomic particles possess a kind of proto-consciousness, which provides the foundation for the consciousness that pervades the universe.

I will not use that argument in this chapter, because:

(A) I reject premise (1) because, like Peirce, I believe that the universe is fundamentally a process, which shapes itself into items of varying sizes, none of which is more causally fundamental than the others. There are no *fundamental* particles, because the continuous process described by synechism is more fundamental than any of the particulate forms that process takes on. Large items are not just abstract patterns that supervene on the smaller particles, and do not derive their causal powers from those particles. Subatomic particles are real, but larger items are equally real. Consequently, the whole problem of emergence presupposed in premise (2) doesn't come up. (See Rockwell 2008).

(B) I reject premise (3) because I feel there are other equally effective ways of accounting for the explanatory gap. (See Rockwell 2005, pp.118–133). My arguments in this chapter will therefore define consciousness in ways that do not rely on the alleged existence of the explanatory gap.

4 Consciousness in the World

You don't have to be a dualist philosopher to acknowledge that we make distinctions between conscious and unconscious beings all the time. This natural intuitive ability is reliable often enough for most of our daily social interactions, and to at least serve as a starting point for a scientific study of consciousness. We know that *Homo sapiens* are conscious, and rocks are not. We know that frogs are more likely to be conscious than sea slugs. What are the concepts that make those judgments possible? Unlike the consciousness discussed by dualists and mysterians like Chalmers and McGinn, the kind of consciousness we will be considering here is intersubjective. It is, in fact, the concept which make intersubjectivity possible. It is defined entirely in terms of behavior, not as some kind of Cartesian "mysterious glow that only I can feel." This does not mean that consciousness is reducible to a list of individual acts of behavior, such as stimulus–response connections. We attribute consciousness to creatures on account of the overarching pattern of their behavior, and this pattern is not perceptible when you break behavior up into discrete steps. We judge an item to be conscious because that is the best theoretical explanation for its behavior as a whole, just as the existence of electrons is the best explanation for the behavior of macroscopic inanimate objects. Judgments about which items are conscious are often imperfect, but unless we had concepts for making such judgments, we could not survive in the social world (in fact, we would not have a social world at all). My goal is to explicate those concepts and see whether we can use them to make judgments about the relative plausibility of panpsychism vs. Dawkins' atheist reductionism.

Chalmers would not consider these behavioral criteria for identifying consciousness to be part of what he calls the hard problem. Nevertheless, this behavioral problem has powerful and unique challenges of its own. I am not referring here to old skeptical complaints that the process is sometimes fooled by robots, Teddy bears, and complicated devices that don't exist but could. It's not just that we don't always know *who* is conscious. We also don't know *how* we know who is conscious. We make these judgments instinctively, with no real understanding of the inferences involved. This is why the first major scientific attempt to determine the presence of consciousness – the Turing Test – is only a public opinion poll, and makes no attempt to explain the decision-making process of those who are polled. This is also why it is very difficult to retool our natural consciousness-detector to answer questions it wasn't designed to answer. We probably have some chance of success when we extrapolate from ourselves to other medium-sized biological creatures. However, we have no reason to believe that these instincts will ever be reliable in classifying an evolutionary process that takes place over millions of years, of which we can only observe a tiny part. There is no theological equivalent to the Turing Test that can be applied to millennium-long natural processes to determine whether or not they are conscious.

Our close-up view of biological history does not limit our ability to observe and classify mechanical processes. On the contrary, that is where the analytical tools of mechanical thinking do their best work. However, our goals and purposes are invisible to this kind of inquiry because they are high-level properties of a much larger system. You can't see the purposes of a purposeful organism if you analyze it into neural firings and muscle contractions. But that doesn't prove that we don't have goals and purposes, any more than the fact that chairs are made of molecules proves that they aren't really chairs, or the fact that Oxford University is made up of buildings and people proves that there is no Oxford University. This was Gilbert Ryle's reply to both reductionism and dualism, and it works as well for theology as it did for philosophy of mind.

Nevertheless, although I may be attempting a doomed enterprise, I am going to try to outline some of the fundamental principles we use to distinguish conscious from unconscious beings. My goal is to express those principles at a level of abstraction that will hopefully make them applicable to the metaphysical and theological controversies that separate the panpsychists from the reductionist atheists. Most of the time, we tell conscious beings from unconscious ones by relying on accidental, rather than essential, properties of conscious beings. If I look out at a roomful of students during a lecture, I assume that those items in the room which most resemble *Homo sapiens* are conscious, and those which don't (the desks, light fixtures, etc.) are not conscious. I make this assumption even during those times (such as early Monday mornings) when there is little significant difference in behavior between the students and their desks. This assumption is legitimate because we have a set of observational predicates that enable us to identify individuals who are members of species that are normally conscious (four limbs with five digits each, two of which are usually covered by shoes, etc.). These kinds of assumptions will not do, however, when we are trying to make judgments about whole categories of entities whose consciousness is in doubt, such as frogs, Martians, or galaxies. Instead, we will need to formulate some general principles that will enable us to justify judgments where our intuition and/or prejudices cannot be relied on. How do we begin to explore such unfamiliar philosophical territory?

We may have phrased this question backward. Perhaps our fundamental presupposition is that the world is filled with persons, and the question we need to ask is "How are we able to distinguish mechanical unconscious objects from persons?" By fundamental, I do not mean foundational. I am not claiming that our awareness of persons is somehow more direct than our awareness of mechanisms, or that this proves that persons are more real than mechanisms. On the contrary, precisely because persons are prior in our order of knowing, they are secondary in the order of being. This is what I think Wilfrid Sellars meant when he said that "the original image of man-in-the-world" was "a framework in which *all* the objects are persons" (Sellars 1963, p. 10). Judea Pearl made a similar point:

> The agents of causal forces in the ancient world were either deities…or human beings and animals, who possess free will…When machines had to be constructed to do useful jobs… systems consisting of many pulleys and wheels, one driving another, were needed.…Once people started building multistage systems…*physical objects began acquiring causal character.* (Pearl 2009, p. 403 italics in original)

The development of science and engineering led to the discovery that there were certain items in our world which were not like us persons in two diametrically opposed ways:

Items whose behavior is completely predictable are nonpersons. We doubt the personhood of insects because their behavior is far more predictable than the behavior of vertebrates. They do what they do out of instinct, which means when you take their behavior out of the context for which evolution designed it, they will continue that behavior even when it doesn't achieve its goal, rather than spontaneously adapt to the new circumstances. If we found out that insect behavior was not this rigid and inflexible, we would be less willing to deny that insects are persons. We are even more sure that windup toys are not persons because they are even less adaptive and more predictable than insects, and surer still that rocks are nonpersons for the same reason. And we are surest of all that the machines we built are not conscious, because they are designed to be totally predictable means of fulfilling our desires and purposes. When they start to lose that predictablility, we are tempted to attribute conscious personhood to them, which is why we swear at cars that refuse to start and computers that crash. This is part of the joke in Scoop Nisker's comment that a smart bomb would be one that refused to go off.

Items whose behavior is completely unpredictable are nonpersons. We are not willing to grant personhood to objects or states of affairs that are completely random and chaotic. Hume's argument against free will works because of this intuition. By identifying free will with chaos, he created a strong case for the compatibilist position that free will had nothing to do with what made us conscious human beings. We assume that someone who is going mad is losing consciousness as her behavior is becoming more chaotic. Completely chaotic behavior would indicate complete lack of consciousness. This is also why we would never attribute consciousness to a random aggregate of items such as three spoons, a pencil, and a cup of coffee. If there is no systematic coherence at all linking a set of items together, we acknowledge that these items do not constitute a conscious being.

Consciousness is therefore a property we attribute to those items that dwell in a twilight zone between the comprehensible and the incomprehensible. Their behavior is predictable, but only in rough qualitative ways, not precise quantitative ways. This is the relationship that Dennett describes between the intentional stance and the physical stance: the patterns we discover in conscious intentional systems can describe the broad outlines of the system's behavior, but cannot predict the exact details:

> The intentional strategy…is notoriously unable to predict the exact purchase and sell decisions of stock traders, for instance, or the exact sequence of words a politician will utter when making a scheduled speech. But one's confidence can be very high indeed about less specific predictions: that the particular trader *will not buy utilities today* or that the politician *will side with the unions against his party.* (Reprinted in Haugeland 1997, p. 67 italics in original)

> There are patterns that impose themselves, not quite inexorably but with great vigor, absorbing physical perturbations and variations that might as well be considered random; these are the patterns that we characterize in terms of the beliefs, desires, and intentions of rational agents. (Ibid., p. 70)

Dwelling in this causal twilight zone is a necessary characteristic of conscious systems, but I don't think it is sufficient. We also need to add that:

The behavior of conscious beings is explained by final causes, not efficient causes. "Efficient causes" is Aristotle's term for what I have been calling mechanical causes. Mechanical causes explain events by reference to other events that came before them. The rock rolls down the hill now because I kicked it a few seconds earlier. Final causes explain events by reference to events that come after them, i.e., my students come to class because they want to graduate. These final causes are also called goals and purposes, and no being that completely lacks them will be considered conscious.

These three characteristics seem to be both abstract enough to avoid provincial prejudices and yet concrete enough to account for our common sense ideas about the mental. I think they can be made the basis of a theory of mind that renders panpsychism plausible, and perhaps even testable (at least in principle).

5 Consciousness, Predictability, and Strange Attractors

Let us first consider the issue of predictability. Is there anything in nature other than large-brained animals that occupies this twilight zone between determinism and randomness? Can the soft predictability of what Dennett calls intentional systems be quantified in new yet legitimate ways? I think it can. There is a dynamic pattern in nonlinear chaotic systems called a *strange attractor* which fits this description quite well. Strange attractors vary within certain regions that can be described verbally (e.g., when mapped onto Cartesian coordinates, the changes in this system create a pattern of a torus, or a butterfly with three wings). But the exact path through those regions does not repeat, even though it is mathematically predictable. Consequently it cannot be described exactly in geometrical terms, although one can describe the general outlines of the state space that it travels through. Port and Van Gelder (1995) describes this situation by saying that even though a system that contains chaotic patterns is unpredictable, it is not unfathomable (p. 576). David

Skrbina argues that there is a relationship between consciousness and those systems that include this kind of strange attractor:

> The brain is like all dynamic systems – chaotic and unpredictable in detail. This is, at least, consistent with our common sense view of human thought, and of human action. Thoughts and actions are not predictable in detail... However, we know that there is a sense in which thoughts and behavior are predictable, and this is through the concept of human personality. A personality is a quasi-stable entity. In people, it represents the range of typical and expected behavior. For most people, barring injury or severe disruption, it tends to be consistent over time, usually from childhood through old age.
>
> The concept of personality corresponds very closely with the concept of the strange attractor. Recall the Lorenz attractor: a consistent, recognizable, semi-stable pattern, which, in a fuzzy sense, identifies the bounds of the possible states of the system. If the brain is seen as a chaotic system, accompanied by a quasi-attractor pattern in phase space, then a personality can be seen as a logical and necessary consequence. ...So: why do people have personalities? The answer seems the same as: why do real chaotic systems follow quasi-attractor patterns in phase space? (Skrbina 2001, p. 105–106)

The nonlinear neurodynamics of researchers like Walter Freeman provides evidence that the cognitive functions of brains are best understood as fluctuations in systems with strange attractors (see Rockwell 2005, Chapter 9). If this (admittedly controversial) theory of brain function turns out to be right, this would mean that the any system that ran by similar enough dynamic principles would have to be considered conscious in some sense, even if it contained no neurons. This follows from the basic assumption of artificial intelligence (AI): that it is possible to make thinking machines out of something other than protein, such as silicon or galaxies. The success of contemporary AI (or lack thereof) is irrelevant here. The fundamental assumption AI is an essential implication of naturalism, i.e., the rejection of the possibility of "magic meat" which is uniquely capable of generating consciousness. Large-brained organisms are conscious because of the patterns embodied in their nervous systems, etc., not because of any intrinsic properties magically lurking in the meat itself. It is contingently possible that animal protoplasm is the only physical substance capable of embodying these patterns. But there is no reason whatsoever to believe this is true, which is why both panpsychism and AI are possibilities that deserve to be taken seriously. There would be no reason to deny consciousness to a complex system, even a galaxy-sized one, just because it didn't look like us, speak our language, or fluctuate on a time frame we could make sense out of.

Right now we have insufficient evidence to determine whether any of these macro-patterns are either conscious or merely mechanical. Until such evidence is discovered, or if it is never discovered, panpsychists can still legitimately object that Dawkins' Blind Watchmaker theology requires us to deny that such patterns could ever exist. This denial is a mistake. Even if such patterns are not actual, they are surely both logically and physically possible. Science tells us that stable physical systems are either deterministically mechanical or have the quasi-stability of strange attractors. If my criteria for categorizing conscious systems are correct, Blind Watchmaker theology requires us to choose the first alternative, but gives us no legitimate reasons for doing so.

6 Mechanical vs. Final Causality

Much of the plausibility of Blind Watchmaker theology comes from the fact that what Aristotle called final causes seems like childish superstition today. The opening Dawkins quotation implies not only that reality is divided up into fundamental parts, but also that those parts are connected by a chain through which the causal power travels from the past into the future. The idea that a cause can reach back into the past from the future seems magical and preposterous. This is what gives plausibility to Dawkins' claims that "Natural Selection…has no purpose in mind… It does not plan for the future" (Ibid.). Evolution can in principle explain the origin of life entirely with mechanical causes, and mechanical causes by definition make no use of a person-based ontology of final causes. Consequently, atheism appears to be necessarily true. Any occurrence that evolution cannot explain with mechanical causes now, it can in principle explain in the future. "God of the Gaps" theories like Intelligent Design are pseudoscience because they reject this necessary truth.

Nevertheless, the incoherence of Intelligent Design arguments does not imply the truth of atheism. As I mentioned earlier, the Blind Watchmaker argument suffers from the fact that it implies that we ourselves are not conscious. All systems, both conscious and unconscious, are in principle analyzable into mechanical causes, including those (like us) which do have goals and purposes. The lack of gaps provide no evidence one way or the other as to whether a system is purposive. Another problem, however, is that this distinction between final and mechanical causes is not applicable to the science of dynamic systems. The parts of such systems are each comprehensible as past-driven causes, but the system as a whole is not. Complex dynamic systems contain feedback loops called attractor spaces which, like all loops, have neither beginnings nor ends. Consequently, the distinction between past and future causes is not applicable to them. We do refer to certain looped systems as being mechanical, such as ticking clocks. This is because the loops they follow are simple and repetitive. However, if a loop is complex enough that it can only be described by a strange attractor, and/or a system of strange attractors, it seems a likely candidate for being a system which manifests purposive activity. We would probably be hesitant to describe some strange attractor systems as being conscious or purposive, such as waterfalls or cyclones. But although such ideas are problematic at this point in the history of science, they should not be dismissed as unthinkable. There are many dwellers on the ontological borderlines between the animate and the inanimate, such as viruses and crystals. As we get a better understanding of the dynamic patterns which constitute consciousness, we will inevitably reshape the borders between the conscious and the nonconscious, and waterfalls may very well occupy a similar place along those new borders.

The purposes of such a system of strange attractors may be incomprehensible to us, but that does not mean those purposes are nonexistent. As long as the attractor includes a cycle of destabilization and restabilization, and this cycle is itself spiraling toward some other sort of metastability, there is no reason to refrain from describing this cycle with terms like "striving," "fulfillment," and "conditions of satisfaction." We may not think of certain kinds of stability as worth striving for, but that is irrelevant. I find the satisfaction derived from ice fishing and cockfighting to be incomprehensible, but I do not infer from this that the practitioners of these activities are not conscious. The only difference between these examples and other less anthropocentric strange attractors is that all of us have some limited ability to empathize with the former. But this is a difference in degree, and surely striving for the specific things we strive for can't be an essential property of consciousness. The only essential property is that such a system is striving to maintain some state or other. Which state the system settles into would be a matter of the system's taste.

7 Peirce and the Big Bang

Another possible objection would be that the loops in these complex dynamic systems were themselves the result of mechanical causes in their past. Both premises of this argument are wrong. Even if this were true, according to Blind Watchmaker theology, this is equally true of us, and this does not stop us from being conscious. There are, furthermore, plausible interpretations of modern physics that give us reason to believe it is not true. As I understand the Big Bang theory, the patterns in the universe did not emerge by throwing together a bunch of atoms. Instead, the atoms, and the laws that govern them, emerged from the chaos that followed the Big Bang:

> Even if there were events before the big bang, one could not use them to determine what would happen afterward, because predictability would break down with the big bang. Correspondingly, if as is the case, we know only what has happened since the big bang, we could not determine what happened beforehand. As far as we are concerned, events before the big bang have no consequences, so they should not form a part of the scientific model of the Universe. (Hawking 1988, p. 49)

I am hesitant to make extensive metaphysical inferences from a scientific popularization like Hawking's *A Brief History of Time*. Nevertheless, the above quote does seem to imply something like Peirce's claim that the universe began in chaos, and that deterministic laws emerged as chaos settled into spontaneity, which in turn settled into mechanical habits. If this interpretation of Hawking's science is correct, this implies that mechanism emerged from consciousness, rather than the other way around. The time before the Big Bang would be a period of complete randomness, when there would be no predictability at all, and the universe became more predictable after the big bang until it eventually obeyed the laws of physics. It is certainly possible that the intermediately chaotic systems that came into existence in the middle of this process contained systems of strange attractors that should be classified as conscious. If something like my definition of consciousness is correct, then those systems at the midpoint between randomness and determinism could be conscious in some sense, and thus consciousness would have come into existence before mechanical determinism.

There is also no reason to assume that this macro-consciousness ever disappeared. We conscious systems always have habitual parts that behave according to mechanical laws. The bigger such systems are, the harder it is for smaller creatures to see anything but their mechanical subsystems. If there are proton-sized rational scientists who are studying us using Dawkin's methods, they would no doubt be equally convinced that we are not conscious. This is why there are misleading procedural implications to Dawkins' assumption that the puzzle of biogony must be framed as "How could unordered atoms group themselves into even more complex patterns until they ended up manufacturing people?" Perhaps the question might be better framed as something like "How did massive spontaneous systems splinter into medium-sized purposive agents who now interact with a mechanical deterministic environment?"[4]

Is there any difference in cash value between these two different ways of phrasing the question? At this point, it's too early to tell. However, I think it likely that keeping both descriptions in mind would generate more avenues of research and increase our chances of finding the best solutions to this puzzle or puzzles. Walter Freeman once said that trying to understand the mind by studying neurons is like trying to understand a thunderstorm by studying the molecular structure of water (personal communication). It seems to me that trying to understand the emergence of life in terms of unordered atoms grouping themselves into patterns is similarly narrow in focus. We should at least consider the possibility that the patterns are grouping the atoms together, rather than assuming that all the causal power is stored inside the atoms. And if this is the best description, we are not that far away from saying that the patterns might have reasons and purposes, and not just causes, for doing what they do. It's preposterous to speak of microparticles having reasons and purpose, but not so preposterous to attribute these qualities to macro-patterns. After all, we are macro-patterns who have reasons and purposes. The Big Bang theory did not exist in Peirce's time, so he was in no position to cite it in his defense. We are, however, in no position to dismiss the possibility that the big bang theory, or whatever succeeds it, could prove that Peirce was right, and the Blind Watchmaker theory wrong, about which came first, the conscious or the mechanical.

8 Strange Attractors in the Modern Universe

We should also avoid the Deist mistake of inferring unconsciousness from the predictability of the law-like habits into which the tychistic universe has currently settled. If we discovered two interacting systems in today's universe, one of which impinges on the other in such a way as to create deterministic causal networks that surround a sufficiently sophisticated central system of strange attractors, we could legitimately describe the inner system of strange attractors as a mind, and the outer

[4] Even though he makes no specific comments about metaphysics or cosmogeny, I believe that Judea Pearl's new mathematical formulation for causal laws strongly supports this framing of the question. Pearl says that mechanical causality, in which the cause follows the effect, occurs only when one system interacts with another. Although his theory can in principle accommodate situations in which two mechanical systems interact, most of his examples involve purposive agents whose actions give rise to a mechanical set of causal laws. This is one reason he refers to his causal mathematics as an "Algebra of Doing."

deterministic system as that mind's environment. Science and technology are made possible by our ability to create closed systems in our laboratories and/or in the bowels of our machines. Within the context of these closed systems, there are such things as mechanical causes. We trigger a cause, and an effect immediately follows. There is no part of the universe which is in principle immune to the reductive analysis that makes this power possible. But just because *each* part of the universe can be reduced to a chain of mechanical causes, it does not follow that *all* of the universe is so reducible. The reductive materialist takes as an article of faith the claim that we could explain both organic and nonorganic systems purely mechanically if we had a big enough laboratory.

I can't prove this is wrong, but I see no reason to share this faith. Because I can see no reason to deny consciousness to sufficiently sophisticated dynamic systems containing strange attractors, and because there is no reason to believe that we are the only such systems in the universe, panpsychism seems to be a genuinely live option. We might not have as much reason to believe in *pantheism*, i.e., the claim that everything in the universe has a single unified consciousness. As I mentioned earlier, however, my form of panpsychism occupies a middle ground between pantheism and atheist reductionism. Pantheism seems to me to be possible, but a more cautious position would be what we might call a polytheist pantheism, i.e., the belief that there is at least one conscious macro-pattern in the universe, but not necessarily at most one conscious pattern.

Outside of the laboratory, where scientists must rely on observations rather than experiments, we find complex dynamic systems that loop back on each other in such a way as to dissolve the distinction between purposive and mechanical causes. Weather formations, economic booms and busts, and the behaviors of galaxies and solar systems all appear to be probabilistic systems with strange attractors whose behavior is predictable only qualitatively, not quantitatively. Am I saying that thunderstorms and economic depressions are conscious? That seems unlikely, but this surface implausibility tells us little about whether these phenomena are part of a larger conscious system. Our livers are probably not conscious, but like all organic tissue, they possess a quasi-chaotic stability that makes them capable of participating in a conscious system. This relative instability is why laboratory biology has never achieved the replicability of laboratory physics. Peirce was aware of this even in 1892, when he argued that "protoplasm is in an excessively unstable condition" (Peirce 1892, p. 348). He also argued that protoplasm occupied a twilight zone between chaos and determinism he called "unstable equilibrium" (Ibid.), and this was his main reason for arguing that "protoplasm feels" (Ibid., p. 343).

There are nonorganic macro-systems that appear to possess a similar kind of semi-stability, and I would argue that this makes them plausible candidates for being parts of conscious systems. Admittedly, we do not yet have a sophisticated set of principles that could explain conscious behavior in purely dynamic terms. Such a theory would not rely on accidental properties such as the ability to smile and wave, or the possession of neurons, but rather on the truly essential properties that would be possessed by any conscious being regardless of what it looked like or was made of. My short list of properties does not provide such a theory, any more than Democritus provided us with Newtonian physics. I have used vague qualifiers like "sufficiently sophisticated" to indicate how we might eventually distinguish conscious and unconscious dynamic systems. Only careful interdisciplinary communication between neurodynamics and other branches of Dynamic Systems Research would reveal the puzzles such a theory of consciousness would need to solve. My hope is that acknowledging such a theory is possible and would give worthwhile reasons for asking whether Dawkins-style reductionism provides the outer boundaries for all legitimately possible answers to the question "how and why did organic mindedness come to exist in the natural world?" Perhaps it emerged from inorganic mindedness, rather than from mindless mechanism.

References

Clark, D. (2004). *Panpsychism: Past and recent selected readings*. Albany: State University of New York Press.
Dawkins, R. (1976). *The selfish gene*. Oxford: Oxford University Press.
Dawkins, R. (1986). *The blind watchmaker*. New York: Norton.
Dawkins, R. (2006). *The God delusion*. Boston: Houghton Miflin.
Dennett, D. (1979). *True believers: The intentional strategy and why it works*. (Reprinted in *Mind Design II*, by J. Haugeland, Ed., 1997, Cambridge, MA: MIT Press)
Hartshorne, C. (1962). *The logic of perfection*. Lasalle: Open Court.
Haugeland, J. (Ed.). (1997). *Mind design II*. Cambridge, MA: MIT Press.
Hawking, S. (1988). *A brief history of time*. New York, NY: Bantam Books.
Pearl, J. (2009). *Causality: Models, reasoning and inference*. Cambridge, UK: Cambridge University Press.
Peirce, C. S. (1892, October 1–22). Man's glassy essence. *The Monist* (Reprinted in *Essential Peirce*, Vol 1, by N. Houser & C. Kloesel, Eds., 1992, Bloomington: Indiana University Press)
Peirce, C. S. (1940). *Philosophical writings of Peirce* (J. Buchler, Ed.). New York: Dover Publications.
Peirce, C. S. (1958a). *Peirce: Collected papers* (vols. I–VII, P. Hartshorne & P. Weiss, Eds.). Cambridge, MA: Harvard University Press. (Original work published 1931)
Peirce, C. S. (1958b) *Charles S. Peirce: Selected writings* (P. P. Weiner, Ed.). New York: Dover Publications.
Port, R. F., & Van Gelder, T. (Eds.). (1995). *Mind as motion: Explorations in the dynamics of cognition*. Cambridge, MA: MIT Press.
Rockwell, T. (2005). *Neither brain nor ghost: A nondualist alternative to the mind/brain identity theory*. Cambridge, MA: Bradford Books, MIT Press.
Rockwell, T. (2008). Processes and particles: The impact of classical pragmatism on contemporary metaphysics. *Philosophical Topics, 36*(1), 239–258.
Rosenberg, G. (2005). *A place for consciousness: Probing the deep structure of the natural world*. Oxford: Oxford University Press.
Royce, J. (1901). *The world and the individual*. New York: Macmillan.
Seager, W. (2004). The generation problem restated. In D. Clark (Ed.), *Panpsychism: Past and recent selected readings*. Albany: State University of New York Press.
Sellars, W. (1963). Philosophy and the scientific image of man. In *Science, perception, and reality*. London: Routledge and Kegan Paul.
Skrbina, D. (2001). *Participation, organization, and mind: Toward a participatory worldview*. Doctoral thesis, University of Bath, Bath, UK. http://people.bath.ac.uk/mnspwr/doc_theses_links/pdf/dt_ds_chapter4.pdf
Skrbina, D. (2005). *Panpsychism in the West*. Cambridge, MA: MIT Press.
Skrbina, D. (Ed.). (2009). *Mind that abides*. Amsterdam: John Benjamins.
Strawson, G. (2006). In A. Freeman (Ed.), *Consciousness and its place in nature: Does physicalism entail panpsychism?* Exeter: Imprint Academic.

Origins of the Qualitative Aspects of Consciousness: Evolutionary Answers to Chalmers' Hard Problem

Jonathan Y. Tsou

Abstract According to David Chalmers, the hard problem of consciousness consists of explaining how and why qualitative experience arises from physical states. Moreover, Chalmers argues that materialist and reductive explanations of mentality are incapable of addressing the hard problem. In this chapter, I suggest that Chalmers' hard problem can be usefully distinguished into a "how question" and "why question," and I argue that evolutionary biology has the resources to address the question of why qualitative experience arises from brain states. From this perspective, I discuss the different kinds of evolutionary explanations (e.g., adaptationist, exaptationist, spandrel) that can explain the origins of the qualitative aspects of various conscious states. This argument is intended to clarify which parts of Chalmers' hard problem are amenable to scientific analysis.

1 Introduction

In several works, David Chalmers (1995, 1996, 2003) has formulated the hard problem of consciousness in terms of various "why questions": Why does subjective experience arise from a physical basis? Why should the physical processing of the brain give rise to a rich qualitative inner life? Why is the performance of brain functions accompanied by experience? Chalmers suggests that these questions are mysterious and that science cannot satisfactorily answer them. In this chapter, I argue that either Chalmers' why questions do not fall within the proper purview of science or there are evolutionary answers to them. With respect to the latter issue, I discuss evolutionary explanations of the subjective aspects of various conscious states. While these evolutionary explanations can address Chalmers' why questions,

J.Y. Tsou (✉)
Department of Philosophy and Religious Studies,
Iowa State University, Ames, IA, USA
e-mail: jtsou@iastate.edu

L. Swan (ed.), *Origins of Mind*, Biosemiotics 8, DOI 10.1007/978-94-007-5419-5_13,

they do not provide the kind of *global philosophical answer* that his questions demand. I suggest that such a global demand is an unreasonable constraint to place on a satisfactory theory of consciousness.

The main argument of this chapter is that evolutionary explanations can address Chalmers' why questions. The chapter proceeds as follows. In the second section, I explicate Chalmers' presentation of the hard problem as a challenge for reductive explanations of consciousness. Part of Chalmers' challenge for the reductionist is to explain *why* the qualitative aspects of experience (i.e., "qualia") accompany brain states. In the third section, I suggest that Chalmers' challenge is misguided insofar as his why questions either place an unreasonable constraint on what counts as a satisfactory explanation of consciousness or there are evolutionary explanations that can address them. In the fourth section, I discuss evolutionary explanations for the origin of the subjective aspects of various conscious states (e.g., pain, color vision, orgasms). The different kinds of evolutionary explanations that can be given reveal the sense in which Chalmers' demand for a global philosophical answer to his why question (and hence, the hard problem) is misguided.

At the outset, it should be stated that the argument of this chapter does not address Chalmers' hard problem *in its own terms*. Chalmers' formulation of hard problem is a request for a *causal or proximal explanation* that can explain how and why consciousness is produced by the brain. The analysis of this chapter will not address this question. A fundamental assumption of this chapter is that Chalmers' formulation of the hard problem is ill posed and in order to make steps toward addressing it, it is first necessary to reformulate Chalmers' general formulation of the hard problem into a set of more narrowly defined questions. The analysis of this chapter focuses on how science can address why questions related to the origins of the qualitative aspects of consciousness. In engaging in this task, my aim is to clarify which parts of Chalmers' hard problem are capable of being addressed through empirical and scientific means.

2 Chalmers' Hard Problem and Why Questions

Chalmers' hard problem is intended to pose a challenge for physicalist explanations of consciousness and, more generally, reductive explanations that aim to reduce the subjective aspects of consciousness to something more objective (e.g., brain states or functional states). In this regard, Chalmers' analysis augments Thomas Nagel's (1974) argument that any satisfactory explanation of consciousness must capture its qualitative aspects or "what it is like" to be an organism. Like Nagel, Chalmers contends that the subjective aspects of consciousness should not be neglected or eliminated in scientific explanations. Indeed, for Chalmers, explaining the subjective aspects of consciousness ("experience") constitutes the hard problem of consciousness:

> The really hard problem of consciousness is the problem of *experience*. When we think and perceive, there is a whir of information-processing, but there is also a subjective aspect. As Nagel (1974) has put it, there is *something it is like* to be a conscious organism.

> This subjective aspect is experience. ...*It is widely agreed that experience arises from a physical basis, but we have no good explanation of why and how it so arises.* Why should physical processing give rise to a rich inner life at all? (Chalmers 1995, p. 201, emphasis added)

Here, Chalmers presents the hard problem as the task of explaining *how and why experience arises from a physical basis*. On this formulation, neither physicalist nor functionalist explanations can adequately address the hard problem since these explanations proceed precisely by reducing the subjective features of mentality (qualia) to objective (physical or functional) states, thereby circumventing the hard problem altogether (Chalmers 2003, pp. 104–105).

Chalmers' formulation of the hard problem can be distinguished into the following questions:

(1) How does experience (qualia) arise from a physical basis?
(2) Why does experience (qualia) arise from a physical basis?

Distinguishing the hard problem in this manner deviates from the spirit of Chalmers' analysis; however, there are good philosophical reasons for distinguishing Chalmers' how question from his why question (cf. Flanagan and Polger 1995, p. 321). Chalmers (personal communication) has indicated that what his hard problem is intended to solicit is a *proximal or causal explanation*, i.e., what I present in this chapter as the "how question" of (1). With Chalmers, I agree that (1) is a mysterious question, and science has made surprisingly very little progress in addressing this question. At present, we lack a strong scientific understanding of how our qualitative experiences (e.g., the felt quality of an emotion, the subjective experience of blue) arise from brain states. While I think that Chalmers' how question is a hard problem that science cannot address, I will concede this point and not pursue the issue further in this chapter.[1]

This chapter focuses on critically examining Chalmers' hard problem as formulated in (2), which will clarify which aspects of Chalmers' hard problem are amenable to scientific analysis. While I maintain that the how question of (1) is not answerable by scientific or empirical means, I suggest that the why question of (2) is. Chalmers' presentation of the hard problem as a why question is somewhat ambiguous, but at the

[1] It should be noted, however, that from the perspective of materialists, (1) begs the question on behalf of the dualist. If "mental states" simply *are* brain states (as in identity theory), then the question of how mental states *arise from* brain states is a pseudo-question for which there is no meaningful answer. Other materialists would reject Chalmers' (and Nagel's) methodological assumption that a satisfactory theory of consciousness *must* explain the phenomena of experience (or qualia). Some materialists object that this controversial assumption has not been sufficiently argued for, that it rests on a set of flimsy intuitions, or that it ultimately relies on a fallacious appeal to ignorance (Churchland 1996; Dennett 1996; cf. Chalmers 1997). Moreover, some eliminativists argue that the class of things regarded as "qualia" are too poorly defined to constitute a proper explanandum, and hence, qualia should be eliminated (rather than explained) in a theory of consciousness (Dennett 1988; Churchland 1996).

very least this question asks *why*—in addition to the functional aspects of mentality—does consciousness include a qualitative experiential component. As Chalmers puts it:

> What makes the hard problem hard and almost unique is that it goes *beyond* problems about the performance of functions. To see this, note that even when we have explained the performance of all the cognitive and behavioural functions in the vicinity of experience – perceptual discrimination, categorization, internal access, verbal report – there may still remain a further unanswered question: *Why is the performance of these functions accompanied by experience?* (Chalmers 1995, p. 203, emphasis in original)

Chalmers suggests that explaining the performance of particular cognitive functions (e.g., the integration of informational contents) by specifying a physical mechanism (e.g., 35–75 Hz neural oscillations in the cerebral cortex) constitutes the "easy problems" of consciousness, and cognitive science is well equipped to address these problems. However, the *further* question of why the performance of various cognitive functions is accompanied by experience is a hard problem:

> This further question is the key question in the problem of consciousness. *Why doesn't all this information-processing go on in the dark, free of any inner feel?* Why is it that when electromagnetic waveforms impinge on a retina and are discriminated and categorized by a visual system, this discrimination and categorization is experienced as a sensation of vivid red? We know that conscious experience *does* arise when these functions are performed, but the very fact that it arises is the central mystery. (Chalmers 1995, p. 203, emphasis added)

For the purposes of this chapter, it is useful to distinguish Chalmers' why question into a more general and more specific formulation:

(a) Why are neural states accompanied by subjective experience?
(b) Why are particular neural states accompanied by subjective experience?

These two questions pose different kinds of challenges for reductive explanations of consciousness.[2] The more specific question in (b) demands that an adequate explanation of a neural state (e.g., associated with pain or color perception) must—in addition to specifying a physical mechanism—explain why it is associated with a particular subjective experience. The more general question in (a) is more demanding insofar as it requires that a satisfactory theory of consciousness must explain why the subjective aspects of experience (in addition to its physical and functional aspects) exist at all. Neither of these demands is adequately met by materialist (or functionalist) analyses of consciousness.

---

[2] Although I have distinguished Chalmers' why question into a more general and specific formulation, these two questions are clearly related. In the conclusion of this chapter, I suggest that evolutionary answers to (b) will help to make progress on answering the more general question asked in (a). With respect to (a), I maintain that neural states are accompanied by qualitative experience because of evolutionary history; however, I resist drawing the stronger (*adaptationist*) conclusion that qualitative experience exists *because it was adaptive*. While the origins of the qualitative aspects of consciousness can often be explained in terms of their adaptive function (e.g., pain states or hunger states), I maintain that some conscious states are better explained by non-adaptationist explanations.

3 A Dilemma for Chalmers

In this chapter, I argue that Chalmers' presentation of the hard problem as a why question does not provide a grave challenge to materialist (or reductive) explanations of consciousness.[3] More specifically, I maintain that in its more general formulation, Chalmers' why question falls outside the proper domain of science (and hence, an adequate scientific explanation of consciousness is not required to answer it) and that there are evolutionary answers for its more specific formulation. This argument can be formulated as a dilemma:

1. If Chalmers' why question is (a), then there is an answer to this question, but it is not a question that science is required to address.
2. If Chalmers' why question is (b), then there will be evolutionary answers for different mental states, but one can only expect to find answers for particular mental states on a case-by-case basis.
3. Thus, either Chalmers' why question is not a question that science is obligated to answer or there are evolutionary answers to it.

This dilemma suggests that Chalmers' hard problem—formulated as a why question—should not be regarded as an intractable problem for materialists.

The more general interpretation of Chalmers' why question asks the following: (a) why is subjective experience conjoined to neural states at all? Put in this form, this question is a query into why neural activity is accompanied by subjective experience (over and above its functional aspects). While I believe that there is an answer to this question, it is not the kind of question that science is obligated to answer. From this perspective, explaining *why*—for humans (and many animals)—neural activity is accompanied by qualitative aspects would appeal to contingent facts about the kinds of sensory organs and nervous systems that humans (and animals) have evolved to possess. As such, the answer to (a) would appeal to evolutionary history and explain *what it is like* to be a human (or bat, bee, dog, or shark) in terms of the sensory organs and nervous system possessed by that species. Accordingly, there is an answer to be given to (a); however, this answer might not be very interesting from a scientific perspective. At the very least, science would not provide the *specific global kind of answer* to (a) that Chalmers' question solicits.

By analogy, consider the question "why is the sky blue?" To answer this question, one would appeal to facts such as the kinds of eyes that humans have evolved to possess and the kinds of wavelengths of visible light that normal human eyes can detect. If after being told these facts, Ruth thought that there was a *further fact* required to provide an *adequate scientific explanation*, then Ruth is making a conceptual error about what constitutes a satisfactory explanation. Similarly, if

[3] The analysis of this chapter is intended to be neutral on metaphysical issues concerning dualism versus materialism. The main goal of the chapter is to show that there are scientific explanations available for the reductionist and materialist to address Chalmers' why question.

Tom is told that consciousness is accompanied by experience because of the kinds of sensory organs and nervous system that humans have evolved to possess, and he protested that there is a further fact needed to provide an adequate scientific explanation, we should conclude that he is confused. This analogy highlights some characteristics of (a). First, there is an answer for (a), but the proffered explanation would not fall within the class of questions that science normally addresses. Second, addressing (a) would appeal to contingent facts. Finally, it is simply confused to think that there is a *deeper explanation* to be given for such questions beyond pointing to various contingent facts (cf. Chalmers 1996, p. 111). Thus, a reductive answer can be given for (a); however, it is not the illuminating sort of explanation that Chalmers is seeking when he asks "Why *should* physical processing give rise to a rich inner life at all?" (Chalmers 1995, p. 201, emphasis added)

The more specific interpretation of Chalmers' why question asks the following: (b) why are particular neural states accompanied by subjective experience? I think that there are evolutionary answers that can address this question. Chalmers alludes to this kind of response when he writes:

> There is an *explanatory gap* (a term due to Levine 1983) between ... functions and experience, and we need an explanatory bridge to cross it. A mere account of the functions stays on one side of the gap, so the materials for the bridge must be found elsewhere. This is not to say that experience *has* no function. Perhaps it will turn out to play an important cognitive role. But for any role it might play, there will be more to the explanation of experience than a simple explanation of the function. Perhaps it will even turn out that in the course of explaining a function, we will be led to the key insight that allows an explanation of experience. If this happens, though, the discovery will be an *extra* explanatory reward. There is no cognitive function such that we can say in advance that explanation of that function will *automatically* explain experience. (Chalmers 1995, pp. 203–204, emphasis in original)

Chalmers maintains that the explanatory methods of cognitive science and neuroscience are insufficient to address (b). In this chapter, I argue that evolutionary biology has the resources to help to bridge the apparent gap between functions and experience. In articulating this view, I assume that the *kinds of why questions* that evolutionary explanations can address take the form: "why is there any subjective aspect (as opposed to no subjective aspect) attached to a particular neural state?" This captures the thrust of Chalmers' (1995) question: "Why doesn't all this information processing go on in the dark, free of any inner feel?" (p. 203). *If* the kind of explanation that Chalmers is seeking is an answer to the question "why is a particular subjective experience attached to a neural state *rather than another subjective experience*?" (cf. Chalmers 1996, pp. 99–101); then, I think that this places the standard of explanation too high. I assume that humans *could have* evolved such that *some other subjective experience* accompanies a brain state (e.g., a pain state); however, it is a contingent fact that *this subjective experience* has evolved (which is the relevant explanandum that evolutionary explanations can explain). Since it is a contingent evolutionary fact, the demand to explain why *this subjective experience rather than some other* (functionally equivalent) *subjective experience arose*, in my view, sets the bar of explanation too high (far higher than is set in science).

While I believe that evolutionary explanations can address the question of why particular neural states are accompanied by subjective experience, we must be cautious about our expectations regarding what this research can tell us with respect to (b). If Chalmers wants to discover a ubiquitous kind of answer to (b) that tells us what *the function* of experience is (*in general*), then I think that no meaningful answer is forthcoming (cf. Chalmers 1996, pp. 120–121). At best, evolutionary research can provide explanations of why particular neural states are accompanied by specific subjective experiential aspects.

4 Evolutionary Explanations of Qualia

The kinds of evolutionary answers that can be given for (b) are discussed in William James' analysis of consciousness in his *Principles of Psychology* (1890, Chapters 5–6). In the context of an argument (against epiphenomenalist theories) that consciousness has causal efficacy (cf. Robinson 2007), James points out that there is a certain correspondence between (1) beneficial and detrimental conscious states and (2) the subjective experiences appended to such states:

> *It is a well-known fact that pleasures are generally associated with beneficial, pains with detrimental, experiences.* All the fundamental vital processes illustrate this law. Starvation, suffocation, privation of food, drink and sleep, work when exhausted, burns, wounds, inflammation, the effects of poison, are as disagreeable as filling the hungry stomach, enjoying rest and sleep after fatigue, ... are pleasant. Mr. Spencer [1855] and others have suggested that these coincidences are due ... to the ... action of natural selection which would certainly kill off in the long-run any breed of creatures to whom the fundamentally noxious experience seemed enjoyable. ... [I]f pleasures and pains have no efficacy, one does not see ... why most noxious acts, such as burning, might not give thrills of delight, and the most necessary ones, such as breathing, cause agony. The exceptions to the law are ... numerous, but related to experiences [e.g., drunkenness] that are either not vital or not universal. (James 1890, pp. 143–144, emphasis in original)

In this passage, James suggests that there are *good evolutionary reasons* for why certain conscious states are accompanied by particular subjective experiences. In particular, evolutionarily detrimental states (e.g., starving, being wounded, sickness) are associated with painful experiences, whereas evolutionarily beneficial states (e.g., being nourished, rested, or healthy) are associated with pleasurable experiences because these subjective experiential states themselves play a vital (causal) role in helping organisms survive and reproduce.

The Jamesian framework outlined above provides a beginning of an answer to (b): certain neural states are accompanied by qualia because these qualitative experiences play an important role in facilitating some function (e.g., seeking sustenance, avoiding physical damage) that promoted a species' survival and reproduction (cf. Cole 2002, p. 43). For conscious states that fall in this class, *adaptationist explanations* can explain the origins of the qualitative aspects of these states. For example, the qualitative experience of acute pain states (i.e., hurting) is evolutionarily adaptive insofar as these qualitative states helped teach organisms to avoid stimuli and situations (e.g., fire) that can damage their bodies (Polger and Flanagan 2002, p. 21). A creature that lacked qualitative pain states would be evolutionarily disadvantaged (see Puccetti 1975), and we can explain the origins of the qualitative aspects of pain states in terms of their evolutionary benefits. Hence, adaptationist explanations can provide answers to the question of why *some* conscious states (e.g., pain states, states of fatigue) are accompanied by particular qualitative experiences (e.g., hurting, feeling tired).

While it is tempting to think that the qualitative aspects of consciousness can always be explained in terms of their evolutionary benefits (e.g., see Tye 1996; Gray 2004), this assumption is mistaken (cf. Chalmers 1996, pp. 120–121). In this chapter, I take a pluralist stance, which assumes that there are different kinds of evolutionary explanations (besides adaptationist ones) that can explain the origins of the qualitative aspects of various conscious states (cf. Polger and Flanagan 2002). This follows the recommendation of philosophers of biology (e.g., Gould and Lewontin 1979; Gould and Vrba 1982; Gould 1991; Lewontin 1979; Lloyd 1999) who have warned against the adaptationist ("Panglossian") tendency to view all traits that organisms presently possess as *invariably* being naturally selected because they served some adaptive function. These philosophers emphasize that there are multiple evolutionary reasons for why various traits have arisen. Besides adaptationist explanations, other evolutionary explanations that can explain why a trait (e.g., qualia) exists include (1) a trait emerged due to random factors (e.g., genetic drift, demographic events), (2) a trait exists because of developmental effects (e.g., pleiotropy, allometry), (3) a trait was once adaptive but is no longer so, (4) a trait is itself not adaptive but a by-product of an adaptive trait (i.e., "spandrels"), and (5) a trait is an evolutionary by-product but subsequently acquired adaptive value (i.e., "exaptations").

As an example of a qualitative aspect of experience that was once adaptive but is no longer adaptive, consider the question of why humans have the particular qualitative experience of colors (e.g., red) when we perceive objects. Human color vision is trichromatic insofar as it is based on three photopigments contained in different retinal cones, which allows humans to distinguish over two million colors (Gray 2004, pp. 85). Most mammals are dichromats, and trichromacy is thought to have evolved 30 million years ago with the evolution of Old World primates. An explanation for why trichromacy evolved is that trichromacy allowed Old World primates to distinguish more sharply between colors in the red to blue range and their diets consisted largely of fruits that were yellow, orange, or red (Nathans 1999; Gray 2004, pp. 85–86). From this perspective, humans have a particular experience of the color red because we have descended from a species whose color vision conferred upon them an evolutionary advantage. While these qualitative aspects of color experience may have been adaptive in the past for early Homo sapiens, they are not necessarily adaptive in current evolutionary niches (e.g., where color blindness will not significantly compromise an individual's inclusive fitness).

As an example of a qualitative experience that has a less obvious evolutionary history, consider the example of female orgasm. Among evolutionary biologists, it is widely agreed that the qualitative aspects of male orgasm (i.e., pleasure and ecstasy) evolved because it promoted reproductive success. However, this adaptationist answer cannot adequately explain female orgasm since females can become pregnant without experiencing orgasms. In a Chalmersian spirit, one could ask: why are female orgasms accompanied by a particular subjective experience? Elisabeth Lloyd (2005) has examined various competing answers to this question, including the following theories:

(1) Female orgasm evolved because it promoted an enduring attachment between males and females (i.e., pair-bonding).
(2) Female orgasm evolved to stimulate male orgasm.
(3) Female orgasm evolved because it promoted a higher rate of intercourse for females.
(4) Female orgasm evolved because it increased the likelihood of fertilization by facilitating a suction mechanism of the uterus.
(5) Female orgasm evolved as an evolutionary by-product of male orgasm.

Lloyd argues that the scientific evidence favors (5), which maintains that female orgasm did not emerge because it was evolutionarily adaptive, but as a by-product of male orgasm (i.e., as a spandrel). On this account, the evolutionary history of female orgasm is similar to that of male nipples. Male nipples exist because female nipples are adaptive and both sexes go through similar stages in embryological development. Analogously, female orgasm exists because male orgasm is adaptive and both sexes share the same embryological developmental history (such that the penis and clitoris share the same embryological origins).

The examples of color vision and female orgasm illustrate the *different kinds* of evolutionary reasons why conscious states might be associated with particular subjective features. While the reason why these subjective aspects exist can *sometimes* be explained in terms of the adaptive function of such experiences (e.g., pain states, states of nourishment), sometimes the subjective aspect of particular conscious states (e.g., female orgasm) will be explained as contingent evolutionary accidents (e.g., spandrels, exaptations). For this chapter, what is important is not what the correct explanations are, but the fact that there are evolutionary answers that can be given for (b). If this view is correct, then there are respectable reductive (and materialist) explanations that can be given for (b).

5 Conclusion

In this chapter, I argued that evolutionary biology has the resources to address aspects of Chalmers' hard problem and, in particular, the question of why particular neural states are accompanied by specific qualitative features. In its more general interpretation, I argued there is an answer to the question of why neural activity is accompanied by subjective experience (which would appeal to contingent facts about the sensory organs and nervous systems that humans and other species possess through evolution), but it is not very scientifically illuminating. In its more specific interpretation, I argued that there are evolutionary answers to the question of why particular neural states are accompanied by subjective experience, but there will be a multitude of explanations. With respect to the relationship between these two questions, evolutionary explanations of why particular neural states are accompanied by qualia can be helpful for formulating a more precise answer to the more general question of why neural activity is accompanied by qualia (see footnote 2). The analysis of this chapter suggests that brain activity is accompanied by qualia because these qualitative aspects either are themselves adaptive insofar as they help organisms survive and reproduce or they are (sometimes accidental) consequences of other adaptations. To assume that there is a *deeper philosophical explanation* to be given to these questions, however, is to commit a conceptual error.

In offering a naturalistic analysis of the hard problem, my discussion has shifted away from Chalmers' focus on identifying a causal mechanism that connects brain states to subjective experience (the "how question" of this chapter). This neglect was intentional as I think that this issue is ultimately a metaphysical question that science does not have the resources to answer (and arguably, for which no meaningful answer can be given). By reframing Chalmers' why question into a narrower question concerning why particular conscious states have specific qualitative aspects, my aim has been to show that there are reductive explanations (viz., evolutionary explanations) available to account for the origins of qualia. While reframing Chalmers' hard problem in this way will be unsatisfactory to some insofar as it deflates the ambitions of Chalmers' challenge, I contend that the most promising route to progress on understanding the phenomenon of consciousness is by addressing modest questions in a naturalistic manner, rather than trying to answer ambitious questions via conceptual analysis.

Acknowledgements I am grateful to David Chalmers, Stephen Biggs, William Robinson, David Alexander, Liz Stillwaggon Swan, Curtis Metcalfe, John Koolage, Heimir Geirsson, Gordon Knight, and Murat Aydede for very helpful comments and suggestions on earlier drafts of this chapter.

References

Chalmers, D. J. (1995). Facing up to the problem of consciousness. *Journal of Consciousness Studies, 2*(3), 200–219.

Chalmers, D. J. (1996). *The conscious mind: In search of a fundamental theory*. New York: Oxford University Press.

Chalmers, D. J. (1997). Moving forward on the problem of consciousness. *Journal of Consciousness Studies, 4*(1), 3–46.

Chalmers, D. J. (2003). Consciousness and its place in nature. In S. P. Stich & T. A. Warfield (Eds.), *Blackwell guide to the philosophy of mind* (pp. 102–142). Malden: Blackwell.

Churchland, P. S. (1996). The Hornswoggle problem. *Journal of Consciousness Studies, 2*(5–6), 402–408.

Cole, D. (2002). The functions of consciousness. In J. H. Fetzer (Ed.), *Consciousness evolving* (pp. 43–62). Amsterdam: John Benjamins.

Dennett, D. C. (1988). Quining qualia. In A. J. Marcel & E. Bisiach (Eds.), *Consciousness in contemporary science* (pp. 42–77). New York: Oxford University Press.

Dennett, D. C. (1996). Facing backwards on the problem of consciousness. *Journal of Consciousness Studies, 3*(1), 4–6.

Flanagan, O., & Polger, T. (1995). Zombies and the function of consciousness. *Journal of Consciousness Studies, 2*(4), 313–321.

Gould, S. J. (1991). Exaptation: A crucial tool for evolutionary analysis. *Journal of Social Issues, 47*(3), 43–65.

Gould, S. J., & Lewontin, R. C. (1979). The spandrels of San Marco and the Panglossian paradigm: A critique of the adaptationist programme. *Proceedings of the Royal Society, London, Series B, 205*(1161), 581–598.

Gould, S. J., & Vrba, E. S. (1982). Exaptation: A missing term in the science of form. *Paleobiology, 8*(1), 4–15.

Gray, J. (2004). *Consciousness: Creeping up on the hard problem*. Oxford: Oxford University Press.

James, W. (1890). *The principles of psychology* (Vol. 1). New York: Henry Holt.

Levine, J. (1983). Materialism and qualia: The explanatory gap. *Pacific Philosophical Quarterly, 64*(October), 354–361.

Lewontin, R. C. (1979). Sociobiology as an adaptationist program. *Behavioral Sciences, 24*(1), 5–14.

Lloyd, E. A. (1999). Evolutionary psychology: The burdens of proof. *Biology & Philosophy, 14*(2), 211–233.

Lloyd, E. A. (2005). *The case of the female orgasm: Bias in the science of evolution*. Cambridge, MA: Harvard University Press.

Nagel, T. (1974). What is it like to be a bat? *Philosophical Review, 83*(4), 435–450.

Nathans, J. (1999). The evolution and physiology of human color vision: Insights from molecular genetic studies of visual pigments. *Neuron, 24*(2), 299–312.

Polger, T., & Flanagan, O. (2002). Consciousness, adaptation and epiphenomenalism. In J. H. Fetzer (Ed.), *Consciousness evolving* (pp. 21–42). Amsterdam: John Benjamins.

Puccetti, R. (1975). Is pain necessary? *Philosophy, 50*(July), 259–269.

Robinson, W. S. (2007). Evolution and epiphenomenalism. *Journal of Consciousness Studies, 14*(11), 27–42.

Spencer, H. (1855). *The principles of psychology*. London: Longman, Brown, Green, and Longmans.

Tye, M. (1996). The function of consciousness. *Noûs, 30*(3), 287–305.

Part IV
Philosophy of Mind

Neuropragmatism on the Origins of Conscious Minding

Tibor Solymosi

Abstract The philosophy of pragmatism has much to offer mind and life scientists in their thinking about the origins and nature of experience. In this chapter, I provide an introduction to neurophilosophical pragmatism by reviewing how classical pragmatists, such as John Dewey, reconceived concepts like experience, mind, and consciousness in light of the advances ushered forth by Darwinism. I then elaborate on a recent debate in cognitive science and neurophilosophy over how to think about conscious mental activity. In doing so, I draw on and modify the pragmatist framework sketched in the first part of the chapter.

After several decades of animosity between philosophy and science, philosophers and scientists are beginning to value the contributions each discipline brings to understanding and explaining the world.[1] Historically, philosophy and science were not separate enterprises. Only very recently has a strong distinction been made between them. In broad strokes, typically under the banner of *naturalism*, the distinction is being rejected by many. While this confluence of philosophy and science is showing promise, its nature is multifaceted and problematic, for there is no consensus on what the nature of philosophy is, even among self-proclaimed naturalists.

---

[1] Indeed, it is very much a beginning. Massimo Pigliucci (2008) offers an excellent description of what he calls the "borderlands between science and philosophy." In it, he notes physicist Steven Weinberg's essay "Against Philosophy" (1992) as an exemplar of anti-philosophy coming from scientists. This hostility from science toward philosophy recently gained attention when another physicist, Lawrence Krauss, gave an interview in *The Atlantic* in which he contended that physics has made philosophy irrelevant (Andersen 2012). His mockery and apparent contempt for philosophy—particularly when it came to a philosophical critique of his recent book—received so much criticism that Krauss quickly offered an apology (Krauss 2012). Some might see this apology as half-hearted; regardless, I see it as a bit of progress over the last 20 years.

T. Solymosi (✉)
Department of Philosophy and Religious Studies, Allegheny College, Meadville, PA, USA
e-mail: tibor@neuropragmatism.com

L. Swan (ed.), *Origins of Mind*, Biosemiotics 8, DOI 10.1007/978-94-007-5419-5_14,

Subsequently, the nature of science is also unclear. Consequently, the relationship between philosophy and science remains undefined. The need for greater self-reflection, mutual understanding, and clearer conceptions of philosophy and science is particularly strong when we—philosophers, scientists, artists, and laypersons alike—aim to understand and explain the origins of mind in nature. Among the first philosophers to consider the origins of mind in nature in light of Charles Darwin's theory of evolution were the American pragmatists: Charles Sanders Peirce, William James, John Dewey, and George Herbert Mead. Their views on the nature of philosophy, of science, of their interrelationship, and of the origins of mind are not only pertinent to these issues today but are also gaining new support from advances in the sciences of life and mind.

In this chapter, I aim to introduce philosophical pragmatism to those unfamiliar with it. In doing so, I also offer something to those who are familiar with pragmatism, namely, a reconstruction of conscious activity in consideration of the origins of mind in nature. To be sure, this chapter does not aim to review the main ideas and themes of each of the pragmatists listed above (though I do draw from them), for there is enough disparity among them that to provide an overarching view on a particular issue like the origins of mind is anathema not only to the originality of these pragmatists' thoughts but to the spirit of pragmatism as well. The central aim of this chapter, then, is to show how pragmatism offers an empirically responsible, scientifically pluralistic, and critically constructive philosophy. With respect to the question of the origins of mind in nature, pragmatists recognize not only a deep continuity between mind and nature but also the necessity of bringing multiple scientific perspectives to the question. Lastly, as a philosophy, pragmatism offers more than describing how the world is or how it works; pragmatism offers imaginative possibilities for how to improve human experience in light of what our best science tells us about the workings of the world. In addressing the question of the origins of mind from a pragmatist standpoint, I hope to offer a vision of how this question can not only be addressed scientifically but also philosophically in the sense just now described.

What I aim to accomplish in this chapter is threefold. First, through an introduction to pragmatism, I suggest that the job of philosophers is distinct from but dependent upon the work of scientists. Second, through the advocation of my neurophilosophical pragmatism, I follow up on Dewey's inclination to consider organic activities in terms of adjectives, adverbs, or gerunds instead of substantive nouns, that is to say, as conscious or minding rather than consciousness or mind. This is not to say that we must eliminate concepts like mind or consciousness. The third aim of this essay is to follow through on the view of pragmatism that I advocate by elaborating on a metaphor for thinking about consciousness and mind, which I have introduced elsewhere (Solymosi 2011). This metaphor—that conscious activity is like cooking, that is, that consciousness is to the brain, body, and world as cooking is to brain, body, and world—may initially come across as counterintuitive. However, if I am successful in this chapter, readers should be sufficiently provoked to take up the challenge of either fleshing out the details of the proposed view or criticizing it with stronger evidence and an alternative perspective.

1 Pragmatism, Naturalism, and Falliblism

Pragmatism is America's most original contribution to the Western philosophical tradition. It emerged in the aftermath of the American Civil War, in the midst of industrialization, and in the fire of Darwin's theory of evolution by means of natural selection. With intellectual roots in the idealism of George Berkeley, Immanuel Kant, and Georg Hegel in Europe, the transcendentalism and romanticism of Ralph Waldo Emerson, and the democratic spirit of Walt Whitman in America, the classical pragmatists Peirce, James, Dewey, and Mead turned philosophical tradition on its head. This rejection of traditional philosophical practice is perhaps best evidenced in the attempt to define philosophical pragmatism itself. The early twentieth-century Italian pragmatist Giovanni Papini wrote that "Pragmatism cannot be defined. To offer a brief definition of pragmatism is to do the most antipragmatic thing possible" (Weiner 1973, 552). The philosopher and historian of ideas Arthur O. Lovejoy added support to Papini's claim (though with less enthusiasm than Papini) in his "Thirteen Pragmatisms I and II" (Lovejoy 1908a, b). The plethora of pragmatisms illustrates its core anti-essentialism. That pragmatism resists philosophical definition, especially in terms of necessary and sufficient conditions, does not imply that pragmatism cannot be characterized. Among its anti-essentialism are attitudes of anti-skepticism and anti-dualism.

Of course, such negative characterizations are not typically satisfying or especially useful on their own. The contemporary pragmatist and Dewey scholar Larry Hickman has offered the following characterization of pragmatism, which will guide my further elaborations. Hickman works chronologically through Peirce, James, and Dewey:

> Here is Peirce in 1878: "Consider what effects, that might conceivably have practical bearings we conceive the object of our conception to have. Then, our conception of these effects is the whole of our conception of the object." Here is James, twenty years later, in 1898: "The effective meaning of any philosophic proposition can always be brought down to some particular consequence, in our future practical experience, whether active or passive; the point lying rather in the fact that the experience must be particular, than in the fact that it must be active." And here is Dewey in 1938, sixty years after Peirce's statement: "The proper interpretation of 'pragmatic,' [involves] namely the function of consequences as necessary tests of the validity of propositions, *provided* these consequences are operationally instituted and are such as to resolve the specific problem evoking the operations."
>
> Put succinctly, the Pragmatic theory of meaning insists that we treat the whole meaning of a concept not just in terms of its use in a language game, as Wittgenstein urged us to do, but in terms that are overtly experimental and behavioral and in ways that *transcend* particular language games: the meaning of a concept is the difference it will make within and for our future experience. (Hickman 2007b, 36)

Hickman concludes, "Another way of putting this is that the Pragmatic method is experimental at its core" (2007b, 36).

This experimentalism ties directly to the Darwinian naturalism of pragmatism, especially with regard to pragmatism's anti-essentialism, anti-skepticism, and anti-dualism. Prior to Darwin, philosophy and science were both fixated on finding the fixed universals of nature. Philosophers sought the final cause of nature, viz., its grand purpose based on its underlying reality behind the appearances of experience. Scientists (though they were not called scientists at the time but "natural philosophers") aimed at uncovering the laws of nature. These laws served as the logical basis of further observations and experiments. In short, philosophy and science aimed at finding the essences of nature. These essences were expressed in the language of mathematics. The mathematics at the time immediately prior to Darwin's contribution, however, was only beginning to be developed into probabilistic and statistical methods that were effective in scientific work. These methods were at the core of the Darwinian revolution in science and philosophy as Peirce and Dewey readily recognized (Peirce 1877; Dewey 1976–1988 [1910/MW4]).

The significance of this shift is not easily overstated. If the products of scientific inquiry—indeed all inquiries—are probabilistic, the ancient and modern criteria that knowledge be absolute, universal, final, indubitable, and unchanging are no longer appropriate from a scientific perspective. Moreover, if humans and everything humans do are the products of evolution, then our scientific activity, the products of that activity, and what we call knowledge are also products of evolution. The pragmatists took this evolutionary fact seriously in their understanding of the relationship between humans and nature.

If inquiry produces provisional beliefs to habitually guide action of a human in a world, as Peirce first suggested, then there must be evolutionary ancestors to inquiry and habit formation more generally. Indeed, as the contemporary pragmatist Daniel Dennett has illustrated, the process that is recapitulated ontogenetically and phylogenetically is a process of generating things (e.g., acts, skills, ideas, hypotheses) and testing them (sometimes in the world without reflection, other times in imaginative reflection, and also at times in the world after such reflection is done). As Larry Hickman notes about Dewey's view of technology, it is a process of generate and test.[2]

This evolutionary continuity is significant for a pragmatist conception of the origin of mind in nature, for the naturalism of Dewey is not necessarily the naturalism of many analytic philosophers today, who contend that bridge laws and other reductive principles and tools can express or explain higher-level phenomena, such as mentation, in lower-level terms (whether it is neural, chemical, or physical, depending on the reductionist). Dewey's naturalism shares a rejection of the supernatural with these reductive naturalists. However, this pragmatic naturalism—Peirce and James largely agree with Dewey on this point—recognizes the continuity between the living and the nonliving, between the human and the animal, and between experience and nature.

This continuity between the human and the natural is important for explaining not only the origins of mind and experience in nature but also for understanding the means by which we gain that explanation. To conclude this section, I provide a general and brief statement of the nature of inquiry, science, and philosophy, as

[2] On the details of this evolutionary view of inquiry in Peirce, Dewey, Hickman, and Dennett, see Solymosi (2012a).

pragmatists conceive it, in order to frame the rest of this chapter. The next section elaborates on the pragmatist reconstruction of experience, which is a broader category than mind or intelligence. I then turn to the consequences for our conception of mind and intelligence that follow from the pragmatist reconstruction of experience. Finally, with this conceptual scaffolding in place, I draw on recent scientific and philosophical work to discuss my new metaphor for thinking about consciousness.

Following Darwin, living organisms are adaptive to environments that are both precarious and stable. Since these environments are often changing—sometimes with regularity that can be anticipated, sometimes not—organisms that are more capable of adjusting to changes have a greater likelihood of survival. Of the organisms capable of such adjustment, only those who can pass on these traits to their progeny are likely to continue their evolutionary line. The adjustments that organisms make are both to themselves and to their environments. Adaptive changes are the ones that continue the living process of the individual organism and, especially, its progeny.

As problems of survival and viability are solved through trial and error, through a process of generate and test, some organisms evolve that can better attend to their environment than others. The attentive behavior of these organisms relies on habits formed that afford two related activities. The first is automatic response to a specific set of conditions (e.g., a frog's snapping its tongue upon seeing an object of a certain size in a certain part of its visual field). The second activity is the slowing down of automaticity in order for further information processing to occur. The information processed comes from three sources: the immediate environment, recollection of previous interactions with other environments, and anticipation of various courses of action. This processing occurs simultaneously in a dynamic circuit and not in a reflex arc of stimulus–response mechanisms.[3] The dynamic circuit of a nervous system becomes amplified in social organisms capable of communication. Not only are alerts communicated but also other suggestions for action are made, from predator warnings to a request for help. In time, problems were being solved more and more deliberately, due in large part to the development of tools.

The emergence of tools indicates many important developments in hominin evolution. What is worth noting here is that tools illustrate the deliberate modification of an environment for multifarious purposes. Among the consequences of tool use is sophisticated symbolism in language and art. With the rise of human culture, inquiry becomes symbolized, deliberate, and institutionalized. The ability of humans to adaptively adjust to their environments, argue the pragmatists, is indicative of our evolutionary trajectory. We become fallibilists in recognizing that we are part of an evolutionary process in which tools are useful for certain purposes before realizing they are useful for other purposes too or are detrimental to larger aims.

[3] See Dewey (1969–1972, [1896/EW5]), Rockwell (2005), Chemero (2009), and Solymosi (2011) on Dewey's critique of the reflex arc concept and the significance of it for contemporary dynamic systems theory.

Our knowledge claims are tools that are open to further modifications, revisions, and abandonment as are any other tools.

As scientific inquiry benefits from industrial society, the need to critique older knowledge claims in light of newer scientific claims becomes ever greater. The job of taking old beliefs and putting them to new uses and of taking new beliefs to achieve older aims now reconsidered is the job of philosophy, according to Dewey. This is what Dewey called the project of reconstruction. It is the philosophical inquiry into how the claims of science provide the means for achieving our ideals. On this view, as we continue to inquire in order to solve the problems we perceive, our beliefs about how the world is *and* how the world could be in light of what we know about it are provisional. The world continues to change, due in no small part to our own interactions with it. Because our actions often have unforeseen consequences, we must be open to revising what we believe and know about the world and its possibilities.

To conceive of knowledge in this way, James and Dewey realized that a new view of experience was necessary. Proponents of science regularly refer to the empirical component of scientific activity as a cornerstone of its success in informing us about the workings of the world.[4] Traditional empiricism held that experience was a passive affair, in which the mind received sense data from behind a veil of ideas that kept the external world from being (easily) known. The pragmatic reconstruction of experience in light of Darwin is a significant departure from the sensationalism of the moderns. For this reason, James called it *radical empiricism.*

2 Reconstructing Experience

When philosophers discuss experience, what they mean is not necessarily what people think of when they talk about experience. For most philosophers, experience is sensationalistic. This is a view that Dewey called the spectator theory of mind or what Dennett has referred to as the Cartesian Theater.[5] The idea is that the mind is a passive receiver of data provided by the bodily senses about the world, but is not in direct contact with the world. These data are viewed on a screen or stage by the mind.

[4] One of the significant characteristics of modern science as opposed to the science or *scientia* (i.e., systematic knowledge) of antiquity is its emphasis on empirical observation in experimentation. In the next section, I distinguish between a passive sense of experience and an experimental one. For now, it is worth emphasizing that the science with which I am concerned is empirical, that it gains a significant part of its authority from its empirical component, and that, most controversially, all fields, which consider themselves scientific, are empirical *even if they insist otherwise*. The most obvious example of this would be mathematics. However, as pragmatists have long argued (see Dewey 1981–1991 [1938/LW12]), and as Lakoff and Núñez (2001) have further corroborated, mathematics is based in bodily experience and metaphors and is therefore empirically based. For more details on the empirical nature of scientific activity, see Godfrey-Smith 2003.

[5] See Dewey (1981–1991 [1925/LW1]), Dennett (1991), and Solymosi (2011).

To use philosophical parlance, there is a veil of ideas (our thoughts, conceptions, appearances, and illusions) that separates our mental life from the external world. This dualism is just what Dewey sought to reject in his efforts to reconstruct experience in light of science and Darwin.

Dewey recognized that the evolutionary process was a continuous one of adjustment of an organism to its environment. This adjustment could be one of the organism's modifying some aspect of itself to better fit the environment (adaptation) or of the organism's modifying its environment (alteration) to better fit the organism—these are not mutually exclusive processes and often dynamically occur (see Hickman 2007b). From an evolutionary perspective, there is no organism without an environment and no environment without an organism.[6] The two are entangled. So great is this entanglement that contemporary thinkers have corroborated Dewey's insight in suggesting that organism and environment should be treated as the single evolutionary unit, symbolized by *Œ* (see Griffiths and Gray 2001). This view is very much in line with Dewey's insight into how to reconstruct experience.

Experience for Dewey was the interaction or transaction between organism and environment (1981–1991 [1925/LW1: 12; 1939/LW14: 16]). This is a radically different view from the spectator theory. For one, the spectator was passively taking it in, whereas the organism-environment transaction is dynamic and active. The dynamism of this transaction also opens experience up to scientific investigation because it does not posit a distinct ontological substance inaccessible to scientific methods.[7] It does, however, raise questions about how to talk about experience of this sort, particularly as it relates to mind and culture.

This transactive conception of experience may seem odd to some, but the general idea is not unfamiliar. The contemporary neopragmatist Robert Brandom illustrates the difference between sensationalistic experience and transactional experience by appealing to German. He writes of the classical pragmatists,

> In the service of a renovated empiricism to go methodologically with that naturalism in ontology [as influenced by Darwinian evolution], they developed a concept of *experience* as *Erfahrung* rather than *Erlebnis*: as situated, embodied, transactional, and structured as *learning*, a process rather than a state or episode. Its slogan might be 'No experience without experiment'. Representing and intervening were for them two sides of one conceptual coin—or less imagistically, reciprocally sense dependent concepts concerning aspects of processes exhibiting the selectional, adaptational structure common to evolution and learning. (Brandom 2004, 14)[8]

---

[6] For Dewey, the contextual whole, what he called a "situation," is prior to any distinction between organism and environment. If there is difficulty in conceiving of an environment's dependence on an organism, consider its etymology. Without something to environ—to surround—there can be no environments (no surroundings). See Dewey 1981–1991 [1938/LW12].

[7] Rockwell (2005) is the first application of dynamical systems theory to Dewey's conception of experience. See also Chemero (2009) and Solymosi (2011).

[8] Despite Brandom's useful discernment here, readers should be alerted to the unfortunate misunderstandings Brandom makes in the second half of this article, in which he criticizes classical pragmatism for making semantic mistakes for which there is no warrant as Hickman (2007a) illustrates.

When a person has experience with something—an object, an event, an activity—we say that the person has familiarity with it. Experiential learning, on this view, is the means of acquiring knowledge through familiarization. To become familiar with a thing is to interact with it, to play with it, to try it out—to *experiment* with it. Depending on how these interactions go, more practice is needed, or the proposed object of learning is discarded. Successful interaction yields new skills through which more familiarizations can occur. From an evolutionary perspective, experience evolves as a developmental process of pattern production through natural selection—through the trials and errors of patterns of organism-environment interaction, patterns that are generated and tested.

3 Reconstructing Mind and Culture

Through iteration upon iteration of generating and testing, evolutionary experience cumulated to the point at which social animal life was not only communicative but deliberately so. Symbolic use of environmental manipulation, particularly in tool construction and modification, signifies the emergence of culture. Whether it was through the locutions of animal calls, the gestures of body language or emotional releases, hominins were creating new patterns of interaction in an environment that was not solely physical or biological. It was also social. The environment of this social organism was an environment filled with other social organisms like itself.

Since the evolutionary process is one in which problems of survivability and viability arise, creatures that can solve these problems are more likely to persist and thus have the opportunity to pass on their means of problem solving. Prior to social interactions, the best means of passing on problem-solving traits was largely genetic. Genes, after all, are patterns themselves that interact with cellular mechanisms, which interact with each other to operate the cell, which interacts with other cells of its type to operate tissues, which operate organs, which operate organic systems, which operate the body. Regulatory processes occurred but were not executed deliberately on the strictly biological level.[9] Once communication between animals developed to the point where problems could be solved through group communication instead of waiting for genetic mutation and selection to occur, problem solving was something that individuals could share with each other and with progeny.

This sharing was done through communication. This transaction between multiple individuals, particularly as the goods permeated through time thanks to verbal and written stories, is integral to the pragmatic reconstruction of mind. For Dewey, mind was not the introspective first person we traditionally conceive it to be.

[9] On the dynamics of regulatory processes from a neuroscientific and pragmatist perspective, see Schulkin 2003, 2009, 2011a, b. Of particular importance is Schulkin's distinction between the regulatory processes of homeostasis (which is passive and resistant to change) and allostasis (which is dynamic and anticipates change).

Instead of the mind as an entity that one is or a body has, pragmatists such as Dewey conceived of it as an operation or a process. To James's provocative question, "Does Consciousness Exist?" (1904/1977), the pragmatist must answer no, for there is no *thing* that is mind or consciousness, especially in the sense that its existence is distinct from the things that are conscious or minded. Dewey's view became even more radical in its reconstruction. Not only did Dewey emphasize using the gerund—that is a person is minding as opposed to a person's mind—he also emphasized the environmental conditions that made the conscious or minding activity of an organism possible.

The environment in which a conscious organism develops or is *cultivated* is culture. Later in his life, Dewey lamented over his attempts to reconstruct experience and thought that he should have used the word *culture* instead to denote what he was after (Dewey 1981–1991 [1925/LW1]: 361). For Dewey, culture is social human transaction. That is, the human organism is interacting with other human organisms in a social medium of shared symbols, values, and facts. Dewey's dissatisfaction with *experience* comes from the numerous misunderstandings his contemporaries had of his dynamic view (i.e., they kept confusing *Erfahrung* for *Erlebnis*). One of the consequences of his view is that the mind is not something unique to an individual organism. Rather, it is not something found within the organism at all. It is at once something the organism does and something that affords the organism that activity.

Just as I run with my legs, I mind with my brain and body. Consider how I run. I do not run with my legs alone: I require an environment conducive to that activity. I cannot run in a deep lake, on ice, or in the air. I need not only ground that is appropriate to the activity, I need muscles and feet appropriate to the activity too. Furthermore, since I am not one for running barefoot, I either need a very specialized environment to protect my bare feet, or I need a good pair of running shoes. Through attempting to run, through running rather poorly, then mediocrely, I eventually develop into running fairly well. That is, I become familiar with the activity, viz., I develop experience. Such experience is not just with the moving of the legs, it is with the environment in which I run: an environment that I or others have modified for the purpose of running (i.e., I run on a treadmill, or a track, and not on the Interstate).

In a similar fashion, I do not mind independently of my environment, social and biological. The culture affords me the opportunities and means of acting toward the ends I seek. Of course, not any ends are permissible since taboos, cultural norms, and laws (both natural and social[10]) serve to limit my ability to do whatever I please.

[10] Social laws are constructed to manage the behavior of individual persons; there are consequences for violating them. Natural laws are regularities that one ignores at one's own peril: no matter how hard I try I cannot walk on the ceiling—unless, of course, I learn how to manipulate the natural regularities to work in my favor by substantial experimentation. In which case, what it means to walk on a ceiling has been reconstructed in light of the possibilities created through imaginative scientific activity.

Nevertheless, so much of the culture, of the mental environment, viz., the symbolic scaffolding that makes meaningful action possible, is the dynamic product of eons of evolutionary experience that has only recently (in geological time) been more deliberately adjusted to serve human ends, including but not limited to survivability and viability. On this pragmatist view, mind and culture are interchangeable. There are no individual minds without a culture in which to cultivate the activity of mental living; there is no culture without the richly symbolic but nevertheless organism-environment transactions of individual human organisms in a human environment.

Experience is not a passive affair in which sense data that somehow represent the external world are received. Rather it is an active and dynamic affair in which the transactions between organisms and their environments co-regulate and co-constitute the patterned activities of problem solving. These problems of survivability and viability are not deliberately solved; they are not even recognized by the organisms. Yet these experiential events accumulate and develop through iterations. Eventually, the pattern of transaction of social organisms is generated and tested. Some of the successful patterns produce groups of organisms that communicate and cooperate to solve common problems. Some of these problem solvers happen upon new solutions to old problems that grow out of but do not require the community to be solved at all times. The abilities to talk to oneself and then to think to oneself are integral features of a mindful culture. This culture has developed into a rich and intricate scaffolding that provides both the stability needed for effective action and the flexibility for innovation in problem solving. The problems of survivability and viability remain, but new problems emerge with symbolic culture. When deliberate or mindful activity is undertaken, individuals find myriad possibilities open to them. The selection of which trajectory to take, however, is no easy task. One may choose carefully or poorly—that is to say one can be more or less intelligent.

4 On the Origins of Intelligence: Cooking as Consciousness

> To see the organism *in* nature, the nervous system in the organism, the brain in the nervous system, the cortex in the brain is the answer to the problems which haunt philosophy. And when thus seen they will be seen to be *in*, not as marbles are in a box but as events are in history, in a moving, growing never finished process. (John Dewey (1981–1991 [1925/LW1]: 224))

In light of my reconstructions of experience, mind, and culture, it is appropriate to ask where intelligence fits in. Specifically, we can ask, where does intelligence originate? To be sure, there is no precise moment when experience, mind, culture, or any other biological trait first appeared. All products of evolution slowly emerge from other products and processes. As with speciation, there may be no clear speciation mark as it is happening, but once it has happened we can distinguish *retrospectively* between two separate species. I submit that intelligence originates with the ability to *retrospect* on and to *evaluate* one's experience in order to ameliorate one's transactions with one's environment.

Retrospection fits with my conception of experience in that it results from earlier non-retrospective transactions. From these non-retrospective transactions, social organisms evolved to begin cooperating to solve their problems. As familiarity with specific problems grew in these social groups, the means for looking back on how each instance of a problem came about were being laid. That is, in order to ask why or how some event came about, one must have ready at hand the details about the event's sequence. Being able to articulate such a sequence requires not only communication of the details but also symbolization of the sequence and the sub-events. The symbolization affords comparison to other tokens of a similar type. Comparison and reflection are retrospective. This is the first step in intelligent action.

The second step is to look forward. To get stuck in the past is detrimental to successful activity in the present. The similarities and continuities between past and present are important sources of information. They afford us opportunities to act. But without consideration of the present situation and how present or near future action will adjust the organism-environmental situation—i.e., how action could change our experience—no evaluation of which course would be better or worse is possible. What guide such an evaluation are the ideals the decision-making entity (the community or the individual) holds. Intelligent activity, then, requires both the recognition of what has been and is the case and the imagination of what could be the case. In both recognition and imagination, however, the mental life, while reflective, is not inert: it is tied to activities undergone, undergoing, and to be undertaken.

This pragmatist view I have put forth has so far drawn heavily on classical sources, especially the writings of John Dewey. His view, particularly, anticipated much of the current work in dynamical systems, and in enactive, embodied, and extended mind theories. The pragmatist standpoint, however, has more to offer than these historical roots. By way of concluding, I draw on contemporary scientific research to distinguish a neuropragmatic perspective on the nature of consciousness.

As Alva Noë has rightly argued (2009), the orthodoxy of cognitive science today is that the mind is the brain. This is expressed by an analogy with digestion. That is, just as digestion is what the gut does, the mind is what the brain does. Noë appeals to impressive data that such an analogy between mind and digestion (as functions of brain or gut, respectively) is misleading. To account for the richness of mental life, Noë rightly argues that the body is as important as the brain, especially when we consider the interactions between the two. Finally, Noë notes that just as brains cannot be so easily extracted from their bodies, bodies are not so easily detached from their environments. To aid in distinguishing his position from the cognitive science orthodoxy, Noë offers the metaphor that consciousness is like dancing. The core of this metaphor is that dancing is something we do and that consciousness is no different. Digestion is something that just happens inside of us when we happen to ingest food. There is a degree of automaticity that Noë finds problematic for thinking about consciousness in this way. On Noë's account, consciousness is not the sort of thing that just happens. It takes work that involves the "nexus of brain, body, and world," as so many of the enactive, embodied, and extended mind theorists like to say.

Despite the initial affinities between what is conveyed by Noë's metaphor and the pragmatist view I have sketched above, there are points of disagreement. First, the emphasis on dancing seems to neglect the important role the brain does play in conscious activity. Second, dancing is something that requires a minimum of an environment. A person can dance wherever there is a floor or something on which or with which to dance. The aesthetic and bodily aspects of dancing are well appreciated from a pragmatist standpoint. However, there is a delicate balance between brain, body, and world that the dancing metaphor fails to capture.

I propose that a better metaphor for thinking about conscious activity is cooking (Solymosi 2011). Unfortunately, this metaphor, like the other two, is a bad metaphor because it needs explaining. The digestion metaphor has a couple of kernels of truth to it. These include the recognition that a specific bodily system is primarily, though not exclusively, focused on the process and that this specific process is biologically adaptive. Of course, the limitations of the digestion model are clear: not only is conscious activity far more dynamic than the digestion model suggests, so is digestion, which is a dynamically active and complex process. Another limitation is that the digestion model implies that conscious activity is strictly biologically adaptive. This simply is not the whole story, for conscious activity is also culturally adaptive. Dancing is an improvement because it brings in the body and the world, albeit minimally. Cooking, I believe, captures these positive aspects of digestion and dancing because cooking, from an evolutionary and ecological perspective, is the bodily extension of digestion into the environment.

This way of thinking about consciousness, moreover, fits the empirical data quite well. The biological and subsequent cultural changes that cooking had on our brains and bodies are substantial (Laland et al. 2000: 140a; and Power and Schulkin 2009: 68–89). Notably, as our brains grew larger, the caloric demands required by a larger brain came at a cost to digestive tissue. That is, as our brains grew larger, our gastrointestinal tracts grew smaller. The nutritional requirements once provided by the work of a longer GI tract were nevertheless met. This work seems also to have been done through the advancements in tools and fire maintenance, i.e., by cooking. In breaking down animal and plant material before ingestion, human digestion is a process that begins deliberately and actively outside the body. It begins through the bodily activities of many individuals working together in a community.

From the gathering of the raw materials to their preparation, significant communication must go on between people. This communication is not simply for getting the immediate job done; it is also for passing on the skills to the next generation. This technological capacity of problem solving with various degrees of flexibility is at once neural and anthropological. The advances in our growing understanding of mirror neuron systems are detailing the neural means of learning by observing and mimicking what others do (see Cozolino 2006; Franks 2010; and Solymosi 2012b). Anthropologically, what seems to be going on is that learning occurs by apprenticeship. Kim Sterelny has done an impressive job synthesizing several fields of inquiry, from evolutionary biology to archeology and anthropology to cognitive science, to argue that what most distinguishes humans from other primates is how we construct our environments to encourage learning via apprenticeship

(Sterelny 2012).[11] Briefly, what appears to be happening is that mirror neuron systems are at work while the young of a community interacts with their parents and other elders. In short, this sort of transaction—of giving and taking with regard to specific skills and how to improve upon them—is at the heart of experience, just as the classical pragmatists claimed. Or as Brandom suggested as their motto: "No experience without experiment."

That cooking is an ongoing experiment in extending digestion beyond the body is easily apparent to anyone who has tried to cook (success is not required) or to anyone who has tried a novice's attempt at a dish or an avant-garde chef's latest triumph. But we should not underestimate the power of the metaphor because of our contemporary conceptions of cooking as something isolated in a kitchen. For most of human history, the preparation of food was a communal activity that required the participation of many individuals. Through our technological advances, we have created an infrastructure that distributes so much of that work that a person in a first-world nation need only use a microwave, pick up a phone, or walk down the street to easily acquire a meal. Some of us, however, may remember a time from our childhood where learning how to cook a specific meal or a style of cuisine specific to one's ethnic background was simply something a family did. The symbolism, stories, and recipes that are passed down are indicative of tradition that extends beyond nutrition.

Conscious activity is something we do through our brains, bodies, and cultures. Each of us is born into a culture in which numerous affordances are already present to provide opportunities for action (see Gibson 1979, and Chemero 2009). These affordances are not only physical, like a ground suitable for bipedal walking, nor strictly biochemical, like a source of clean water; these affordances are also and emphatically cultural. Viewing cooking in a broad sense affords us an opportunity to consider the origins of our conscious and evaluative activities because we are all familiar with evaluating our food. From simply not liking the taste to unfortunate late nights with stomach pain (and worse) to disapproving of the effects of certain diets on our waistlines, our animals, and our environment, we are able to select better and worse ways to eat. This is the mark of intelligence that Dewey hoped more and more people would strive toward. Since the pragmatists always sought to dissolve dualisms, the question of the origins of mind in nature is perhaps better considered in light of the origins of intelligent behaviors. Such a shift in attention requires us to draw on several scientific perspectives, from the neurobiological to the anthropological. A pluralistic view on cooking is a powerful analog to how we should consider the nature of conscious activity. Such a perspective offers promising and productive answers to a central philosophical question: What are the sorts of things we can deliberately do to effect richer experiences—conceived as educational and experimental organism-environment transaction—for ourselves and others,

[11] Bill Bywater's recent work on synthesizing Dewey and Goethe (see Bywater unpublished manuscript), and pragmatism with the work of Sterelny 2012 (see Bywater 2012), further corroborates the view put forth here.

today and tomorrow?[12] Pragmatists like James and Dewey held that the best hope we have is to reconstruct intelligently our old ideas and beliefs in light of the best science of our day. To achieve such reconstructions, we must not settle for simply experiencing the world in a passive and disinterested fashion. We must engage it experimentally so that we may not only learn about how the world is but also how the world could be.

References

Andersen, R. (2012, April 23). Has physics made philosophy and religion obsolete? *The Atlantic*. Available online at: http://www.theatlantic.com/technology/archive/2012/04/has-physics-made-philosophy-and-religion-obsolete/256203/. Accessed 1 May 2012.

Brandom, R. B. (2004). The pragmatist enlightenment (and its problematic semantics). *European Journal of Philosophy, 12*(1), 1–16.

Bywater, B. (2012, March 15–17). *Neuropragmatism's pedagogy*. Presentation at annual meeting of the society for the advancement of American philosophy. Fordham University, New York.

Bywater, B. *The* Bildung *tradition: From Dewey through Goethe to apprenticeship as a new habit of whiteness*. Unpublished manuscript.

Chemero, A. (2009). *Radical embodied cognitive science*. Cambridge, MA: MIT Press.

Cozolino, L. (2006). *The neuroscience of human relationships: Attachment and the developing social brain*. New York: W. W. Norton.

Dennett, D. C. (1991). *Consciousness explained*. Boston: Little, Brown.

Dewey, J. (1996). *The collected works of John Dewey, 1882–1953: The electronic edition* (L. A. Hickman, Ed.). Charlottesville: InteLex Corporation.

Dewey, J. (1969–1972). *The early works of John Dewey, 1882–1898* (5 vols., Jo. A. Boydston, Ed.). Carbondale: Southern Illinois University Press.

Dewey, J. (1976–1988). *The middle works of John Dewey, 1899–1924* (14 vols., Jo. A. Boydston, Ed.). Carbondale: Southern Illinois University Press.

Dewey, J. (1981–1991). *The later works of John Dewey, 1925–1953* (17 vols., Jo. A. Boydston, Ed.). Carbondale: Southern Illinois University Press.

Franks, D. (2010). *Neurosociology: The nexus between neuroscience and social psychology*. New York: Springer.

Gibson, J. J. (1979). *The ecological approach to vision perception*. Boston: Houghton-Mifflin.

Godfrey-Smith, P. (2003). *Theory and reality: An introduction to the philosophy of science*. Chicago: University of Chicago Press.

Griffiths, P. E., & Gray, R. D. (2001). Darwinism and developmental systems. In S. Oyama, P. E. Griffiths, & R. D. Gray (Eds.), *Cycles of contingency: Developmental systems and evolution* (pp. 195–218). Cambridge, MA: MIT Press.

Hickman, L. A. (2007a). Some strange things they say about pragmatism: Robert Brandom on the pragmatists' semantic 'mistake'. *Cognitio, 8*(1), 93–104.

Hickman, L. A. (2007b). *Pragmatism as post-postmodernism: Lessons from John Dewey*. New York: Forham University Press.

James, W. (1977). Does consciousness exist? In J. McDermott (Ed.), *The writings of William James*. Chicago: University of Chicago Press. (Original work published 1904)

[12] To put the question in traditional philosophical parlance: "How ought we act in order to live a good life (*eudaimonia*)?"

Krauss, L. M. (2012, April 27). The consolation of philosophy. *Scientific American*. Available at: http://www.scientificamerican.com/article.cfm?id=the-consolation-of-philos. Accessed 1 May 2012.

Lakoff, G., & Núñez, R. (2001). *Where mathematics comes from: How the embodied mind brings mathematics into being*. New York: Basic Books.

Laland, K. N., et al. (2000). Niche construction, biological evolution, and cultural change. *Behavioral and Brain Sciences, 23*, 131–175.

Lovejoy, A. O. (1908a). The thirteen pragmatisms I. *The Journal of Philosophy, Psychology and Scientific Methods, 5*(1), 5–12.

Lovejoy, A. O. (1908b). The thirteen pragmatisms II. *The Journal of Philosophy, Psychology and Scientific Methods, 5*(2), 29–39.

Noë, A. (2009). *Out of our heads: Why you are not your brain, and other lessons from the biology of consciousness*. New York: Hill and Wang.

Peirce, C. S. (1992). The fixation of belief. In N. Houser and C. Kloesel (Eds.), *The essential Peirce: Selected philosophical writings, volume 1 (1867–1893)* (pp. 109–123). Bloomington/Indianapolis: Indiana University Press. (Original work published 1877)

Pigliucci, M. (2008). The borderlands between science and philosophy: An introduction. *The Quarterly Review of Biology, 83*(1), 7–15.

Power, M. L., & Schulkin, J. (2009). *The evolution of obesity*. Baltimore: The Johns Hopkins University Press.

Rockwell, W. T. (2005). *Neither brain nor ghost: A nondualist alternative to the mind-brain identity theory*. Cambridge, MA: MIT Press.

Schulkin, J. (2003). *Rethinking homeostasis: Allostatic regulation in physiology and pathophysiology*. Cambridge, MA: MIT Press.

Schulkin, J. (2009). *Cognitive adaptation: A pragmatist perspective*. Cambridge/New York: Cambridge University Press.

Schulkin, J. (2011a, January). Social allostasis: Anticipatory regulation of the internal milieu. *Frontiers in Evolutionary Neuroscience, 2*, 1–15.

Schulkin, J. (2011b). *Adaptation and well-being: Social allostasis*. New York: Cambridge University Press.

Solymosi, T. (2011). Neuropragmatism, old and new. *Phenomenology and the Cognitive Sciences, 10*(3), 347–368. September 2011.

Solymosi, T. (2012a). Pragmatism, inquiry, and design: A dynamic approach. In L. S. Swan, R. Gordon, & J. Seckbach (Eds.), *Origin(s) of design in nature: A fresh, interdisciplinary look at how design emerges in complex systems, especially life* (pp. 143–160). Dordrecht: Springer.

Solymosi, T. (2012b). Can the two cultures reconcile? Reconstruction and neuropragmatism. In J. Turner & D. Franks (Eds.), *The handbook of neurosociology* (pp. 83–98). Dordrecht: Springer.

Sterelny, K. (2012). *The evolved apprentice*. Cambridge, MA: MIT Press.

Weinberg, S. (Ed.). (1992). Against philosophy. In *Dreams of a final theory* (pp. 166–190). New York: Pantheon.

Weiner, P. P. (1973). *Dictionary of the history of ideas, Vol. III* (p. 552). New York: Scribner's Sons.

Not So Exceptional: Away from Chomskian Saltationism and Towards a Naturally Gradual Account of Mindfulness

Andrew M. Winters and Alex Levine

Abstract It is argued that a chief obstacle to a naturalistic explanation of the origins of mind is human exceptionalism, as exemplified in the seventeenth century by René Descartes and in the twentieth century by Noam Chomsky. As an antidote to human exceptionalism, we turn to the account of aesthetic judgment in Charles Darwin's *Descent of Man*, according to which the mental capacities of humans differ from those of lower animals only in degree, and not in kind. Thoroughgoing naturalistic explanation of these capacities is made easier by shifting away from the substance-metaphysical implications of the search for an account of *mind*, toward a dispositional account of the origins of *mindfulness*.

1 Introduction

The term 'naturalism' has been variously used and misused. For most purposes, the provisional definition proposed by Owen Flanagan et al. will serve well enough, enshrining naturalism as "a view of the world, and of man's relation to it, in which only the operation of natural (as opposed to supernatural or spiritual) laws and forces is admitted or assumed" (Flanagan et al. 2007, 1).[1] But of course this definition simply offloads any ambiguity in 'naturalism' onto 'natural'. In the spirit of David Hume's "Of Miracles" (1999, 169–186), we prefer to take naturalism as a methodological "no-miracles" principle. On this principle, we must assume that, for the most part, things do not happen without antecedent. In the absence of some compelling reason to think otherwise, every event or process in the world must be assumed to have an explanation consistent with the natural order of things. When novelty arises,

[1] We are grateful to Jared Kinggard for alerting us to this text. See Kinggard (2010).

A.M. Winters (✉) • A. Levine
University of South Florida, Tampa, FL, USA
e-mail: wintersandrewm@gmail.com; alevine@cas.usf.edu

L. Swan (ed.), *Origins of Mind*, Biosemiotics 8, DOI 10.1007/978-94-007-5419-5_15,

as it occasionally does, novel processes and events must be assumed (again, in the absence of compelling evidence to the contrary) to have antecedents. Nothing arises *ex nihilo*.

This chapter sets out from this same assumption, applied specifically to the origins of mind. Let us suppose that there was a time in the distant past when the universe was devoid of minds, whereas now it is replete with them. When and how did minds come about, and what were their antecedents? A similar question can also fruitfully be posed about any *particular* mind, viz., when and how did *my* mind come about, and what were *its* antecedents? Both questions concern the origins of mind, though on very different timescales. Events on the geological and evolutionary timescales of the first question must set the boundary conditions for addressing the second. Both timescales have been the subject of fruitful philosophical intervention, as has the intersection between the two (see e.g., various contributions to Oyama et al. 2003).

In this chapter, we are specifically concerned with the origins of mind on the evolutionary or geological timescale, as opposed to the historical or developmental. We begin by discussing two related problems that an account of the evolution of mind must overcome: human exceptionalism and dogmatic saltationism. In overcoming these problems, we are guided by the work of Charles Darwin (1859, 2004). Darwin was careful to avoid both of these problems. Like Darwin in the *Descent of Man*, we will focus on the origin of one particular aspect of what organisms with minds are disposed to do—to make aesthetic judgments. Judgment begins with discrimination, the capacity to respond differentially, not to different stimuli so much as to different interactive environments. Whereas stimuli only require a one-way interaction, in which a subject responds not differentially but passively to some causal influence, interactive environments require a two-way interaction between an organism and its environment, which may include other organisms. At some point along what Robert Campbell and Mark Bickhard (1986) call the "macroevolutionary sequence" in the emergence of cognition, this capacity gives rise to what we will call, for lack of a better phrase, *mindfulness*: the organism's further capacity to partition the space of its possible interactive environments and to enact preferences for some potential environments over others (see Levine 2011). For this reason, our discussion will have more to do with the origins of mindfulness than the origins of minds, traditionally conceived.

The diverse implementations of this capacity for aesthetic judgment across the animal kingdom evince numerous differences in degree across multiple dimensions. The macroevolutionary emergence of aesthetic judgment is thus likely to provide a story of the emergence and accumulation of such differences in degree. Such a story challenges deeply held convictions about the uniqueness of human mindfulness. Whatever the merits of these convictions, we argue they have nothing to do with the evolutionary origins of the human mind. In studying the latter, we are drawn to the continuity between human judgment and mindfulness and the capacities of all organisms capable of differentiating and choosing among potential interactive environments.

2 Human Exceptionalism

A standard early modern exemplar of human exceptionalism is the work of René Descartes. Descartes was a pioneer in the naturalistic explanation of many elements of human and animal cognition and perception, formulating mechanistic hypotheses on numerous aspects of human and animal anatomy, physiology, and behavior. Yet, notoriously, he was inclined to resist any analogous explanation of human thought and language. "What brings it about that beasts do not speak," he asserted, "is that they have no thought, and not that they lack the organs for it" (Descartes 2000, 276).[2] Though human eyes are structurally and doubtless functionally similar to bovine eyes, human minds are fundamentally different from bovine minds (if cattle can be said to have minds at all). For someone like Descartes, humans are thus partially removed from nature, and the origins of human minds are removed from the natural order of things. To be fair, it should be noted that the question of the origins of mind or mindfulness did not exist for Descartes in the sense in which it presents itself to us now.

In the contemporary context, advocates of human exceptionalism typically at least attempt to evoke naturalism. A good example is Noam Chomsky, for whom

> ...there is surely no reason today for taking seriously a position that attributes a complex human achievement entirely to months (or at most years) of experience, rather than to millions of years of evolution or to principles of neural organization that may be even more deeply grounded in physical law—a position that would, furthermore, yield the conclusion that man is, apparently, unique among animals in the way in which he acquires knowledge. Such a position is particularly implausible with regard to language.... (Chomsky 1965, 59)

The position that Chomsky is rejecting, which he elsewhere (Chomsky 2009) calls "empiricism," in contradistinction to his own aptly named "Cartesian linguistics," treats a human infant's first language acquisition as a learning process in which general-purpose rules are applied to data. Empiricism fails, Chomsky argues, to account for the rapidity and efficiency of nearly all human language acquisition, especially given the "poverty of the stimulus" the infant has at his or her disposal.

The merits of his arguments need not concern us here. What is of interest is the surprising, or at any rate ironic fact that "the conclusion that man is, apparently, unique among animals in the way in which he acquires knowledge" also falls neatly out of the Chomskian view that Generative Grammar is innate to humans and only humans. In his recent introduction to the third edition of *Cartesian Linguistics*,

---

[2] We are grateful to Christine Wieseler for alerting us to the source of this observation, a letter by Descartes to the Marquis of Newcastle, November 23, 1646. Later in the same text Descartes allows, "if they [animals] thought as we do, they would have an immortal soul as we do" (Descartes 2000, 277). But this conclusion is unacceptable if one aims to provide a purely naturalistic explanation.

James McGilvray acknowledges the Chomskian commitment to a kind of human exceptionalism.

> If much of the mental machinery needed to develop concepts and their combinatory principles is innate and one is going to try to explain how it comes to be in the mind at birth, it won't do to say that God put it there (Descartes) or to construct myths of reincarnation (Plato). The only course open to us is to look to biology and those other natural sciences that can say what an infant human begins with at birth and how what s/he is born with develops. And taking that tack also makes it possible to at least begin to speak to the question of how human beings came to have apparently unique machinery in the first place—to address the issue of evolution. (Chomsky 2009, 18)

The project McGilvray has articulated at first appears to have an eminently naturalistic aim, that of providing a biological explanation of "how human beings came to have apparently unique machinery in the first place." But thus articulated, the project does not offer any support for the uniqueness of human machinery beyond its brute apparentness.

Such an assumption requires justification. To be sure, the animal kingdom is diverse, with the members of every taxon in the Linnean hierarchy exhibiting all sorts of morphological and physiological differences from members of other taxa. But while it is surely true (and trivially so) that only humans speak human language,[3] this does not make the cognitive machinery subtending this fact unique in any especially interesting sense. Alone among Ursids, the Panda possesses an enlarged metacarpal (the Panda's "thumb"; Gould 1992) that allows it to grasp stalks of bamboo; yet this appendage is clearly a *metacarpal*, homologous with every other mammalian metacarpal. Thanks in part to Chomsky, there is a widespread conviction that, as Steven Pinker puts it,

> The discrete combinatorial system called "grammar" makes human language infinite (there is no limit to the number of complex words or sentences in a language), digital (this infinity is achieved by rearranging discrete elements in particular orders and combinations, not by varying some signal along a continuum like the mercury in a thermometer), and compositional (each of the infinite combinations has a different meaning predictable from the meanings of its parts and the rules and principles arranging them). (Pinker 2007, 342)

Inquiring with the requisite degree of care into whether human language actually has all three of these features, and if so, whether they (severally or jointly) are *unique* to human language, would go well beyond the scope of this chapter. Our point here is that the uniqueness of human language thus described is not *self-evident*. As Andy Clark has argued (1992), our willingness to take this uniqueness as given is surely in part an artifact of our experience with *written* language, which clearly involves the explicit, quasi-recursive manipulation of discrete symbol tokens. But by our best estimates, written language is no more than 6,000 years old. This would suggest that written language arose much later than the onset of anatomically

---

[3] This ignores, for the moment, the many fascinating attempts to teach such languages to nonhumans, of which arguably the most successful have involved not primates, but *birds* (see Pepperberg 2002).

modern humans (c. 200,000 years ago). For this reason, written language is better understood as a product of historical or cultural achievement rather than of evolution. Whether, and to what degree, the capacity to become literate is subtended by the same evolved capacities that allow us to acquire spoken language (as opposed, say, to the evolved capacities that make us such prodigious tool users) ought to be an empirical question.

We have no basis for asserting that every variety of human exceptionalism need necessarily violate naturalist strictures. We also take it that the consistency of Chomskian linguistics with the data and theory of human evolution is, or ought to be, an empirical question.[4] But the claim that this approach "makes it possible to at least begin to speak to the question of how human beings came to have apparently unique machinery in the first place" is somewhat misleading. If it could be shown that the cognitive machinery of human language or concept acquisition was *not* unique, or at any rate, that it differed from the machinery available to our nonhuman relatives only in degree, and not in kind, then the task of naturalistic explanation would be enormously simplified. Conversely, by committing himself to human uniqueness, or human exceptionalism, Chomsky has enormously complicated this same task. The resulting complications are especially troublesome when the naturalistic explanation of any biological structure or process requires some sort of evolutionary account. In constructing such an account, the human exceptionalist may be tempted toward *dogmatic saltationism*—to which we now turn.

3 Dogmatic Saltationism

Darwin was an evolutionary *gradualist*, convinced that on the whole the evolutionary process proceeded slowly by small increments. His corpus is replete with expositions of the gradualist doctrine; for our purposes, one classic example will suffice. Of "organs of extreme perfection," such as the mammalian eye, Darwin reasons:

> ...if numerous gradations from a perfect and complex eye to one very imperfect and simple, each grade being useful to its possessor, can be shown to exist; if further, the eye does vary ever so slightly, and the variations be inherited, which is certainly the case; and if any variation or modification in the organ be ever useful to an animal under changing conditions of life, then the difficulty of believing that a perfect and complex eye could be formed by natural selection, though insuperable by our imagination, can hardly be considered real. (Darwin 1859, 186)

In *Descent of Man*, as we shall see, Darwin employed similar arguments in defense of the gradual evolution of human mental faculties. On the modern synthesis in evolutionary theory, still broadly Darwinian in its outlines, very rapid evolutionary

[4] Though we have our doubts about whether it has been treated as an empirical question in the practice of comparative linguistics. If every time a new language is described that appears to violate one or another stricture of Generative Grammar, the community response is to tweak Generative Grammar to accommodate it, one begins to suspect a self-sealing argument.

change is possible when measured on the geological timescale. One way it can occur is by the "founder effect," in which a small (and thus inevitably nonrepresentative) sample of a larger population becomes geographically isolated, and gives rise to a daughter population in which the distribution of traits diverges significantly from that in the ancestor population. Such possibilities are acknowledged in Stephen Gould and Niles Eldredge's account of "punctuated equilibria" (Gould and Eldredge 1977). It must be conceded that these considerations lower the bar for an explanation of human exceptionalism consistent with evolutionary naturalism by allowing the possibility that unique human characters might have arisen suddenly (*saltationally*), but not miraculously.

They do not, however, entirely eliminate the difficulty. First of all, though Gould and Eldredge argue that speciation is often very fast, on the geological timescale, it does not occur overnight, at least not on the shorter "ecological" timescale (Gould and Eldredge 1977). In other words, speciation does not typically occur from one generation to the next.[5] Second, suppose that all of the species in a given clade, save one, lack a particular derived trait. The more complex the novel trait—the greater the number of evolutionary changes necessary to bring it about—the less likely it is to have arisen quickly in the ancestors of the outlier species. Conversely, while simpler derived traits are more likely to arise over shorter spans of geological time, the simpler a derived trait found in a particular species—the smaller the number of evolutionary changes necessary to bring it about—the more likely it is to arise independently in related taxa and to be found throughout the clade in question.

The human exceptionalist who wishes to explain human exceptionalism naturalistically thus faces a dilemma. This dilemma is illustrated by the fate of Generative Grammar in the decades since Chomsky (1965), a trajectory ably summarized by McGilvray. Initially,

> …accommodating a theory of language to biology…looked daunting. It was particularly hard to understand how the human genome could be expected to contain all the information needed to allow for any of a large number of languages while providing too for a way to choose between them. Even the most optimistic account of language universals at the time…would still demand that the genome carry a massive amount of language-specific information, more than any plausible account of evolution could plausibly explain. (Chomsky 2009, 29)

Faced with this challenge, those toiling in the Chomskian fields sought to simplify their task.

Fortunately, in the years following the 1965 publication of *Aspects of the Theory of Syntax*, "Different languages came to look less and less different." This insight led to the "minimalist program in the early 1990s," until finally,

> …very recently it has come to seem as if perhaps the sole 'operation' (rule, principle) needed to explain *both* basic structure and movement is what Chomsky and several others call "Merge." Oversimplifying…Merge is an operation rather like concatenation, putting

[5] Though it *can*—at least in plants, where allopolyploid speciation is possible. This occurs when a hybrid, which is capable of reproduction, is not capable of breeding with either of its parent species. See e.g., Soltis and Soltis (1989).

> items or elements (lexical items) together and creating a new item...Something like that is surely needed for there to be language at all, for all languages 'compose'—they make complexes called "sentences" out of "words." (Chomsky 2009, 29)

Several observations are in order. First, if the innate endowment by virtue of which humans are capable of acquiring language is confined to an operation like "Merge," then language acquisition has come to resemble the kind of learning process an empiricist might well endorse. (Concatenation is a general-purpose tool, after all.) But this is the very sort of position that Chomsky set out to reject.

Second, as noted above, if the emergence of the language faculty was made possible primarily by the evolution of a rudimentary cognitive capacity for concatenation or by the evolutionary refinement of a prior capacity, similar capacities would be likely to be found among our close nonhuman relatives. A simple change that can arise once can also arise more than once when given enough time. But this, too, undermines the uniqueness that Chomskians attribute to human cognition.

Third, it strikes us that the cognitive capacity for putting things together to form novel wholes *is* widespread among our close nonhuman relatives and we would not be surprised to find it widespread throughout much of the animal kingdom. To save human exceptionalism one would have to deny this—on pain of replacing human exceptionalism with mere human speciesism. This forces the human exceptionalist to take recourse to *dogmatic saltationism*:

> ...if...Merge alone is 'contained' in the genome, it becomes much easier...to explain how language could have come about as the result of a single mutation. It need not be a "language specific" mutation; it could, for example, be a side result...It must, though, be 'saltational'—happen in a single jump—for otherwise we would have to suppose that language developed over millennia, and there is no evidence of that. (Chomsky 2009, 34)

McGilvray dates the "single jump" to between 200,000 and 50,000 years ago (between the advent of anatomically modern *H. sapiens* and the migration out of Africa), though not on any especially specific or persuasive grounds. Something more, however, needs to be provided to account for the development of language since other early hominins made it out of Africa for which we lack any evidence suggesting that they developed language.

Following evidence and arguments adduced by Richard Wrangham and others (Carmody and Wrangham 2009; Wrangham 2010), it strikes us as at least as likely that characteristically human language evolved in concert with cooking, perhaps as long as 1.9 million years ago and perhaps over a period of a several hundred thousand years. But were the assumption of evolutionary saltation to be dropped, Chomsky's human exceptionalism would be left without any consistently naturalistic evolutionary ground. This is why we call it *dogmatic saltationism.*

A dogmatic gradualism would be just as bad. But as Darwin was at pains to argue in Ch. 3–5 of *The Descent of Man* (Darwin 2004), every one of the "mental powers" often cited as the sole province of humans may be found among other animals. If he is right, then at least with regard to these traits, gradualism is warranted. We now turn to discuss one of these powers that may at one time have been thought to belong only to humans, thereby further garnering support for gradualism.

4 Darwin on Aesthetic Judgment

Like such contemporaries as Max Müller, Darwin also had a fair bit to say about language. After considering and dismissing a number of ways in which the linguistic faculties of humans might have been said to differ from the communicative faculties of other animals, he concludes, "The lower animals differ from man solely in his almost infinitely larger power of associating together the most diversified sounds and ideas; and this obviously depends on the high development of his mental powers" (Darwin 2004, 107–108). The difference between the mental abilities of humans and nonhuman animals is one of degree, not kind. Language depends on the capacity for association (for Hume and other empiricists, the basis of all reasoning and learning), and while smarter animals form more diverse and complex associations, many animals are capable of forming simple associations, even for purposes of communication. For the remainder of this chapter, however, we focus on a faculty of the mind even more important to understanding its evolutionary origins: the capacity for aesthetic judgment. Communication arises only among social animals. But sociality, in turn, is the prerogative of animals that reproduce sexually. In their reproductive projects, many of them are assisted by aesthetic judgment.

Perhaps the most succinct statement of Darwin's views on aesthetic judgment may be found in Ch. 3 of *The Descent of Man*:

> *Sense of Beauty*—This sense has been declared to be peculiar to man. I refer here only to the pleasure given by certain colors, forms, and sounds, and which may fairly be called a sense of the beautiful…When we behold a male bird elaborately displaying his graceful plumes or splendid colors before the female, whilst other birds, not thus decorated, make no such display, it is impossible to doubt that she admires the beauty of her male partner. As women everywhere deck themselves with these plumes, the beauty of such ornaments cannot be disputed. (Darwin 2004, 114–115)

This passage, occurring in a chapter entitled "Comparison of the Mental Powers of Man and the Lower Animals," is crucial to the whole project of Darwin's book. With its first seven chapters devoted to similarities between humans and other animals, the next 11 to sexual selection in nonhuman animals, and the final two to sexual selection among humans, the conclusion that sexual selection was central to Darwin's conception of "the descent of man" would be inescapable even to a reader content with only browsing the book's table of contents. Aesthetic judgment, or the sense of beauty, is in turn a necessary condition for sexual selection anywhere in the animal kingdom.

5 Implications and Advantages

By focusing on the evolutionary origins of human mindfulness, specifically in regards to the capacity for aesthetic judgment as a necessary condition for sexual selection, we are better able to recognize the continuity between humans and nonhuman animals. Since both humans and nonhuman animals formulate preferences that play a significant role in determining how they will respond to different interactive environments, including the selection of which environments they will respond to, both humans and nonhuman animals exhibit the capacity to partition the space of their possible interactive environments. Among the resources in these possible interactive environments are potential mates. For this reason, mate selection is itself an exhibition of mindfulness, and since aesthetic judgment is necessary for sexual judgment, it follows that there is a strong connection between aesthetic judgment and mindfulness.

In addition to recognizing the connection between aesthetic judgment and mindfulness to better understand the continuity between humans and nonhuman animals, a shift of the discussion of the origins of mind to the origins of mindfulness carries with it many benefits. The first of these has to do with the fact that the problematic character of the question concerning the *origins* of mind has its roots in discussions regarding the *nature* of mind. After all, one is tempted to say, understanding something's origin first requires understanding what that thing is. Discussions of the nature of mind, in turn, have typically focused on identifying the essence of mental *substance* (i.e., as material or immaterial). This approach, however, has fallen short of fulfilling the philosopher's expectations of an account of the nature of mind. We see this in Descartes' writings, in his attempt to explain how the immaterial mind can interact with the physical body. We also see this from the opposing end through attempts to account for how consciousness can arise from material substances (what David Chalmers has called the "Hard Problem"; Chalmers 1997). Without an adequate account of what the mind *is*, philosophers have not had the proper theoretical tools to begin pursuing the problem of the *origin* of mind. This has been a consequence of metaphysical presumptions that the mind is a substance in the first place, which has saddled the theorist with the task of resolving many untenable metaphysical debates for the sake of maintaining the initial presupposition. Rather than attempting to develop a strong metaphysics program around an initial assumption that seems to bring with it more problems than solutions, it may be advisable to recast the initial assumption.

In the case of shifting the focus of the origins of mind to the origins of mindfulness, we assume that the mind should be thought of in processual or dispositional, rather than substance, terms. We take mind to be the capacity to act in particular ways, but, as mentioned above, the term 'mind' is already loaded with substance-based terminology. For this reason, we prefer another term that highlights an organism's capacity for distinguishing among potential interactive environments. So, rather than thinking of mind in terms of something that an organism has, we take mind to be a description of an organism's interactive potential—the behaviors that an organism is disposed to exhibit. This shift from a substance-based view of mind to a dispositional view further highlights the additional benefits of moving the discussion of mind to one of mindfulness.

Specifically, a discussion of mindfulness of the kind we envision is not susceptible to the problems that arise with exceptionalist and saltationalist accounts.

To briefly review, mental exceptionalism is the view that the mental traits possessed by humans are different in kind from any found among nonhuman animals. As shown above, this view is problematic, since positing that humans possess any special trait different in kind from the traits that our nonhuman ancestors possess places a wedge in the naturalist explanations for our traits that evolutionary accounts provide. To suggest that humans possess any special mental trait, though, is to think of the mind in substance-based terms—in terms of Aristotelian essence. Shifting to the dispositional account of mind, in terms of what an organism has the capacity to do, allows us to recognize that the mental capacities exhibited by both humans and our nonhuman relatives exist on the same continuum. This removes the barrier that human exceptionalists place in the way of such naturalistic explanations as evolutionary theory affords.

Similarly, shifting to a dispositional account of mind overcomes the temptation toward saltationism. Since the discussion of mindfulness given here, especially regarding its connection to aesthetic judgment, highlights the continuum that exists between humans and nonhumans, there is no need to posit an account of sudden jumps in evolution to account for the differences in traits between humans and nonhumans. A further upshot for the dispositional account of mind is that rather than having to give up our account of mindfulness when presented with new biological evidence that further demonstrates that there may *not* have been such drastic jumps in the evolutionary chain as the saltationalist insists, which would thereby force the saltationalist to abandon some key features of her account, a proponent of mindfulness as discussed here would be able to use the new biological findings to further elucidate the continuum offered by the gradualist account of evolution. This is an outcome of the saltationalist requiring gaps in the evolutionary story for her position to be tenable, whereas the gradualism endorsed by our dispositional account of mindfulness welcomes the filling in of these gaps.

We believe there is an additional benefit gained by shifting to a dispositional account of mind in considering how the concept of mindfulness avoids both mental exceptionalism and saltationalism. In both cases, there is no need to appeal to anything like a miracle. In the case of the former, rather than believing that humans possess something exceptional beyond their nonhuman counterparts, which requires some additional evidence beyond the current biological data, the discussion of mindfulness allows us to see our abilities as having a similar developmental and evolutionary origin as other species that exhibit similar, although not exact, mental prowess. In the case of the latter, by understanding the differences between animals and nonhuman animals as one of gradation, there is no need to posit sudden developmental ruptures that do not have any antecedents. In other words, the account of mindfulness offered here allows us to offer antecedents for our capacities to differentiate and make judgments regarding potential interactive environments, thereby avoiding any appeals to miracles. For this reason, our account of mindfulness is consistent with naturalism.

References

Campbell, R. L., & Bickhard, M. H. (1986). *Knowing levels and developmental stages* (Contributions to human development). Basel: Karger.
Carmody, R. N., & Wrangham, R. W. (2009). The energetic significance of cooking. *Journal of Human Evolution, 57*, 379–391.
Chalmers, D. (1997). *The conscious mind: In search of a fundamental theory*. Oxford: Oxford University Press.
Chomsky, N. (1965). *Aspects of the theory of syntax*. Cambridge, MA: MIT Press.
Chomsky, N. (2009). *Cartesian linguistics: A chapter in the history of rationalist thought* (3rd ed.). Cambridge: Cambridge University Press.
Clark, A. (1992). *The presence of a symbol* (Reprinted in J. Haugeland (Ed.), *Mind Design II*. Cambridge: MIT Press, 1997). Cambridge, MA: MIT Press.
Darwin, C. (1859). *On the origin of species*. London: John Murray.
Darwin, C. (2004). *The descent of man* (2nd ed.). London: Penguin.
Descartes, R. (2000). *Philosophical essays and correspondence*. Indianapolis: Hackett.
Flanagan, O., Sarkissian, H., & Wong, D. (2007). Naturalizing ethics. In W. Sinnott-Armstrong (Ed.), *Moral psychology*. Cambridge, MA: MIT Press.
Gould, S. J. (1992). *The Panda's thumb: More reflections in natural history*. New York: W.W. Norton.
Gould, S. J., & Eldredge, N. (1977). Punctuated equilibria: The tempo and mode of evolution reconsidered. *Paleobiology, 3*(2), 115–151.
Hume, D. (1999). *An enquiry concerning human understanding*. Oxford: Oxford University Press.
Kinggard, J. (2010). *Rethinking ethical naturalism*. PhD thesis, University of South Florida, Tampa.
Levine, A. (2011). Epistemic objects as interactive loci. *Axiomathes, 21*, 57–66.
Oyama, S., Griffiths, P. E., & Gray, R. D. (Eds.). (2003). *Cycles of contingency: Developmental systems and evolution*. Cambridge, MA: MIT Press.
Pepperberg, I. (2002). *The Alex studies: Cognitive and communicative abilities of African grey parrots*. Cambridge, MA: Harvard University Press.
Pinker, S. (2007). *The language instinct: How the mind creates language* (3rd ed.). New York: Harper.
Soltis, D. E., & Soltis, P. S. (1989). Allopolyploid speciation in Tragopogon: Insights from Chloroplast DNA. *American Journal of Botany, 76*(6), 1119–1124.
Wrangham, R. W. (2010). *Catching fire: How cooking made us human*. New York: Basic.

Mental Organs and the Origins of Mind

Thomas S. Ray

Abstract I introduce a new hypothesis of the origin of complex mind through the emergence of "mental organs," populations of neurons that bear a specific G-protein-coupled receptor (GPCR) on their surface. Mental organs provide a direct connection between mental properties (compassion, comfort, awe, joy, reason, consciousness), and the genes and regulatory elements associated with GPCR. Mental properties associated with mental organs have heritable genetic variation and are thus evolvable. Mental organs evolve by duplication and divergence. Over three hundred different GPCR are expressed in the human brain, providing a genetic and regulatory system that allows evolution to richly sculpt the *mind*.

1 Mental Organs

> There is something fascinating about science. One gets such wholesale returns of conjecture out of such a trifling investment of fact. (Mark Twain – *Life on the Mississippi*)

The human heart, mind, spirit, and soul emerged through the same process that created all of life: evolution by natural selection. In order to understand how the mind evolved, we must understand how it is structured, and how its structure is tied to genes. Here, I propose that "mental organs" (defined as the population of neurons that bear a specific receptor on their surface, such as serotonin-7, histamine-1, alpha-2C) provide the structure and genetic mechanisms that allow evolution to sculpt the *mind*. It should be noted that mental organs currently hold the status of a hypothesis that I am proposing. Their existence remains to be confirmed by rigorous experimental methods. This new hypothesis about a fundamental organizational principle

T.S. Ray (✉)
Department of Biology, University of Oklahoma,
Norman, OK, USA
e-mail: tray@ou.edu

L. Swan (ed.), *Origins of Mind*, Biosemiotics 8, DOI 10.1007/978-94-007-5419-5_16,

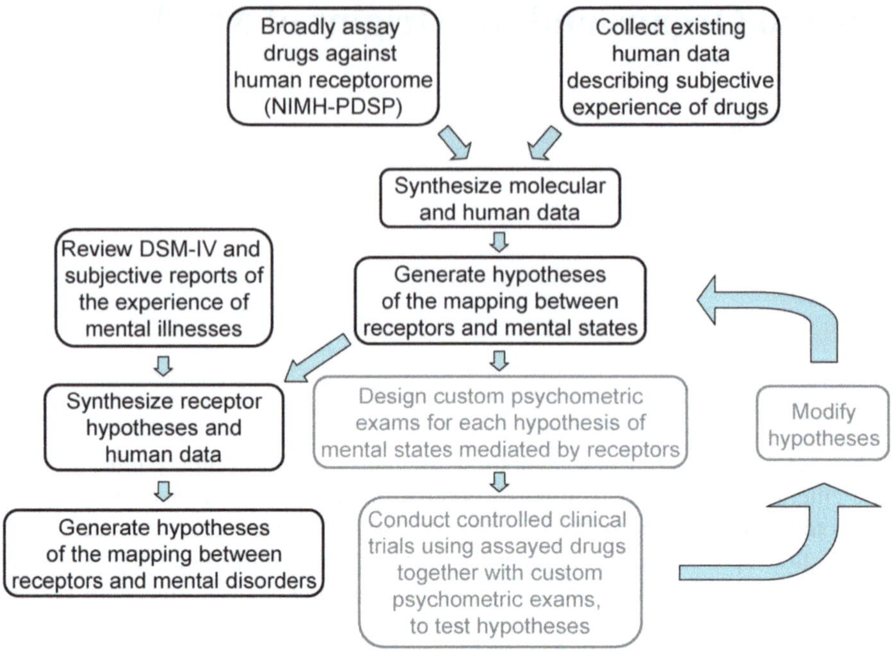

Fig. 1 The overall flow of the research method for discovering, characterizing, and utilizing human mental organs. NIMH-PDSP refers to the National Institute of Mental Health – Psychoactive Drug Screening Program. The three steps in gray in the lower right have not been attempted yet. These three steps are needed to test and refine the hypotheses of the mapping between receptors and mental states

of the mind emerges from my studies of the effects in humans, of drugs that selectively activate neurotransmitter receptors (Fig. 1).

The diverse set of psychoactive drugs collectively represents a rich set of tools for probing the chemical architecture of the human mind. These tools can be used to explore components of the psyche whose discreteness is normally obscured by their being embedded in the complete tapestry of the mind. By activating specific components of the mind, they are made to stand out against the background of the remainder of the psyche. Thus both their discreteness and their specific contribution to the psychic whole can be better appreciated. That the revealed mental elements can be pharmaceutically manipulated suggests that they may be naturally modulated through chemical systems. These receptor mediated mental components are the distinct elements from which the mind has been fashioned through evolution.

In this nontechnical chapter, I will present my findings on the nature of mental organs and the implications of their existence, without doing the heavy lifting of providing the supporting evidence. That technical work will be published elsewhere. Although I will name a dozen receptors, you do not need to know anything about them to follow my arguments. If you have some knowledge of psychopharmacology, then I must ask you to set that knowledge aside, to avoid confusion. The view of psychopharmacology that I present here is new and is not consistent with current paradigms (set aside what you may have heard about serotonin-2 and dopamine).

> It ain't what you don't know that gets you into trouble. It's what you know for sure that just ain't so. – Mark Twain

I ask the reader to suspend disbelief and allow me to present a new view of the mind that has tremendous coherence and explanatory and predictive power. The human mind is populated by mental organs, which play diverse roles within the mind. Some mental organs provide consciousness (in separate adult and childhood forms); others function as gatekeepers to consciousness (in long- and short-time scales); others give salience, meaning, or significance to the contents of consciousness, while others provide content to consciousness. Some mental organs support the facilities of language, logic, and reason, which appear to have arisen in the last hundred thousand years in humans. I will refer to language, logic, and reason simply as cognition. The facilities of cognition appear to be fully developed only in adult humans. The children we develop from and the animals we evolved from lack those facilities and yet have fully functional minds and are capable of making their way in the world. Other mental organs provide affective ways of knowing the world, through feeling alone, which provide the complete archaic mind in our developmental and evolutionary antecedents. Most mental organs have not yet been characterized.

I propose the following list of hypotheses concerning the mental functions mediated by different receptors:

- *Serotonin-7*: adult consciousness and creativity, holds both cognitive (language, logic, reason) and affective (feeling, emotion) content. What we are aware of: the present scene, fantasy, imagination, idea, theory, memory. The spark of creativity. Rather than creating a central theater of consciousness, may bestow the property of consciousness on other mental organs. When strengthened, can create a sense of sumptuousness, sparkle, grandeur, majesty, transcendence, something greater, cosmic, divine, god. As consciousness is strengthened, the contents of consciousness are rendered at higher resolution, become more tangible, and begin to be perceived as if through the five senses. At a critical point, we pass through a mental event horizon, as the contents of consciousness become more salient than actual reality. We mentally exit the actual space and time and enter a space and time created by the mind, within which the mind can create an alternate reality. At this point a mental big bang may occur. Consciousness is a generative system, capable of creating worlds, universes. This creative property may be the basis of free will.
- *Kappa*: childhood consciousness and creativity, holds only affective content. Pretty much everything said of serotonin-7 applies here, except that kappa is a purely affective system, so the contents of consciousness have a very different quality. Kappa consciousness creates a complex, subtle, and richly detailed representation of the world constructed exclusively from feelings.
- *Serotonin-1*: pure cognition: logic, reason, concepts, thought, language. Produces no feeling, can only be detected by engaging in cognitive tasks.
- *Serotonin-2*: dynamic filtering, inhibition, protection. Provides dynamic moment-to-moment selective filtering of access to consciousness, may focus attention.

Activation of serotonin-2 closes the gates to consciousness, while relaxation or inhibition of serotonin-2 opens the gates to consciousness. May be involved in integration.

- *Cannabinoid-1*: long-term filtering, inhibition, protection. The cannabinoid system probably coordinates with serotonin-2, to do on a long time frame, what serotonin-2 does dynamically. The cannabinoid system may operate through long-term potentiation of the filtering function of the serotonin-2 system. A mental immune system, one of whose functions is to provide selective long-term protection against the recurrence of intense mental states, whatever their etiology (spontaneous or drug induced), by selectively blocking access to consciousness. Another function of the mental immune system is to produce an evenly proportioned set of mental organs at maturity, by attenuating access to consciousness of over-expressed mental organs. As we mature, the cannabinoid system gradually, progressively, and permanently (at least for years) blocks access to consciousness of many systems, particularly the affective mental organs.
- *Sigma*: our heart and soul, the core of our being, the core sense of self. Apparently a purely affective domain. The seat of the basic emotions (anger/rage, fear, happiness, sadness, surprise, and disgust). The seat of biographical affective memory. Very sensitive to pleasure and pain. Needs the protection of the serotonin-2 and cannabinoid systems. A strong sense of self. Completely genuine, sees the affectations, façades, and masks that people wear, while putting on none of its own. Manifests innocence, honesty, integrity, and is uncorrupted but also is uncivilized, selfish, hedonistic, and emotional. Intimately connected to the body. May be capable of causing psychosomatic problems such as chronic pain.
- *Mu*: sense of comfort, security, protection; dissipation of pain, hunger, tension, anxiety, frustration, fear, anger, and aggression. A primary role may be the pacification of the fetus and early infant.
- *Beta*: a sense of home, family, community, society, humanity, and human nature that shows as wisdom and may provide a moral compass in human affairs; the sense of happiness, joy, elegance, luxury; the feeling of a fine brandy; the feeling of the season when all the fruits ripen; the feeling of the bustle in the street; the feeling of the smoke from the chimney when dinner is cooking; the joy of cooking. The sense of aesthetics.
- *Imidazoline*: compassion, forgiveness (of others or of one's self; not the concept or gesture of forgiveness, but true letting go in one's heart of anger, grudge, guilt, or shame), healing (letting go of psychological burdens may heal psychosomatic illness), open-hearted tenderness, altruism, empathy, platonic love.
- *Alpha-1*: the sense of place, scene, context. The sense of the unfolding, coherence, continuity, liveliness, and vitality of a scene. The sense that the scene and the entities that populate it extend in space and time, beyond what we directly perceive (it continues behind walls, around corners, and tomorrow). Likely fundamental to the emergence of our sense of reality.
- *Alpha-2*: the sense of the essence or soul of *things* (material objects). Rasa (Sanskrit): "Capturing the very essence, the very spirit of something, in order to evoke a specific mood or emotion in the viewer's brain" (Ramachandran 2007a).

Activation of alpha-2 may provoke recall of (predominantly childhood) memories stored in alpha-2 format. The sense of aesthetics.

- *Histamine*: affective theory of mind (ToM), constructs a persistent representation of the affective domain (heart and soul) of close relations, such as close family members (but also works for nonfamily). ToM is not exclusively constructed on the fly. For each person, we build a model of his or her affective domain, which is stored and refined with each interaction. For close relations, it accumulates a complete detailed model, or representation, of their affective domain. We hold their heart and soul within ours, even after they have died. The more we interact with them, the more completely we hold them. Extraordinary sexual sensibility. The sense of aesthetics.
- *Dopamine*: salience, meaning, significance, insight, integration, deep emotions, and moods (both positive and negative); awe, certainty, religious sentiment; the sense of aesthetics. Establishes the significance of mentation and in this way may modulate the influence of mentation on behavior. Able to associate feeling with thought, making us passionate about ideas.

Each mental organ mediates a domain of human experience with great depth and breadth. I have described each one with a few words, which fall within the domain, but which do not begin to convey the richness, depth or breadth of the mental domain mediated by each organ.

Mental organs are a fundamental organizational property of the human brain and the mind that emerges from it. When we think of brain anatomy, we think of structures like the frontal lobes, cerebral cortex, cerebellum, thalamus, limbic system, pons, and Broca's area. Mental organs are another form of brain anatomy that is less visible to the naked eye but which underlies a no less fundamental relationship to the organization of the mind.

Individual mental organs are real physical entities, just like hearts and lungs, but they have distinctive topological properties because they are composed of populations of neurons woven into networks. The population of cells that make up a mental organ would probably be compatible with definitions of "tissue" based on patterns of gene expression, in that they express the gene for the corresponding receptor. All mental organs identified so far are associated with receptors in a single gene family, the G-protein-coupled receptors (GPCR).

Mental organs do not necessarily have the physical cohesiveness that we associate with conventional organs, such as the liver or kidneys. It is theoretically possible for one neuron to be a component of more than one mental organ or for a mental organ to consist of a dispersed population of neurons, none of which makes any contact with the other neurons of the organ.

On the other hand, the population of neurons composing a mental organ could have all their cell bodies clustered together as is found in the raphe nuclei, a cluster of neurons that release serotonin. However, mental organs are not defined by what neurotransmitter they release, but rather by what kind of neurotransmitter receptor they bear on their surface. We can imagine a different mental organ associated with each of the hundreds of different kinds of modulatory receptors (GPCR). The mental organs associated with different receptors may be anatomically separated or interwoven.

2 Consciousness

2.1 Theater and Gates

Collectively, the mental organs form the apparatus of consciousness. Consciousness renders that which we are aware of. It is a kind of mental space where a representation is created. This might be a representation of the present scene, or it might be a body sensation, a fantasy, a memory, a vision of the future, a feeling, an idea, etc. Consciousness is a complex phenomenon, and the participating mental organs play a variety of roles. Bernard Baars describes consciousness as a theater, with a stage of working memory, a spotlight of attention, context operators (director, spotlight controller, local contexts), players (outer senses, inner senses, ideas), and an unconscious audience (memory systems, motivational systems, interpreting conscious contents, automatisms) (Baars 2001).

Serotonin-7 (or kappa) may provide the stage upon which other mental organs can perform. Serotonin-2 and cannabinoid can be seen as the director. Dopamine can be a spotlight controller. The ways of knowing (serotonin-1, histamine, beta, alpha-1, alpha-2) can be the players. Sigma can be a part of the unconscious audience (Fig. 2).

The theater metaphor suggests that the mentation produced by various mental organs enters into a mental space produced by the organs of consciousness. Let us examine this for the specific example of beta and serotonin-7. Beta produces the sense of home, family, community, and the joy of life. However, activation of beta does not cause a subject to experience these feelings unless they enter consciousness. In order to enter consciousness, the feelings produced by beta must pass through the gates mediated by serotonin-2 and cannabinoid (Fig. 3).

For some subjects, activation of beta alone will not produce a conscious experience of the joy of life, because the gates to consciousness are permanently blocked closed by the cannabinoid receptors. For these subjects, the experience of beta can only occur if at the same time that beta is activated, the cannabinoid blocks are also removed. Then the effects of beta can get past the gates and enter consciousness (Fig. 4).

2.2 Central vs. Distributed

Now we need to consider an interesting observation: the expansion of consciousness itself can be permanently blocked by the cannabinoid receptors. For subjects that have such blocks, consciousness can only expand if the cannabinoid blocks are removed. A possible implication of this is that the gates of consciousness do not mediate the access of the mental organ (e.g., beta) to the organ of consciousness (serotonin-7); rather, the gates mediate the access of the organ of consciousness to the other mental organ(s) (e.g., beta).

This possibility suggests a fairly different view than that suggested by the theater metaphor. In the theater metaphor, the mental organs that act as players (e.g., beta) enter onto the central stage of consciousness (e.g., serotonin-7). This presents some practical and conceptual difficulties. If a specialized mental organ is required to

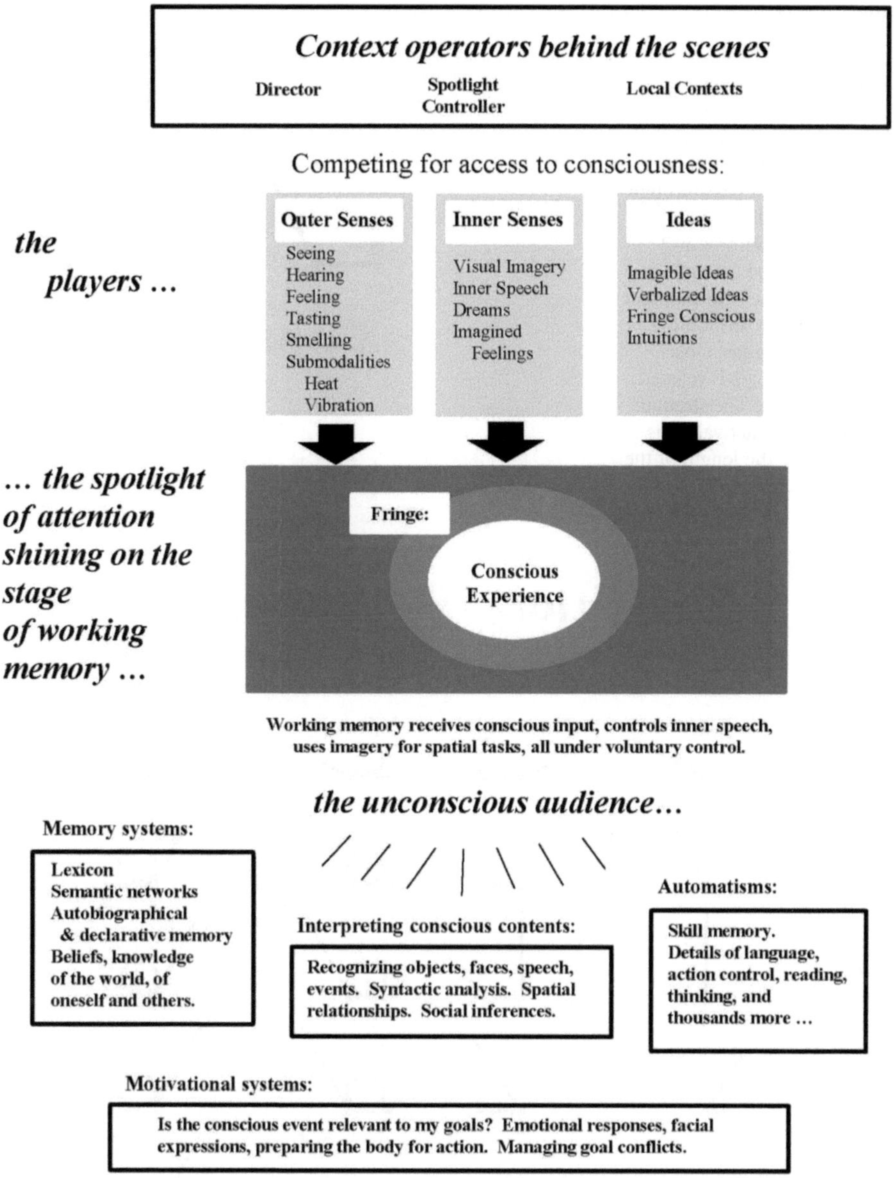

Fig. 2 Fig. 2-1 redrawn from Bernard Baars "In the Theater of Consciousness." His caption: "*A theater metaphor for conscious experience*. All unified theories of cognition today involve theater metaphors. In this version, conscious contents are limited to a brightly lit spot of attention onstage, while the rest of the stage corresponds to immediate working memory. Behind the scenes are executive processes, including a director, and a great variety of contextual operators that shape conscious experience without themselves becoming conscious. In the audience are a vast array of intelligent unconscious mechanisms. Some audience members are automatic routines, such as the brain mechanisms that guide eye movements, speaking, or hand and finger movements. Others involve autobiographical memory, semantic networks representing our knowledge of the world, declarative memory for beliefs and facts, and the implicit memories that maintain attitudes, skills, and social interaction. Elements of working memory – on stage, but not in the spotlight of attention – are unconscious. Notice that different inputs to the stage can work together to place an actor in the conscious bright spot, a process of *convergence*, but once on stage, conscious information *diverges*, as it is widely disseminated to members of the audience. By far, the most detailed functions are carried on outside of awareness

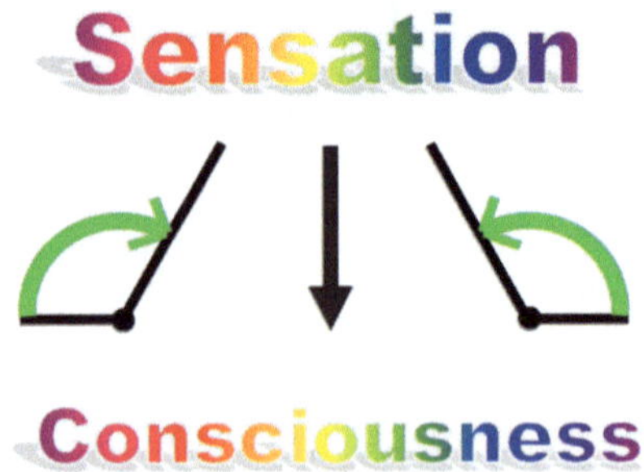

Fig. 3 In order to enter consciousness (mediated by serotonin-7), sensation (mediated by a variety of receptors) must pass through the gates (mediated by serotonin-2). Serotonin-2 manipulates the gate in a dynamic moment-to-moment fashion. When serotonin-2 is activated, the gates close, when serotonin-2 is relaxed, the gates open. The strength of serotonin-2 activation is indicated by the length of the curved green arrows

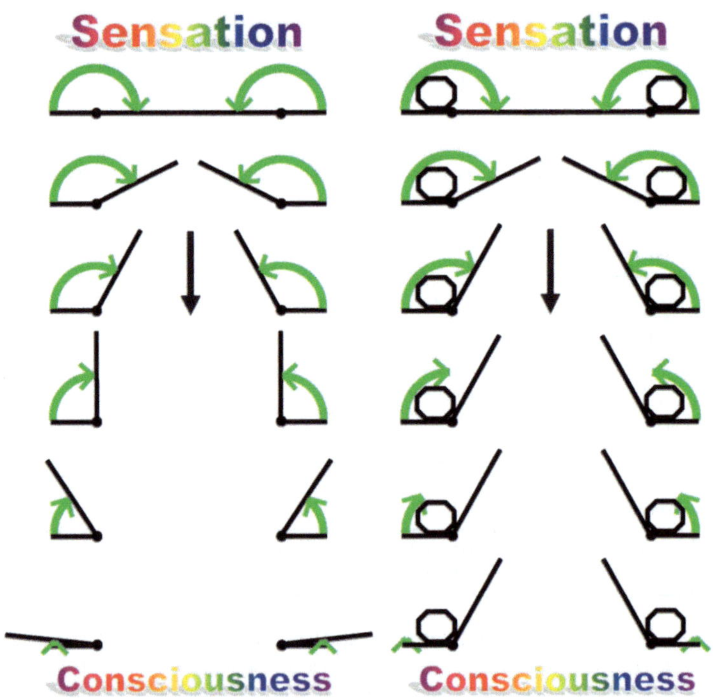

Fig. 4 *Left*: Illustrates that serotonin-2 operates the gates along a spectrum from relaxed/open (*bottom*) to activated/closed (*top*). *Right*: Illustrates the interaction of the serotonin-2 gates with the cannabinoid blocks. Note that the *green arrows* are the same length at each level of the spectrum in the diagrams on the *right* and *left*. On the *left*, the positions of the gates are determined by the strength of serotonin-2 (length of the *green arrows*). However, on the *right*, the ability of the gates to open is limited by the blocks imposed by the cannabinoid receptors. In the figure on the *right*, the cannabinoid blocks are all at the same position, allowing the gates to open partially. However, cannabinoid also operates on a spectrum so that the blocks may allow the gates to open most of the way, partially, barely, or not at all

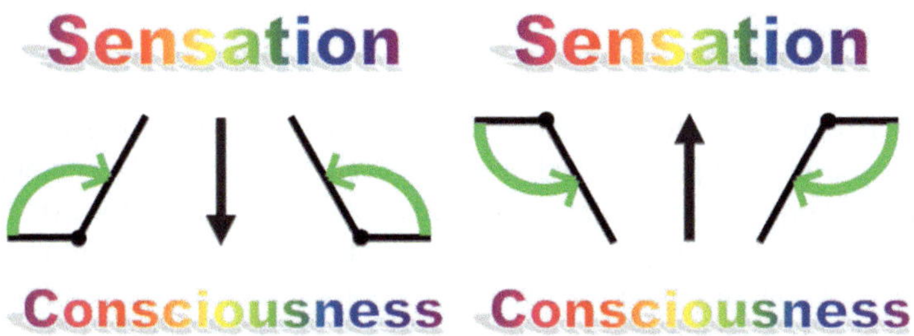

Fig. 5 Two alternate hypotheses of the relationship between sensation, consciousness, and the gates. *Left*: The conventional view is consistent with the theater metaphor, in which sensation must pass through the gates in order to play in the spotlight on the stage of the theater. This view implies that the organ of consciousness (serotonin-7 or kappa) is a central theater where consciousness manifests. *Right*: An alternate view suggests that consciousness is distributed among the sources of sensation (various mental organs). Rather than the other mental organs having to send their sensation through the gates into the theater of consciousness, the organ of consciousness must pass through the gates in order to bestow the property of consciousness on the organs that generate sensation. In this alternate view, consciousness is not centralized in any one organ, and the theater is not an appropriate metaphor

produce a specific domain of feeling (e.g., the sense of home, family, community, and the joy of life), then could there be a general purpose organ of consciousness capable of rendering the experience generated by each of the many different kinds of mental organs? And how is the feeling communicated from the source mental organ (e.g., beta) in all of its richness and detail, to the organ of consciousness?

In the alternate view, the organ of consciousness does not provide a mental space into which other mental organs enter; rather, the organ of consciousness performs the function of making other mental organs conscious. In this view, the mental space is distributed across mental organs, not centralized in one, and the above-mentioned conceptual and practical issues evaporate (Fig. 5).

2.3 *Sense of Self*

When serotonin-7 is strongly activated without simultaneously activating serotonin-2, the subject is very likely to experience a loss of the sense of self, ego-loss, the fully non-dual state. This is a curious observation, because it occurs without any actual inhibition of the serotonin-2 system but with only a strong activation of serotonin-7. It appears that if serotonin-7 is strongly activated while serotonin-2 is not altered, the serotonin-2 system is overwhelmed, and consciousness floods through the gates, with the gatekeeping function of serotonin-2 effectively completely disabled. The loss of the sense of self in this situation suggests that an important component of the egoic sense of self is the *act* of the serotonin-2 system manipulating the gates of consciousness. The ability of serotonin-2 to manipulate the gates of consciousness appears to depend on the relative strengths (level of expression) of serotonin-2 and serotonin-7. Balance matters.

2.4 Generative

In the description of serotonin-7 above, I discussed how when sufficiently activated: "At a critical point, we pass through a mental event horizon, as the contents of consciousness become more salient than actual reality. We mentally exit the actual space and time and enter a space and time created by the mind, within which the mind can create an alternate reality. At this point a mental big bang may occur. Consciousness is a generative system, capable of creating worlds, universes. This creative property may be the basis of free will."

It has been suggested that the naturalistic view challenges free will, the idea that human beings are first causes. I would like to suggest that while human beings may not be first causes in the sense of the big bang and while they operate within the fully causal flow of the laws of nature, they none-the-less contain generative mental centers (serotonin-7, kappa) that contribute novel input into this flow. Human creativity (art, music, and literature) illustrates this generative property (Fig. 6).

Consciousness is a generative system capable of creating worlds in the mental plane, and capable of influencing the body in the physical plane. The generative system of consciousness introduces original causal input while completely obeying the laws of nature. Thus the causal creativity of the human mind coexists peacefully with the causality of the laws of nature.

However, this generative property does not occur when serotonin-7 is activated alone. When activated almost alone (together with serotonin-1), it produces an empty state of non-duality. It is only when affective mental organs are simultaneously activated that the generative property becomes apparent. Thus the generative process is not a property of serotonin-7 alone, but of affective mental organs when they are brought very strongly into consciousness by serotonin-7.

They are transformed by serotonin-7, a process I call "serotonin-7ization." There seem to be common themes to its effects: adds a creative exuberance; takes it to a higher level; makes connections; comprehends the bigger picture; creates sumptuousness, sparkle, grandeur, majesty, transcendence; intangible becomes tangible; and thoughts, feelings, motivations originating from within may be perceived to originate from without. In her novel *Jane Eyre*, Charlotte Brontë describes the natural process in ordinary life:

> Won in youth to religion, she has cultivated my original qualities thus:—From the minute germ, natural affection, she has developed the overshadowing tree, philanthropy. From the wild stringy root of human uprightness, she has reared a due sense of the Divine justice. Of the ambition to win power and renown for my wretched self, she has formed the ambition to spread my Master's kingdom; to achieve victories for the standard of the cross. So much has religion done for me; turning the original materials to the best account; pruning and training nature. (Bronte 2009)

This creative process is not limited to religion. Simple curiosity could be cultivated, developed, reared, formed, and turned into a Nobel Prize winning insight. Serotonin-7ization is a fundamentally creative process that may form the basis of free will.

Fig. 6 When consciousness is expanded by activating serotonin-7, the contents of the mind become more tangible and may be experienced as if perceived through the five senses. The mind becomes increasingly creative as consciousness expands. Photo by: LSD-photos Marco Casale – Paolo Dall'Ara (http://lsd.eu/, http://lsd.eu/index.php?gallery/show/adv1)

2.5 *Balance*

With serotonin-7 strongly activated, the contents of consciousness are more richly rendered. The ratio of expression of consciousness (serotonin-7) and the other mental organ will influence the quality of the expression of the mentation. If the ratio leans toward the other mental organ, the expression of the mental organ (joy, compassion, comfort), will be more grounded in actual reality. If the ratio leans more toward consciousness (serotonin-7), the expression will be more invented, more generative, more creative, and able to go beyond actual reality. The organs of consciousness are organs of creation. Balance matters. Outside of a certain range of balance, mental difficulties are likely.

3 Ways of Knowing

As adult humans, we largely know and understand the world through reason, and many of us have lost touch with, forgotten, and no longer value other ways of knowing. Here, I will attempt to remind us of what we have lost.

> What Mrs. Coulter was saying seemed to be accompanied by a scent of grownupness, something disturbing but enticing at the same time: it was the smell of glamour. (Philip Pullman, "The Golden Compass" p. 66)

3.1 Flavor

I begin with flavor (odor and taste), because it is a nonrational way of knowing that we retain and value. Most of us know the odor of a rose, the flavor of cinnamon or vanilla, or the rich flavor of a fine curry. It is through odor and taste that we know the flavor of foods and the smells of our world. Flavor is a feeling and a way of knowing that is independent of reason. We generally do not attempt to reason about flavor, and we do not doubt the truths about the world that it reveals to us. We accept flavor for what it is and leave it at that.

While we do not generally intellectualize flavor, the 2004 Nobel Prize in Physiology or Medicine was awarded for unraveling the biological mechanisms of odor (Buck and Axel 1991). Odor and taste receptors are also GPCR. Although about 800–1,200 different functional odor receptors are expressed in the mammalian genome, humans express fewer than 400 (Niimura and Nei 2007). Humans have about a third the number of functional odor receptors as other mammals. The human genome is littered with hundreds of odor receptor pseudogenes (genes that have mutated such that they no longer function).

This suggests that the human experience of odor is relatively impoverished. Dogs are not just more sensitive to odor as a result of having a larger nose; they also have a qualitatively much richer and more subtle and nuanced experience of odor than we do.

When flavor is conveyed from person to person, the language we use takes the form of words like "floral," "minty," "musky," "citrusy," etc. This implies several things. We assume that if we have both experienced a flavor (e.g., vanilla, mint), then we have had a shared experience of the feeling that is that flavor, and by naming a shared tastant or odorant, we can convey the feeling of the flavor. And it may be largely true (except due to variation in expression of the relevant odor or taste receptors). If we had not shared the experience, there would be no language to describe the feeling. Flavor is ineffable.

The same principles apply to feelings in general. There is no language for feelings, other than reference to a shared experience. It could be the odor of a rose, the taste of cinnamon, the feeling of falling in love, the sense of family and humanity mediated by beta, or the sense of the essence or spirit of things mediated by alpha-2. The same applies to any class of feeling, mediated by any mental organ.

If we had never smelled a rose, no one could communicate that sensation to us in any meaningful sense. Similarly, if we had never experienced smell, we could never understand what it feels like. And this is also true for the affective ways of knowing. Those who experience an effective way of knowing cannot communicate the feeling to those who have not. The only way to know feeling is to experience it.

In the description of a dozen mental organs above, I have attempted to identify the feelings associated with them, in ordinary language. But I cannot convey the feelings themselves. What I have attempted to do is to allow us to understand the feelings intellectually, to the extent that is possible.

3.2 Emotion

When we think of feelings, most of us think of emotions, such as anger/rage, fear, happiness, sadness, surprise, and disgust. Emotions play a role in determining motivational states. When strong, emotions can take control of us and dramatically affect our behavior. Many people rightfully feel that emotions are something that needs to be controlled and dominated, lest they take over and cause us to do things we regret or cause us to suffer. While emotions and ways of knowing are both in the affective domain, ways of knowing are not as tightly linked to motivation. Feelings that fall into the category of ways of knowing play the role of painting the world for us, just as flavors do. Affective ways of knowing are a way of truthfully representing the world in our minds and do not have the troublesome motivational properties of emotions. As adults, we are barely aware of and have largely forgotten these ways of knowing through feelings.

3.3 Cognitive and Affective

At the Tofukuji Buddhist temple in Kyoto Japan, there is a large rock, which is 10–15 ft tall, 3–4 ft wide, and about a foot thick (Fig. 7). On this rock is carved, in beautiful flowing vertical Japanese script, a haiku. The haiku reads: "Furuike ya kawazu tobikomu mizu no oto." This translates into English as "old pond, frog jump, sound of water." This is perhaps the most famous haiku, written by Matsuo Bashō (1644–1694). The book "One Hundred Frogs" (Sato 1995) is a collection of nearly 150 different translations into English of this simple haiku. There is a joke about a haiku vendor with a sign that reads "Haiku 100 yen. With frog, 25 extra."

There are fundamentally two ways of knowing this haiku. We can know the haiku with our rational mind. In this case, well, if a frog jumps into water, it will make a splash and that will cause vibrations in the air, so of course there will be a sound, which we can hear. If we know the haiku this way, it is kind of silly and pointless. Or we may rationally interpret it as a metaphor, in which case we may be able to find symbolic meaning in it.

Fig. 7 Rock at the Tofukuji Buddhist temple in Kyoto Japan, with a carving of the famous haiku by Matsuo Bashō (1644–1694). The haiku reads: "Furuike ya kawazu tobikomu mizu no oto." This translates into English as "old pond, frog jump, sound of water." (Photo by Tom Ray)

The other way to know the haiku is with our heart. If we know it this way, it paints a moment, a beautiful and timeless scene of an ancient pond, with a frog jumping in and splashing, as frogs have jumped in for millions of years. While we may not have a visual image of the scene, we can feel it. We paint the scene with feelings. It may even be better not to visualize it, because then its representation is purely affective. When we know the haiku in this way, we can understand why it is so famous.

There are, broadly speaking, two fundamental ways of knowing, the cognitive and the affective, the head and the heart, reason and feeling, modern and archaic. The cognitive domain understands the world in terms of language, reason, ideas, symbols, and concepts, while the affective domain understands the world in terms of feelings. Both domains, cognitive and affective, are capable of "knowing" and "understanding" the world, each in its own way. And each domain is able to construct a "model" of the world in consciousness, a rich, subtle, and complex representation of the world.

It appears that children are dominated by the affective domain, while adults are largely dominated by the cognitive mind, at the expense of emotions, feelings, and intuition. When we mature into adults, we find ourselves knowing the world largely through language, logic, and reason. We tend to lose touch with the way we knew the world as children, the archaic way of knowing, through feelings, through our heart.

Reason as a way of knowing and understanding is evolutionarily new and appears to be fully developed only in adult humans. However, before the emergence of reason, we still knew and understood our world and ourselves through feelings, and adult humans retain this capacity (even if it is not exercised). Our developmental and evolutionary antecedents (children and nonhuman higher animals) have a fully developed affective mind and still know the world exclusively in this way. The affective mind of humans predates the cognitive mind (developmentally and evolutionarily) and is ancient, complex, subtle, rich, and capable of knowing and understanding the world, based on feelings alone. The ineffability of many kinds of mystical experiences arises from this affective way of knowing.

While reason has emerged in the last hundred thousand years of our evolution, the affective ways of knowing have been elaborating through evolution for hundreds of millions of years. This archaic way of knowing has great evolutionary depth, and like flavor, remains profoundly valid today, revealing truths about the world. Perceiving truth can be a matter of life or death (i.e., natural selection). Multiply this by many millions of generations (iterations). If truth can be found, evolution can find it.

While the faculties of language, logic and reason seem to be mediated by one or a few mental organs based on serotonin receptors, the affective mental organs are numerous and diverse, mediated by a wide variety of receptors (among the dozen mental organs that I have characterized: alpha-1, alpha-2, beta, histamine, imidazoline, dopamine, sigma, mu, kappa). Thus the affective systems do not represent a single, alternate, way of knowing, but rather a multiplicity of ways of knowing.

We might suppose then that the cognitive way of knowing is relatively monolithic in part due to being evolutionarily young. Perhaps over evolutionary time, the cognitive way of knowing will mature, diversifying and differentiating across widely varying mental organs, as have the affective ways of knowing.

3.4 Ontological Categories

The ways of knowing collectively represent a set of natural ontological categories which evolution has settled upon for representing the world in the mind:

- Laws and patterns of nature – serotonin-1
- Things – alpha-2
- Place, scene – alpha-1
- Home, family, community – beta
- Beings – histamine

3.5 Traditions of Knowing

It may be that each of our great teachers and spiritual leaders achieved their unique insights as a result of the exceptional blooming of a particular mental organ. In each case, this was a great achievement, and often religions or major philosophical or secular traditions formed around them. It should be possible to identify the mental organ(s) associated with each tradition.

Socrates taught how to think rationally, at a time when it was not done, and is credited with the origin of the "concept" (Jaspers 1962). From Socrates and others, ultimately flowed the age of reason and the age of enlightenment. Socrates experienced an exceptional bloom of the mental organ of reason, built from the five serotonin-1 receptors.

Siddhartha Gautama (Buddha) experienced an expansion of consciousness by a blooming, through meditative practices, of the mental organ of adult consciousness, defined by serotonin-7.

Confucius displayed the deep understanding of humanity and human nature that shows as wisdom (Jaspers 1962), which likely resulted from an exceptional bloom of the mental organ defined by beta.

Jesus Christ had absolute faith in God and absolute faith in the immanent end of the world and coming of the kingdom of heaven (Jaspers 1962). This suggests an exceptional bloom of dopamine. Also, his reputation for open-hearted tenderness, compassion, forgiveness, healing, and love suggests an exceptional bloom of imidazoline.

Among the affective ways of knowing that I have characterized, alpha-2 may be the most ineffable. Ramachandran discusses a word from Sanskrit, "rasa": "Capturing the very essence, the very spirit of something, in order to evoke a specific mood or emotion in the viewer's brain"(Ramachandran and Hirstein 1999; Ramachandran 2004, 2007a, b), which precisely describes the way of knowing mediated by alpha-2.

Alpha-2 appears to provide the basis for several philosophical and religious traditions. The Shinto religion "teaches that everything contains a kami ("spiritual essence", commonly translated as god or spirit)." "There are natural places considered to have an unusually sacred spirit about them, and are objects of worship. They are frequently mountains, trees, unusual rocks, rivers, waterfalls, and other natural edifices" (Wikipedia 2010b). This is also characteristic of animistic religions in general and of alpha-2.

In Taoism, attributed to Laozi, the goal is to attain a mental state in which is revealed "the soft and invisible power within all things." It is a state in which "everything is seen as it is, without preconceptions or illusion." "It is believed to be the true nature of the mind, unburdened by knowledge or experiences" (Wikipedia 2010a). While this may represent the entire affective domain, facets of Taoism clearly manifest alpha-2 and beta.

3.6 The Full Bouquet

The paradigmatic individuals that I described above, each of which achieved a unique insight through the exceptional blooming of a particular mental organ, would have also had an exceptionally well-developed consciousness (serotonin-7) to render the key mental organ in rich resolution, and an exceptionally well-developed cognitive organ (serotonin-1) to be able to articulate their insights. Thus, their unique insight and teaching requires a triple bloom of consciousness and cognition with another mental organ(s).

But this picture is not whole, because each of our teachers or spiritual leaders, while endowed with the full bouquet of mental organs, experienced the exceptional bloom of only one or perhaps a few mental organs (other than serotonin-1 and serotonin-7). Each of these traditions celebrates only a narrow domain of human potential. The discovery and characterization of a significant set of mental organs opens the possibility of a new tradition of knowing. We now have the potential to experience the blooming of the full bouquet of mental organs, resulting in the realization of our full human potential (Ajaya 2009). This full bouquet of mental organs is what is great in us. This is our humanity, this is our evolutionary heritage. This is what makes us rich. It should be cultivated in its wholeness, not only narrowly selected parts of it, chosen by the historical accident of our birth into a particular religious, philosophical, secular, or ethnic tradition.

Recognizing and valuing the full bouquet has the potential, at least theoretically, to unify the competing traditions, by showing the contribution of each one to the richness of the human spirit. We see how, taken together, they form the beautiful bouquet of the human heart, mind, soul, and spirit. Each mental organ is like a unique flower, contributing to the floral arrangement that evolution has left us, here a rose, there an iris, and there a daisy... Only when all are taken together are we *fully* human.

Can scientists, naturalists, materialists, and rationalists of various sorts acknowledge that their way of knowing, reason, is only one of many ways of knowing with which our ancient evolutionary heritage has endowed us? Are not all of these ways of knowing equally valid? Can the followers of any tradition, religious or secular, acknowledge that their particular tradition is not exclusive and above all others?

3.7 Loss of Affective Ways of Knowing

An existential risk that my work identifies is loss of neurotransmitter receptor (mental organ) diversity as a result of the aggressive spread of a cognitive monoculture. The mind is populated with mental organs. To persist, each mental organ must contribute to fitness. The cognitive mental organs caused such a jump in Darwinian fitness (witness the population explosion and the elaboration of war technology) that fitness variation among affective mental organs is *relatively* negligible, as is their contribution to fitness *relative* to the cognitive organs (Fig. 8).

At present, the affective mental organs appear to be fully active in childhood, but by adulthood, the mental immune system has largely converted them into vestigial organs. For now, they may be preserved by their critical roles in childhood, and their unconscious activity in adulthood may still influence our judgment. However, the affective neurotransmitter receptors may be in danger of becoming pseudogenes as have most of our odor receptors. Preservation of human neurotransmitter receptor diversity deserves a place alongside preservation of biological species diversity.

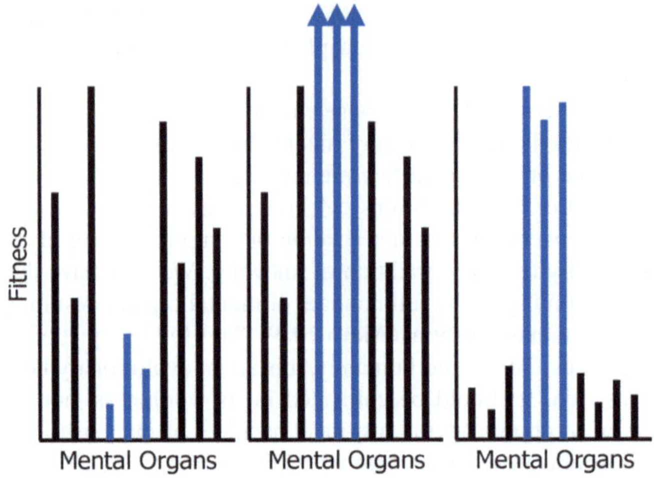

Fig. 8 Patterns of fitness of the mental organs that populate the mind. *Left*: To persist, each mental organ must contribute to fitness. Let the fourth, fifth, and sixth bars (*in blue*) represent the cognitive mental organs, and the other bars (*in black*) represent the affective mental organs. *Center*: By providing logic and reason, cognition gave us science and technology that produced the human population explosion, an extraordinary payoff in Darwinian fitness that dwarfs the fitness contribution of the affective mental organs. The fitness of the cognitive mental organs is off the chart. *Right*: Adjusting the vertical fitness axis, we see that the fitness contribution of the affective mental organs is slight *relative* to the fitness contribution of the cognitive mental organs

The worlds of feeling and reason need to recognize one another, reconcile, learn mutual respect, and merge, because only then can we truly be whole.

> we have lost the way… Our knowledge has made us cynical, our cleverness hard and unkind. We think too much and feel too little: More than machinery we need humanity; More than cleverness we need kindness and gentleness. – Charlie Chaplin, 1940, *The Great Dictator* (Chaplin 2011)

There is another side of this issue. What kind of mind can be comfortable with, or ignore, or willingly participate in the destruction of our planet, each other, or ourselves? The shutting off of the affective domain in adults can be a contributing factor to such a mentality. Our history of warfare may have selected for the shutting off of the affective domain more completely in adult males (through more aggressive serotonin-2 and cannabinoid systems). The degree to which this shutting off occurs is highly variable within the population and varies between individuals, ages, genders, cultures, and mental organs.

The various ways of knowing do not compete. They blend together to form a perceptual whole, like the flavors in a rich stew. Each mental organ adds spice to our lives. Reason coevolved with a preexisting affective domain and is designed to be informed by affective input. Various authors have suggested that the cognitive mind is built on top of and remains fully dependent on the affective mind and that without the underpinning of affect, humans are not able to make sound judgments

(Damasio 2005; Pham et al. 2012). The cognitive domain alone can produce reason, intelligence, and knowledge, but wisdom requires a healthy unity of both the cognitive and affective domains (Hall 2010). When reason reigns at the cost of losing touch with the other ways of knowing, we retain the ability to manipulate nature, but we do not understand its essence, and cannot make wise judgments. The accumulation of material goods and power over nature cannot make us wealthy if we lack feelings. It is the rich experience of the flavors and feelings of life that makes us wealthy.

The affective ways of knowing are the means by which children grow into adults who understand the world. Alison Gopnik describes the way that children explore the world through fantasy, imagining scenarios, in order to translate an understanding of chains of causality into an understanding of the nature of the world and how to flourish in it (Gopnik 2009). But how do children obtain such understanding before the emergence or maturation of cognition? It is in large part through the affective ways of knowing, which are more obviously active and dominant during childhood.

4 Mind of the Dog

My interpretation of human mental organs is based on the synthesis of molecular data with reports of subjective experience. Unfortunately the same methodology cannot be applied to nonhuman animals, because they cannot tell us about their experience. However, there is one animal with which humans have an intimate enough relationship that humans have generated detailed descriptions of the animal's mind: dogs.

Dogs are the first animal that humans domesticated, from wolves, about 15,000 years ago. The ancestral dog has evolved through artificial selection into hundreds of different breeds. Wolves were preadapted to evolve into our best friend, by virtue of their social nature. The American Kennel Club provides temperament data for 161 breeds of dog (AKC 2012). Here are a few examples:

Sloughi – The Sloughi is a dog with class and grace. The attitude is noble and somewhat aloof.
Bichon Frise – Gentle mannered, sensitive, playful and affectionate. A cheerful attitude is the hallmark of the breed, and one should settle for nothing less.
Briard – A dog of heart, with spirit and initiative, wise and fearless, and no trace of timidity. Intelligent, easily trained, faithful, gentle, and obedient, excellent memory, ardent desire to please his master.
Irish Water Spaniel – Very alert, inquisitive, and active. Stable in temperament with an endearing sense of humor. May be reserved with strangers but never aggressive or shy.
Pekingese – A combination of regal dignity, intelligence, and self-importance make for a good natured, opinionated, and affectionate companion to those who have earned its respect.

Pomeranian – The Pomeranian is an extrovert, exhibiting great intelligence and a vivacious spirit.
Toy Fox Terrier – Intelligent, alert and friendly, and loyal to its owners. He learns new tasks quickly, is eager to please, and adapts to almost any situation. Self-possessed, spirited, determined, and not easily intimidated. He is a highly animated toy dog that is comical, entertaining, and playful all of his life.
Newfoundland – Sweetness of temperament is the hallmark of the Newfoundland; this is the most important single characteristic of the breed.

The mental properties of dogs are characteristic of their breed and vary between breeds. These mental properties of dogs are clearly genetically based and heritable. I believe that dog and human personalities are constructed from the same kinds of elements: mental organs. Not the same two sets of elements, but the same kinds of elements. However, I also believe that the individual properties that are shared by the two (e.g., sense of humor) are examples of convergence, not homology. Neither the wolf nor the common ancestor of the dog and human had a sense of humor. The ability of such a great diversity of distinct mental properties to emerge rapidly through selective breeding is indicative of the speed with which mental organs are able to evolve.

It appears that fully developed cognition is unique to humans. Thus, the minds of dogs and other nonhuman animals are purely affective minds. In order to understand the animal mind, we need to understand the affective mind.

5 Modulatory Personality

Just as individuals vary in the size and proportioning of features such as ears, noses, breasts, and hands, the degree of development and expression of individual mental organs varies dramatically between individual persons (Borg et al. 2003). Thus each individual person has a unique pattern of expression, or proportioning, of the full set of perhaps 100 or more mental organs. I call this individual pattern the "modulatory personality." Modulatory personalities are as unique and variable as human faces, perhaps more so, and probably underlie much of what we refer to as character, temperament, and personality. Extreme modulatory personalities may produce exceptional individuals but also may be pathological.

Each mental organ plays its role in the mind along a spectrum of degree of expression from low to high (Fig. 9). We would generally expect that the mean of the distribution would correspond to the normal and healthy condition, while the two extremes of the distribution may correspond to exceptional individuals or pathological conditions. Thus, each human mind represents a configuration in a space of hundreds of dimensions, in which each axis represents the level of expression of a single receptor or mental organ (see Fig. 10) and each point in the space represents the modulatory configuration of all receptors in an individual human.

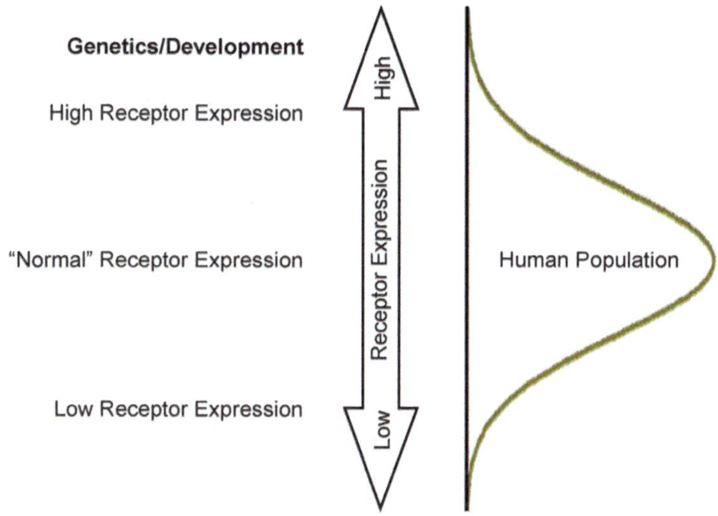

Fig. 9 Each mental organ expresses itself along a spectrum. Each individual human will at any point in time be at some position on the spectrum of each mental organ. The human population will be distributed along the spectrum, perhaps in a normal distribution

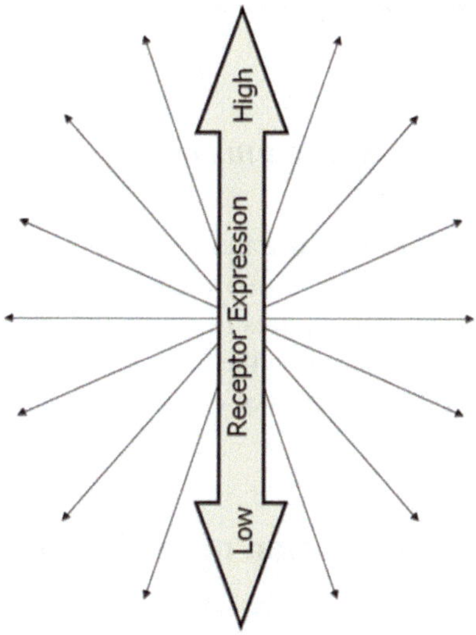

Fig. 10 One spectrum will exist for each mental organ, of which there are likely hundreds. Collectively, the population of mental organs could be represented by a high-dimensional space, with one axis per mental organ or receptor

We would expect a cloud of points representing the human population, with the highest density centered around the point representing the median values of all the hundreds of receptor distributions, and the density of the cloud decreasing as we move away from this global mean, in any direction in the receptor space.

In evolutionary terms, we would expect selection to shape the population variation in receptor expression such that the mean of the distribution would maximize fitness, while the extremes would tend to be less fit. When selection is strong, it

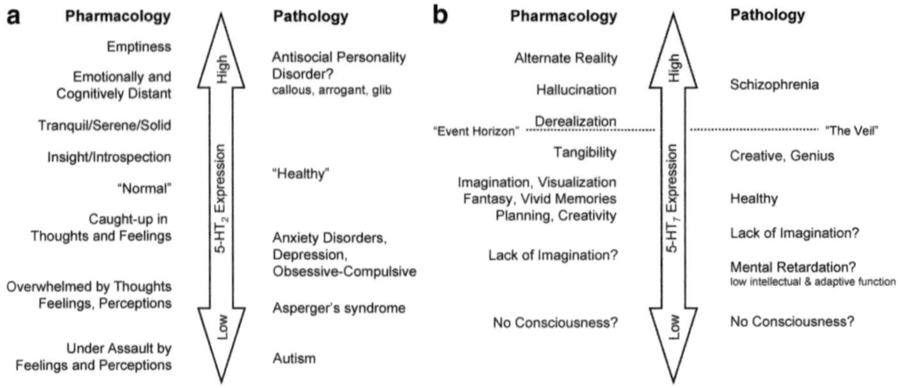

Fig. 11 A spectrum representing a hypothesis of the mental continuum associated with the range of expression (from low to high) of the mental organ defined by the serotonin-2 receptors (*left*) and serotonin-7 receptors (*right*)

would likely maintain a narrow distribution, when weak a broader distribution. The overall configuration of the relative levels of expression of the many mental organs is the "modulatory personality" and is highly variable within the human population.

5.1 *Spectrums of Expression*

Figure 11 illustrates hypotheses of the spectra relating mental states and the level of expression of two mental organs, serotonin-2 on the left and serotonin-7 on the right. The figures illustrate the variations in mental properties found along the spectrum, including the central healthy range, as well as pathologies that might be found at the extremes of expression of these two mental organs (Fig. 11).

6 Evolution of the Mind

Mental organs are all tied to a single gene family, the G-protein-coupled receptors (GPCR), and thus evolve through duplication and divergence of the underlying genes and regulatory elements. The GPCR include receptors for serotonin, dopamine, histamine, and many other neurotransmitters. GPCR genes provide a genetic and regulatory system to richly specify the structure of the *mind*, not just the brain, and thereby make the *mind* highly evolvable. A little more than 300 different GPCR are expressed in the human brain. However, individual mental organs are often made up of groups of closely related receptors. There may be half as many, or fewer, mental organs than receptors.

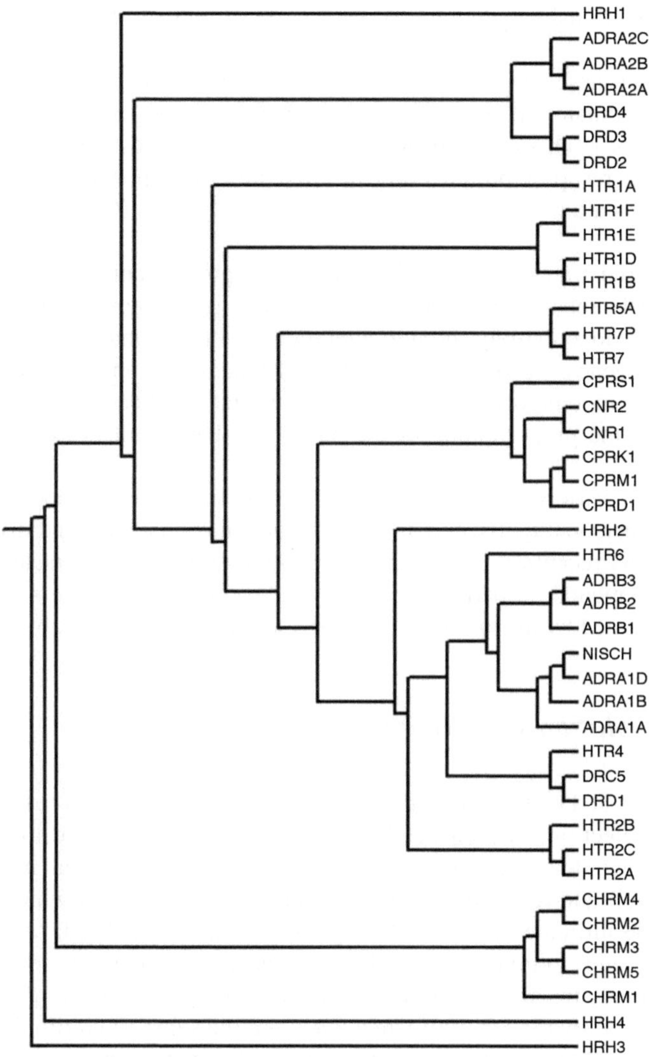

Fig. 12 A tree showing the relationships between 43 G-protein-coupled receptors (out of over 300 expressed in the human brain), based on sequence similarities

6.1 Shaping Mental Organs

If modulatory receptors implement components of the mind, then new components can be created through the process of duplication and divergence of receptor genes. Each individual GPCR corresponds to a single protein encoding gene, whose expression is influenced by many genetic regulatory factors (which largely remain unknown). The GPCR are one of the largest gene families in the human genome and have diversified through the process of duplication and divergence. Figure 12 shows the relationships among a small sample of GPCR (those examined in this study).

On a long timescale, evolution shapes and fine tunes the qualitative properties of individual modulatory components (i.e., do they modulate joy, empathy, consciousness, or reason). This evolutionary process likely involves alteration of both the genes coding receptor proteins and the regulatory components; and in addition likely involves alteration of the second messenger systems (G-proteins) coupled to the receptors.

On a shorter timescale, evolution shapes the proportioning of the modulatory personality (the relative levels of expression of all the receptors). This proportioning likely can be done entirely through alteration of the regulatory elements, without affecting the genes encoding proteins. To understand the importance of the proportioning of the modulatory personality, consider that psychoactive drugs only (transiently) alter the proportioning of the mind, yet they result in radically different mental states.

There are a variety of means of evolutionary shaping of the proportioning of the modulatory personality. Evolution can influence the abundance and distribution of the modulatory receptors, the pattern of activation of these receptors, and the degree to which the mental property mediated by the receptor gains access to consciousness (as mediated by the inhibitory systems, serotonin-2 and cannabinoid).

6.2 Coevolution with Religion

It has been argued that religion provides adaptive benefits and so has been favored by natural selection (Wilson 2003). If so, it would be expected that our innate psychology would have evolved to facilitate religion. Several mental organs appear to facilitate religion. Alpha-2 mediates a sense of soul which some have argued is the ultimate basis of all religions (Tylor 1958). Imidazoline mediates compassion and forgiveness which are central to some religious traditions. Dopamine mediates awe which has been called the distinctive religious emotion, as well as certainty (Smith 2001), meaning, and the sense of spiritual significance (Griffiths et al. 2006). Dopamine appears to be the most quintessentially religious mental organ. Beta may form the basis of Confucianism, which some consider to be a kind of religion. When beta is activated together with serotonin-7, it can produce ecstatic joy which can have a religious quality. Serotonin-7 can produce a sense of transcendence of the body, the cosmic, the infinite, a greater power, and even god.

6.3 Exploration of Mental Space

If a mental organ is relatively well expressed in a population, on average, it will play a prominent role in mental life. Under these circumstances, it will experience more selection than a mental organ that is on average, poorly expressed. As evolutionary time goes by, the highly expressed mental organ will become more richly shaped than the poorly expressed mental organ. At this level of selection, we are not talking about the proportioning of mental organs in the mind, but the qualitative properties of individual organs. A mental organ that is well expressed in a population, and thus experiencing strong selection, can more elaborately evolve the regulatory elements that shape the connection patterns and distribution of the population of neurons that make up the organ. As an element of the mind, this mental organ can become richer, deeper, more subtle, more detailed, more clearly defined, more complex.

If a mental organ decreases in its relative strength of expression in a population, then with weakened selection, the receptor gene may become vulnerable to being converted into a pseudogene, but in addition, the myriad properties of the mental organ may begin to wander and more randomly explore more distant realms of mental space. Under weaker selection, the mental properties can wander through regions of low fitness and may eventually settle on a new function or a new variation of an existing function, and a new mental organ will have been born. But these periods of weak selection correspond to exploration, not refinement. Refinement requires stronger selection than does exploration.

6.4 *Origin of Mind*

In order for evolution to sculpt exquisitely complex, large, multicelled organisms, it needed an evolvable genetic and regulatory mechanism that could specify a developmental program to give rise to such form: the homeotic genes. The evolutionary discovery and elaboration of that genetic and regulatory system was likely one of the key facilitators of the Cambrian explosion and the origin of complex life.

The true evolutionary elaboration of the mind requires a genetic system analogous to the homeotic genes, for shaping *mental* life. In order for the mind to be shaped by evolution, there has to be a genetic and regulatory system that allows heritable genetic variation in coherent mental features. It appears that mental organs, modulatory receptors (GPCR), and the genetic systems that regulate them provide the evolvable genetic keys for the origin and evolution of the mind. I suspect that the association of mental organs with receptors is a matter of evolutionary convenience, related to evolvability.

This is a possible answer to one of the fundamental questions in neurobiology: why are there so many different kinds of modulatory receptors (with over 300 expressed in the mammalian brain)? The mental organ hypothesis suggests a possible answer to this evolutionary question: modulatory receptor diversity is a mechanism to structure and modularize the *mind*, allowing it to be shaped, fine-tuned, and elaborated by evolution. The emergence of mental organs with the genetic systems to regulate and evolve them facilitated the origin of complex minds.

References

Ajaya, A. (2009). *The evolution of human consciousness*. Available via http://www.beingawareness.org/writings/writings_Evolution.htm

AKC. (2012). *American kennel club*. Available via http://www.akc.org/breeds/

Baars, B. J. (2001). *In the theater of consciousness: The workspace of the mind*. Oxford: Oxford University Press.

Borg, J., Andree, B., Soderstrom, H., & Farde, L. (2003). The serotonin system and spiritual experiences. *American Journal of Psychiatry, 160*, 1965–1969.

Bronte, C. (2009). *Jane Eyre*. Radford VA: Wilder Publications.

Buck, L., & Axel, R. (1991). A novel multigene family may encode odorant receptors: a molecular basis for odor recognition. *Cell, 65*, 175–187.

Chaplin, C. (2011). *One of the greatest posts on Youtube so Far!* Available via http://www.youtube.com/watch?v=M8C-qIgbP9o

Damasio, A. (2005). *Descartes' Error: Emotion, reason, and the human brain*. New York: Penguin.

Gopnik, A. (2009). *The philosophical baby: What Children's minds tell us about truth, love, and the meaning of life*. New York: Farrar, Straus and Giroux.

Griffiths, R. R., Richards, W. A., McCann, U., & Jesse, R. (2006). Psilocybin can occasion mystical-type experiences having substantial and sustained personal meaning and spiritual significance. *Psychopharmacology (Berl), 187*, 268–283.

Hall, S. S. (2010). *Wisdom: From philosophy to neuroscience*. New York: Knopf.

Jaspers, K. (1962). *Socrates, Buddha, Confucius, Jesus: The paradigmatic individuals (A Harvest book, HB 99)*. New York: Harcourt.

Niimura, Y., & Nei, M. (2007). Extensive gains and losses of olfactory receptor genes in mammalian evolution. *PLoS One, 2*, e708.

Pham, M. T., Lee, L., & Stephen, A. T. (2012). Feeling the future: The emotional oracle effect. *Journal of Consumer Research*. Available via http://www.jstor.org/stable/10.1086/663823

Ramachandran, V. S. (2004). *The neurological basis of artistic universals*. Available via https://notes.utk.edu/Bio/greenberg.nsf/0/7222777efe4b2d2885256e2c007d85f8

Ramachandran, V. S. (2007a). *The artful brain*. New York: Pi Press.

Ramachandran, V. S. (2007, December 3). *Podcast 118 – "What neurology can tell us about human nature, synesthesia and art"*. Available via http://www.matrixmasters.net/salon/?p=146

Ramachandran, V. S., & Hirstein, W. (1999). The science of art: A neurological theory of aesthetic experience. *Journal of Consciousness Studies, 6*, 15–51.

Sato, H. (1995). *One hundred frogs*. New York: Weatherhill.

Smith, H. (2001). Do drugs have religious import? A thirty-five-year retrospective. In T. B. Roberts (Ed.), *Psychoactive sacramentals, essays on entheogens and religion* (pp. 11–16). San Francisco: Council on Spiritual Practices.

Tylor, E. B. (1958). *Religion in primitive culture*. New York: Harper & Brothers.

Wikipedia. (2010b). *Shinto*. Available via http://en.wikipedia.org/wiki/Shinto

Wikipedia. (2010a). *Taoism*. Available via http://en.wikipedia.org/wiki/Taoism

Wilson, D. S. (2003). *Darwin's Cathedral: Evolution, religion, and the nature of society*. Chicago: University of Chicago Press.

Mnemo-psychography: The Origin of Mind and the Problem of Biological Memory Storage

Frank Scalambrino

Abstract The internal logic of a semiotic view of life suggests memory is the origin of mind. Interpreting the meaning of "sign" by way of Charles S. Peirce, the object of this chapter is to provide a response to the biosemiotic problem of the origin of mind in respect to both its general and specific formulations, i.e., as evolutionary emergence and as human environmental experience. As such, I hope for this chapter to express the biosemiotic view of mind and function heuristically for future research regarding memory and mind. "Mnemo-psychography" means that the mind writes itself out of memory. In regard to biosemiotics, the thesis of mnemo-psychography suggests that the mind originates out of interaction between the environment and the biological capacity for memory. By providing a biosemiotic reading of the results of contemporary memory research, specifically the work of Eric Kandel, Daniel Schacter, and Miguel Nicolelis et al., I argue for the thesis of mnemo-psychography, over a biosemiotic version of identity theory, as the solution to the problem of the origin of mind.

> In heaven, learning is seeing; on earth, it is remembering.
> Happy are those who have experienced the Mysteries.
> They know the beginning and the end of life.
> –Pindar (c.518–438 BCE)

In this chapter, I advocate for my thesis that the internal logic of the biosemiotic paradigm commits biosemiotics to the view that memory is the origin of mind. The "problem of the origin of mind" asks: out of what does mind originate? And, since biosemiotics takes life as its context, this question can be general or specific. Generally, "origin" is limited to the evolutionary emergence of mind, and specifically, it is limited to a living system or a particular organic being. I provide examples regarding mind at varying levels of complexity to address the problem both generally and specifically. This chapter is divided into four sections.

F. Scalambrino, Ph.D (✉)
The Chicago School of Professional Psychology, Chicago, USA
e-mail: fscalambrino@thechicagoschool.edu

L. Swan (ed.), *Origins of Mind*, Biosemiotics 8, DOI 10.1007/978-94-007-5419-5_17,

In the first section, I explain what I mean by the "internal logic" of a semiotic view of life by discussing biosemiotic research in conjunction with the philosopher Charles S. Peirce's notion of "sign." In the second section, I argue the internal logic of Peirce's notion of sign commits the discipline of biosemiotics to the claim that memory is the origin of mind. I refer to this claim as "the thesis of mnemo-psychography." In both the third and fourth sections, I argue biosemioticians should affirm this thesis as the biosemiotic solution to the problem of the origin of mind. I invoke the biological problem of memory storage to frame the specific problem of the origin of animal mind for the sake of critically examining a biosemiotic version of identity theory of mind as an alternative to mnemo-psychography. Since the alternative thesis claims what I take to be the only other logically viable biosemiotic solution to the origin of mind problem, by process of elimination, I argue for mnemo-psychography. Finally, I further clarify and support my thesis in regard to easily recognizable features of memory.

1 Introduction: Memory as the Biosemiotic Origin of Mind

"Biosemiotics" refers to the paradigm-shifting idea (Anderson, et al. 1984; Hoffmeyer and Emmeche 1991; Eder and Rembold 1992) that "life is based on semiosis" (Barbieri 2008a, p. 577). In other words, "If signs (rather than molecules) are taken as fundamental units for the study of life, biology becomes a semiotic discipline" (Hoffmeyer 1995, p. 16). Further, biosemiotics is characterized by two principles. First, "semiosis is unique to life, i.e., that it does not exist in inanimate matter" (Barbieri 2008d, p. 1, 2009, p. 230). Second, "semiosis and meaning are *natural* entities," i.e., the "origin of life on Earth" is not supposed to be the result of a supernatural cause (Barbieri 2008d, p. 1, 2009, p. 230). Hence, it is suggested that "Biosemiotics is necessary in order to make explicit those manifold assumptions imported into biology by such unanalyzed teleological concepts as function, adaptation, information, code, signal, cue, etc. and to provide a theoretical grounding for these concepts" (Pain 2007, p. 121).

Asking the question of the origin of mind within the biosemiotic paradigm, then, takes all living creatures as its context, since accordingly, all terrestrial life engages in "The process of message exchanges, or semiosis" (Sebeok 1991, p. 22). So, on the one hand, "Plant semiosis, for example, is distinct from animal semiosis and both of them from the semiosis of fungi, protists and bacteria" (Barbieri 2008c, p. 46). Yet, on the other hand, despite differences, "they are all semiotic processes, and allow us to conclude that semiosis exists in all living systems" (Barbieri 2008c, p. 46). Hence, the question of the origin of mind may refer to a *specific* living system, or it may refer *generally* to the evolutionary emergence of mind.

In this chapter, I argue that the biosemiotic answer to the problem of the origin of mind, in both its general and specific formulations, is memory. Biosemiotics seems committed to the position that memory is necessary both to perpetuate life, i.e., for the transmission of information, and to develop higher levels of complexity, i.e., for the interpretation of information into meaning. Memory as the origin of mind does not deny differences across what may be meant by mind in different living systems. Rather, life itself entails mind of varying complexity, and as the condition for the possibility of ever higher mind, memory must dwell copresent with the very transmission of life.

In so far, then, as signs "are taken as fundamental units for the study of life" (Hoffmeyer 1995, p. 16; cf. Marvell 2007), a brief examination of how biosemioticians understand "sign" will reveal memory as the origin of mind. Biosemioticians (cf. Barbieri 2009; cf. Kull, et al. 2009; cf. Favareau 2007; cf. Hoffmeyer 2006; cf. Emmeche 1991) often turn to the philosopher Charles S. Peirce's notion of a sign as a triadic relation between "representamen, object and interpretant" (Nöth 1990, p. 44; cf. Peirce 1998, p. 290). And, the thesis of mnemo-psychography suggests that the interpretant, whether we are discussing T cells, plants, embryos, animals, or humans, is a form of memory.

Though an extensive discussion of Peirce's philosophy is beyond the current scope, according to Peirce, a sign "is something which stands to somebody for something in some respect or capacity" (Peirce 2011, p. 99). Further,

> A *Representamen* is the First Correlate of a triadic relation, the Second Correlate being termed its *Object*, and the possible Third Correlate being termed its Interpretant, by which triadic relation the possible *Interpretant* is determined to be the First Correlate of the same triadic relation to the same Object. (Peirce 1998, p. 290)

Notice, the triadic relation involves *first* a cognition of a sign, *second* a recognition, and *third* the uncovering of the recognizer. And, as the second part of Peirce's quote indicates, despite the "thirdness" of the interpretant's discovery here as recognizer, to move from cognition to recognition, the recognizer is required and therefore must already be present. In other words, in order for the information, i.e., the representamen, to stand for something, i.e., to be interpreted as meaningful, the interpretant was necessarily present, though its manner of interpreting was not yet discovered until its interpretation was performed in the recognition of the object. In this way, the interpretation of a sign is always already the creation of a new sign which once interpreted creates a new sign, etc., and this process is semiosis.

Complexity develops out of the interpretant, since to recognize the representamen as an object is to retain the representamen through a transmission to a higher level of complexity by interpreting it as meaningful. Further, once an interpretant becomes a representamen in a subsequent sign, the semiotic process has increased in complexity as if it were developing the habit of interpreting as it just had. Notice, then, the key to the construction of the higher complexity is the *retention* of the more complicated structure being composed, and this complicated structure exists nowhere, while being constructed, other than in the memory, i.e., the transmission of the complexity, of the process itself. Hence, the retention through transmission of the representamen as meaningful object is a form of memory, and the higher structure of which the object is a part is constructed out of this memory.

Explicitly, then, here are the three ways in which the interpretant is a form of memory: (1) that the interpretant is present in the recognition of the representamen as the object shows the interpretant had the capacity to recognize the representamen as the object; (2) the interpretant's recognition of the representamen as object retains the representamen through a present transmission into a different complexity; (3) since the interpretant subsequently functions as a representamen, both a tendency to recognize, or habit of interpreting, as it just had and the complexity such tendencies, or habits, construct are transmitted into the future. The following concrete examples should help further clarify this thesis.

2 Interpretant as Form of Memory: Examples Across Living Systems

In their chapter titled "T Cell Memory" in *From Innate Immunology to Immunological Memory*, J.T. Tan and Charles Surh (2006) suggest, "Memory T cells develop in response to a progressive set of cues," and "T Cell memory induced by prior infection or vaccination provides enhanced protection against subsequent microbial infections" (Tan and Surh 2006, p. 85). The three successive phases, then, of what Tan and Surh refer to as the "T cell response to an acute infection," i.e., "expansion, contraction, and maintenance," can be viewed through the biosemiotic paradigm. Recognition of the pathogen invokes the expansion of T cells followed by a contraction upon the elimination of the infectious representamen. The maintenance phase indicates the interpretant's subsequent presence as representamen of the system now in a higher complexity (Tan and Surh 2006, p. 86; cf. Tough and Sprent 1994). As a result, "Upon re-exposure … memory T cells respond faster and stronger than naïve T cells" (Tan and Surh 2006, p. 87). Hence, the interpretant when discussing (memory) T cells can be seen as a form of memory.

In his chapter titled, "Plant Communication," Günther Witzany explains how viewing a plant's interaction with its environment through the biosemiotic paradigm reveals the interpretant as a form of memory. According to Witzany, when "Chemical molecules are used as signs," "They function as signals, messenger substances, information carriers and memory media in solid, liquid or gaseous form" (Witzany 2010, p. 27). Specifically,

> The detection of resources and their periodic, cyclic availability plays a key role in plant memory, planning, growth and development. When, for example, young trees obtain water only once a year, they learn to adjust to this over the following years and concentrate their entire growth and development precisely in the expected period. (Witzany 2010, p. 32; Hellmeier, et al. 1997)

Witzany explicitly refers to the adaptation as depending on memory. The idea, here again, is that the plant's more complicated comportment to its environment results through a process of communication which depends on, and is constructed out of, memory. Remembering the scarcity, a plant's change of behavior, i.e., concentration of growth, exhibits a more complex relation with its environment.

Barbieri's discussion of embryos provides another concrete example of both the interpretant as a form of memory and this memory as a condition for the possibility of higher complexity. In regard to embryos, Barbieri invokes a celebrated quote from Alberts, et al.'s *Molecular Biology of the Cell* (1989):

> During embryonic development cells must not only become different, they must also 'remain' different ... The differences are maintained because the cells somehow remember the effects of those past influences and pass them on to their descendants ... Cell memory is crucial for both the development and the maintenance of complex patterns of specialization. (Alberts et al. 1989, p. 901; Barbieri 2003, p. 113, entire quote emphasized in original)

What Barbieri and Alberts, et al., refer to as "cell memory" can be seen as the interpretant's subsequent use as a representamen and the capacity for differentiation and complexity which results. Just as Alberts, et al., claim "through cell memory, the final combinatorial specification is built up step by step" (Alberts et al. 2008, p. 466). Barbieri concludes regarding "the overall increase of complexity in the system" that it "is entirely dependent on the memories which are used in a reconstruction, because it is only in the memory space that new information appears" (Barbieri 2003, p. 206). Hence, "memory space" is supposed to refer to the change in complexity transmitted into the future through the development of a tendency, or habit, of recognition regarding the interpretant.

Lastly, in regard to animals, consider Jacob von Uexküll's suggestion that "meaning can and does arise from the interactional 'closure' afforded by the generative functional cycle of perception, action and consequence" (Brier 2010, pp. 699–700). Using such a strategy, Brett Buchanan explicates Uexküll's notion of "embodied anticipatory power" as "evolutionary intentionality" in his book *Onto-Ethologies: The Animal Environments of Uexküll, Heidegger, Merleau-Ponty, and Deleuze*, by invoking Hoffmeyer: "To say that living creatures harbor intentions is tantamount to saying that they can differentiate between phenomena in their surroundings and react to them selectively" (Hoffmeyer 1996, p. 47; cf. Buchanan 2008; cf. Deleuze 1994; cf. Heidegger 1962; cf. Uexküll 2010). Hence, as you can most likely anticipate by now, an animal's capacity for environmental differentiation and selection, viewed from the biosemiotic paradigm, depends on the interpretant's subsequent use as a representamen.

In sum, by interpreting the sign as the fundamental unit for the study of life, biosemiotics seems committed to the claim that mind is transmitted through the interpretant. And, in so far as the interpretant is a form of memory, according to the internal logic of the biosemiotic paradigm, memory is the origin of mind. In fact, biosemioticians Kull, Deacon, Emmeche, Hoffmeyer, and Stjernfelt seem to affirm such a conclusion in stating, "semiotic processes include memory processes in general, which maintain continuity of information and stability of dynamical options" (Kull et al. 2009, p. 172; cf. Jämsä 2008, p. 80). Likewise, Pattee understands "an interpreter as a semiotically closed localized (bounded) system that survives or self-reproduces in an open environment by virtue of its memory-stored constructions and controls" (Pattee 1997, p. 127).

Further, just as Hoffmeyer indicates, "living systems are basically engaged in semiotic interactions, that is, interpretative processes" (Hoffmeyer 2010, p. 367), Barbieri holds, "learning requires a memory where the results of experience are accumulated, which means that interpretation is also a *memory-dependent process*" (Barbieri 2008c, p. 45, emphasis in original). What remains to be shown, then, is an example regarding the human origin of mind in relation to the biological problem of memory storage and an argument showing why biosemioticians should affirm the thesis of mnemo-psychography as the solution to the problem of the origin of mind over biosemiotic versions of identity theory.[1]

3 The Biological Problem of Memory Storage and the Identity Theory of Mind

According to Eric Kandel, the biological problem of memory storage "has a systems and a molecular component" (Kandel and Pittenger 1999, p. 2027). The molecular component pertains to the biological changes necessary for memory to occur, e.g., "the formation of long-term memory requires the synthesis of new protein" (Kandel 2009, p. 12750; cf. Black et al. 1988). The system's component pertains to the biological change involved in the differentiation and determination of spatial configurations external to the organism, e.g., "pattern completion," "pattern separation," and "spatial maps" (Kandel 2009, p. 12750; cf. Buonomano 2007). On the one hand, "These perspectives on the study of memory differ in the questions they ask, their methodology and their conceptual framework" (Kandel and Pittenger 1999, p. 2046). On the other hand, "our understanding of the mechanisms of memory will not be complete until we can unite both perspectives into a single, unified framework" (Kandel and Pittenger 1999, p. 2046).

The framework for which Kandel calls invokes the philosophical question regarding the connection between an organism's central nervous system and its external environment (cf. Place 1956; cf. Feigl 1958; cf. Smart 1959; cf. Chalmers 1995; cf. Sellars 1956; cf. Deacon 2010). Now, the biosemiotic paradigm may provide this unified framework. However, because biosemiotics has yet to solidify an accepted thesis regarding the origin of mind, I will show why choosing (a) the representamen or (b) the object of the sign as the origin of mind is incorrect. In this way, I advocated here for mnemo-psychography by process of elimination.

The framework I have been outlining based on the thesis that the mind originates out of memory, i.e., the thesis of mnemo-psychography, would describe the specific human formulation much like the animal example above. The environmental stimulus would count as the representamen; the central nervous system activity would count as the object; and memory would count as the interpretant. However, as an alternative account, varieties of the identity theory of mind identify nonphysical mind with physical central nervous system activity (cf. Goldberg and Pessin 1997, p. 39). Regarding this framework, then, such a theory would reduce the three terms from environment, brain, and memory to environment and brain.

[1] Elsewhere, I discuss the topic of the relation between mnemo-psychography and the mind-body problem. Here, I am concerned to provide mnemo-psychography as a biosemiotic solution to the origin of mind problem.

William James (1918) provided a famous historical expression of the identity theory of mind:

> however numerous and delicately differentiated the train of ideas may be, the train of brain-events that runs alongside of it must in both respects be exactly its match, and we must postulate a neural machinery that offers a living counterpart for every shading, however fine, of the history of its owner's mind. (James, p. 128)

And, the possibility of just such brain-mind syncing is being raised in conjunction with the success of "brain-machine" interfaces (BMIs), a.k.a. "brain-computer" interfaces (BCIs), by, among others, eminent neuroscientist Miguel A. Nicolelis. According to Nicolelis, "the electrical activity of millions of brain cells (neurons) can be translated into precise sequences of skilled movements" (Nicolelis 2001, p. 403; cf. Nicolelis and Lebedev 2009; cf. O'Doherty, et al. 2011, cf. 2012). Hence, a corresponding syncing between the brain and motor functions of the body seems possible.

However, does this mean that biosemiotics should accept a variety of the identity theory as its thesis regarding the origin of animal mind? Discussing the work of Nicolelis, et al., specifically regarding trained rats, Liz Stillwaggon Swan and Louis J. Goldberg argue that it does. In their article "How is Meaning Grounded in the Organism?" they claim a "short burst of somatosensory neuronal activity (approximately 40 ms in duration) is a spatiotemporal entity that has a specific correspondence to a salient feature of the rat's environment" (Swan and Goldberg 2010, p. 134). They "call this entity a *brain-object*" (ibid). Further, they explain "The Nicolelis' experiments provide a microcosm wherein the external organic world … is linked to the internal organic world of the rat's somatosensory system, and the two are linked by what we are calling brain-objects" (Swan and Goldberg 2010, p. 142). And, they conclude that their model can "be extrapolated to organismic meaning-making in general" (Swan and Goldberg 2010, p. 143). So, how is their thesis a variety of identity theory of mind and how does their thesis relate to the specific animal formulation of the origin of mind problem?

Their thesis suggests that "there is a direct correspondence, an isomorphism really, between the 'objectness' (spatiality and temporality) of the environmental stimulus and the resulting brain-object that represents it" (Swan and Goldberg 2010, p. 141). And, as such, "Brain-objects are the mechanism by which features of the world become features of the brain" (Swan and Goldberg 2010, p. 142). Hence, they have eliminated the presence of interpretation at the level of firstness – their semiotic process begins with "direct correspondence" (cf. Swan and Goldberg 2010, pp. 143–145). This is not eliminative materialism because they are not advocating for eliminating mental representation or mental properties (cf. Churchland 1981; cf. Mach 1897, p.30); rather, if there are mental representations, then what is meant by "brain-object" ultimately encompasses them. As a result, mental representation is *identified* as neural activity, i.e., this is a variety of identity theory of mind. However, by eliminating interpretation from the initiation of the experiential semiotic process, they advance the position that the (brain-) object is the origin of animal mind. In other words, since the interpretant is eliminated and the representamen is environmental, i.e., it is not part of the organism and cannot be the origin of its mind, their only option is the (brain-) object.

Though discussion of Barbieri's concern to advance a noninterpretation based biosemiotics is beyond the current scope, I suggest two justifications for not following the "direct correspondence," i.e., eliminating interpretation at the level of firstness, strategy. And, I invoke Barbieri here because Swan and Goldberg refer to Barbieri (2008b) to justify not following Pierce's notion of sign (cf. Swan and Goldberg 2010, pp. 144–145). First, it seems to me that whatever meaning an organism experientially derives, the organism is always already in its environment such that the derivation of meaning itself constitutes an interpretation of the environment (cf. Kant 1998, p. 110, B xvi; cf. Favareau 2010). Second, it seems Barbieri himself might not advocate for the (brain-) object strategy. According to Barbieri,

> There is no doubt that processes of interpretation take place almost everywhere in the living world, and the Peirce model applies therefore to an impressive range of biological phenomena. There is however *one* outstanding *exception* to that rule. The exception is *the genetic code*. (Barbieri 2008b, p. 180, emphasis added)

Further, "Animals build representations (or internal models) of the world whereas single cells cannot physically do that. This implies the existence of two distinct types of semiosis, one based on interpretation [for animals] and one based on coding [for single cells]" (Barbieri 2009, p. 237). Hence, I advocate for the interpretant, as opposed to the representamen or the object, as the solution to the specific formulation of the origin of mind problem.

4 Mnemo-psychography: Mind Writes Itself Out of Memory

"Mnemo-psychography" etymologically suggests that the mind (ψυχή) writes (γραφή) itself out of memory. As a biosemiotic thesis, mnemo-psychography refers to the process unfolding from firstness in regard to the signs of life. That the interpretant is not initially an object is not a semiotic reason to deny its presence. And, as present, the transmission of information as meaningful and the future complexity derive from the interpretant's interpretation. Hence, in the case of an animal or a human experiencing its environment, memory is the origin of such transmission and derivation. The following concrete examples regarding humans should finalize this chapter as sufficient support, then, for the thesis of mnemo-psychography as the biosemiotic solution to the origin of mind problem.

In 1956, George A. Miller published a now famous paper titled, "The Magical Number Seven, plus or minus two: Some limits on our capacity for processing information" (Miller 1956). I mention Miller's publication to examine two relevant notions that he discusses – recoding and chunking. According to Miller, "The process of memorizing may be simply the formation of chunks, or groups of items that go together" (Miller 1956, p. 95). Further, he distinguishes between chunks and bits; bits compose chunks. And, the distinction is important because it was in this way that he was able to discuss recoding. Miller was concerned with understanding how mnemonists are able to remember and recall items in such large numbers. He notes,

> It is a little dramatic to watch a person get 40 binary digits in a row and then repeat them back without error. However, if you think of this merely as a mnemonic trick for *extending the memory span*, you will miss the more important point that is implicit in nearly all such mnemonic devices. The point is that recoding is an extremely powerful weapon *for increasing the amount of information* that we can deal with. *In one form or another we use recoding constantly in our daily behavior.* (Miller 1956, pp. 94–95, emphasis added)

Recoding, then, amounts to influencing unit formation at the level of firstness. By "chunking" the "bits," transmission of environmental information as meaningful, i.e., recognizing the representamen as object, can be altered. Notice, if the bits are chunked, then the object will be different from the object that results from the representamen's unchunked recognition.

Hence, the interpretant is a form of memory, since object determination itself correlates with memory, and the complexity resulting from the experience of the environment actually derives from the organism's memory. Moreover, memory is at work influencing an organism's nonconscious engagement with its environment in multiple ways. On the one hand, "There is reason to believe … that each sensory system might conceivably be accompanied by a relatively unique memory system" (Spear and Riccio 1994, p. 346). On the other hand, since "automaticity is not driven by stimuli separately from skills" (Jacoby, et al. 1993, p. 261), "chunking may be the primary process that underlies automaticity" (Dehn 2008, p. 122).

Lastly, then, the feature of memory known as "priming" is an excellent example of both the presence of the interpretant as a form of memory prior to the correlate of object recognition in the sign and Uexküll's notion of "anticipatory power." According to Daniel Schacter and Endel Tulving, "Priming is a nonconscious form of human memory, which is concerned with perceptual identification of words and objects" (Tulving and Schacter 1990, p. 301; cf. Schacter and Badgaiyan 2001; cf. Schacter and Buckner 1998). Further, "Priming is a nonconscious form of memory that involves a change in a person's ability to identify, produce or classify an item as a result of a previous encounter with that item *or a related item*" (Schacter, et al. 2004, p. 853, emphasis added; cf. Tulving and Schacter 1990, 1992). In other words, just as it is possible to finish a sentence for someone else based not on clairvoyance but on memory, e.g., sentence structure in general or an interlocutor's tendencies, this feature of memory is at work in one's engagement with the environment. For example, think of "muscle memory" in regard to perception or catching an object that slips from grip without visually seeing it. Hence, priming points to, in humans, what Hoffmeyer calls "Nature's 'taking of habits' [an allusion, of course, to Peirce] – in other words, its tendency to develop new regularities as the result of its own ongoing interactions – has been at work at all times" (Hoffmeyer 2010, p. 602).

5 Conclusion

To conclude, in this chapter, I advocated for the thesis of mnemo-psychography as the biosemiotic solution to the problem of the origin of mind in both its general and specific formulations. To do so, I invoked Peirce's semiotic notion of sign as a triadic relation between representamen, object, and interpretant, and I considered multiple examples at varying levels of complexity. I argued that the interpretant is a form of memory, and I indicated the three functions of the interpretant which reveal it as a form of memory.

I invoked the biological problem of memory storage from Kandel's research and the neuroscientific research of Nicolelis, et al., toward further illustrating the specific formulation of the origin of mind problem. I examined a biosemiotic version of the identity theory of mind as a possible counterargument to my thesis of mnemo-psychography. And, I invoked contemporary memory research – such as Schacter on priming – further evidencing my claim that the interpretant is a form of memory.

Given the central role of memory in the semiotic process, biosemiotics seems necessarily committed to what I call the thesis of mnemo-psychography. Mnemo-psychography solves the origin of mind problem, since the interpretant as a form of memory is both responsible for the transmission of information as meaning and for the derivation of higher levels of complexity. In other words, the mind writes itself out of memory, and memory is the origin of mind.[2]

Bibliography

Alberts, B., Bray, D., Lewis, J., Raff, M., Roberts, K., & Watson, J. D. (1989/2008). *Molecular biology of the cell*. New York: Garland Publishing.

Anderson, M., Deely, J., Krampen, M., Ransdell, J., Sebeok, T., & Uexküll, T. (1984). A semiotic perspective on the sciences: Steps toward a new paradigm. *Semiotica, 52*(1/2), 7–47.

Barbieri, M. (2003). *The organic codes: An introduction to semantic biology*. Cambridge: Cambridge University Press.

Barbieri, M. (2008a). Biosemiotics: A new understanding of life. *Naturwissenschaften, 95*, 577–599.

Barbieri, M. (2008b). Is the cell a semiotic system? In M. Barbieri (Ed.), *Introduction to biosemiotics* (pp. 179–207). Dordricht: Springer.

Barbieri, M. (2008c). Life is semiosis: The biosemiotic view of nature. *Cosmos and History: The Journal of Natural and Social Philosophy, 5*(1/2), 29–51.

Barbieri, M. (2008d). What is biosemiotics? *Biosemiotics, 1*, 1–3.

Barbieri, M. (2009). A short history of biosemiotics. *Biosemiotics, 2*(2), 221–245.

Black, I. B., Adler, J. E., Dreyfus, C. F., Friedman, W. F., LaGamma, E. F., & Roach, A. H. (1988). Experience and the biochemistry of information storage in the nervous system. In M. S. Gazzaniga (Ed.), *Perspectives in memory research* (pp. 3–22). Cambridge, MA: The MIT Press.

[2] I would like to thank Dr. Bernard Baars, Dr. Patrick Reider, and Dr. Stephanie Swales for their helpful comments. And, I would like to especially thank Dr. Liz Stillwaggon Swan for her patience, dedication, and helpful comments.

Brier, S. (2010). The cybersemiotic model of communication: An evolutionary view on the threshold between semiosis and informational exchange. In D. Favareau (Ed.), *Essential readings in biosemiotics* (pp. 697–730). New York: Springer.
Buchanan, B. (2008). *Onto-ethologies: The animal environments of Uexküll, Heidegger, Merleau-Ponty, and Deleuze*. Albany: SUNY Press.
Buonomano, D. V. (2007). The biology of time across different scales. *Nature Chemical Biology, 10*, 594–597.
Chalmers, D. (1995). Facing up to the problem of consciousness. *Journal of Consciousness Studies, 2*(3), 200–219.
Churchland, P. M. (1981). Eliminative materialism and the propositional attitudes. *The Journal of Philosophy, 78*(2), 67–90.
Deacon, T. (2010). Excerpts from the symbolic species. In D. Favareau (Ed.), *Essential readings in biosemiotics* (pp. 541–852). New York: Springer.
Dehn, M. J. (2008). *Working memory and academic learning: Assessment and intervention*. Hoboken: Wiley.
Deleuze, G. (1994). *Difference and repetition*. (P. Patton, Trans.). New York: Columbia University.
Eder, J., & Rembold, H. (1992). Biosemiotics – A paradigm of biology: Biological signaling on the verge of deterministic chaos. *Naturwissenschaften, 79*(2), 60–67.
Emmeche, C. (1991). A semiotical reflection on biology, living signs and artificial life. *Biology and Philosophy, 6*(3), 325–340.
Favareau, D. (2007). How to make Peirce's ideas clear. In G. Witzany (Ed.), *Biosemiotics in transdisciplinary contexts* (pp. 163–177). Helsink: Umweb Press.
Favareau, D. (2010). Introduction: An evolutionary history of biosemiotics. In D. Favareau (Ed.), *Essential readings in biosemiotics* (pp. 1–80). New York: Springer.
Feigl, H. (1958). The "mental" and the "physical.". In H. Feigl, M. Scriven, & G. Maxwell (Eds.), *Concepts, theories and the mind-body problem* (Minnesota studies in the philosophy of science, Vol. II, pp. 370–497). Minneapolis: University of Minnesota Press.
Goldberg, S., & Pessin, A. (1997). *Gray matters: An introduction to the philosophy of mind*. New York: Armonk.
Heidegger, M. (1962). *Being and time*. (J. Macquarrie & E. Robinson, Trans.). New York: Harper and Row.
Hellmeier, H., Erhard, M., & Schulze, E. D. (1997). Biomass accumulation and water use under arid conditions. In F. A. Bazzaz & J. Grace (Eds.), *Plant resource allocation* (pp. 93–113). London: Academic.
Hoffmeyer, J. (1995). The swarming cyberspace of the body. *Cybernetics and Human Knowing, 3*(1), 16–25.
Hoffmeyer, J. (1996). *Signs of meaning in the universe*. (B. J. Haveland, Trans.). Bloomington: Indiana University Press.
Hoffmeyer, J. (2006). Genes, development, and semiosis. In E. Neumann-Held & C. Rehmann-Sutter (Eds.), *Genes in development: Re-reading the molecular paradigm* (pp. 152–174). Durham: Duke University Press.
Hoffmeyer, J. (2010). The semiotics of nature: Code-duality. In D. Favareau (Ed.), *Essential readings in biosemiotics* (pp. 583–628). New York: Springer.
Hoffmeyer, J., & Emmeche, C. (1991). Code-duality and the semiotics of nature. In M. Anderson & F. Merrell (Eds.), *On semiotic modeling* (pp. 117–166). Berlin: Mouton de Gruyter.
Jacoby, L. L., Ste-Marie, D., & Toth, J. P. (1993). Redefining automaticity: Unconscious influences, awareness, and control. In A. D. Baddeley & L. Weiskrantz (Eds.), *Attention, selection, awareness, and control: A tribute to Donald Broadbent* (pp. 261–282). London: Oxford University Press.
James, W. (1918). The automaton-theory. In *The principles of psychology* (Vol. 1). New York: Dover Publications.
Jämsä, T. (2008). Semiosis in evolution. In M. Barbieri (Ed.), *Introduction to biosemiotics* (pp. 69–100). Dordrecht: Springer.

Kandel, E. (2009). The biology of memory: A forty-year perspective. *The Journal of Neuroscience, 29*(41), 12748–12756.
Kandel, E., & Pittenger, C. (1999). The past, the future and the biology of memory storage. *Philosophical Transactions of the Royal Society of London, 354*, 2027–2052.
Kant, I. (1998). *Critique of pure reason*. (P. Guyer & A. W. Wood, Trans.). Cambridge: Cambridge University Press.
Kull, K., Deacon, T., Emmeche, C., Hoffmeyer, J., & Stjernfelt, F. (2009). Theses on biosemiotics: Prolegomena to a theoretical biology. *Biological Theory, 4*(2), 167–173.
Mach, E. (1897). *Contributions to the analysis of the sensations*. (C. M. Williams, Trans.) Chicago: The Open Court Publishing Co.
Marvell, L. (2007). *Transfigured light: Philosophy, cybernetics and the hermetic imaginary*. Bethesda: Academia Press.
Miller, G. A. (1956). The magical number seven, plus or minus two: Some limits on our capacity for processing information. *Psychological Review, 63*(2), 81–97.
Nicolelis, M. A. L. (2001). Actions from thoughts. *Nature, 409*, 403–407.
Nicolelis, M. A. L., & Lebedev, M. A. (2009). Principles of neural ensemble physiology underlying the operation of brain-machine interfaces. *Nature Reviews Neuroscience, 10*, 530–540.
Nöth, W. (1990). *Handbook of semiotics*. Bloomington: Indiana University Press.
O'Doherty, J. E., Lebedev, M. A., Ifft, P., Zhuang, J., Katie, Z., Shokur, S., Bleuler, H., & Nicolelis, M. A. L. (2011). Active tactile exploration using a brain-machine-brain interface. *Nature, 479*, 228–231.
O'Doherty, J. E., Lebedev, M. A., Zeng, L., & Nicolelis, M. A. L. (2012). Virtual active touch using randomly patterned intracortical microstimulation. *Neural Systems and Rehabilitation Engineering, 20*(1), 85–93.
Pain, S. P. (2007). The ant on the kitchen counter. In M. Barbieri (Ed.), *Biosemiotic research trends* (pp. 113–140). New York: Nova Science.
Pattee, H. H. (1997). The physics of symbols and the evolution of semiotic controls. In M. Coombs & M. Sulcoski (Eds.), *Control mechanisms for complex systems: Issues of measurement and semiotic analysis* (pp. 9–25). Albuquerque: University of New Mexico Press.
Peirce, C. S. (1998). Nomenclature and divisions of triadic relations, as far as they are determined. In the Peirce Edition Project (Ed.), *The essential Peirce: Selected philosophical writings* (Vol. 2, pp. 289–299). Bloomington: Indiana University Press.
Peirce, C. S. (2011). Logic as semiotic: The theory of signs. In J. Buchler (Ed.), *Philosophical writings of Peirce* (pp. 98–119). New York: Dover Publications.
Place, U. T. (1956). Is consciousness a brain process? *British Journal of Psychology, 47*, 44–50.
Schacter, D. L., & Badgaiyan, R. D. (2001). Neuroimaging of priming: New perspectives on implicit and explicit memory. *Current Directions in Psychological Science, 10*(1), 1–4.
Schacter, D. L., & Buckner, R. L. (1998). Priming and the brain. *Neuron, 20*, 185.
Schacter, D. L., Dobbins, I. G., & Schnyer, D. M. (2004). Specificity of priming: A cognitive neuroscience perspective. *Nature Reviews Neuroscience, 5*, 853–862.
Sebeok, T. (1991). *A sign is just a sign*. Bloomington: Indiana University Press.
Sellars, W. (1956). Empiricism and the philosophy of mind. In H. Feigl & M. Scriven (Eds.), *Minnesota studies in the philosophy of science,* Vol. 1, (pp. 253–329). Minneapolis, MN: University of Minnesota Press.
Smart, J. J. C. (1959). Sensations and brain processes. *Philosophical Review, 68*(2), 141–156.
Spear, N. E., & Riccio, D. C. (1994). *Memory: Phenomena and principles*. Boston: Allyn and Bacon.
Swan, L. S., & Goldberg, L. J. (2010). How is meaning grounded in the organism? *Biosemiotics, 3*, 131–146.
Tan, J. T., & Surh, C. D. (2006). T cell memory. In B. Pulendran & R. Ahmed (Eds.), *From innate immunity to immunological memory* (pp. 85–115). Berlin: Springer.
Tough, D. F., & Sprent, J. (1994). Turnover of naïve- and memory-phenotype T cells. *The Journal of Experimental Medicine, 179*, 1127–1135.
Tulving, E., & Schacter, D. L. (1990). Priming and human memory systems. *Science, 247*(4940), 301–306.

Tulving, E., & Schacter, D. (1992). Priming and memory systems. In B. Smith & G. Adelman (Eds.), *Neuroscience year: Supplement 2 to the encyclopedia of neuroscience* (pp. 130–133). Boston: Birkhauser.

Uexküll, J. (2010). The theory of meaning. In D. Favareau (Ed.), *Essential readings in biosemiotics* (pp. 81–114). New York: Springer.

Witzany, G. (2010). Excerpts from the logos of the bios. In D. Favareau (Ed.), *Essential readings in biosemiotics* (pp. 731–750). New York: Springer.

Part V
Synthetic Intelligence

Minimal Mind

Alexei A. Sharov

Abstract In contrast to the human standard for mind established by Alan Turing, I search for a "minimal mind," which is present in animals and even lower-level organisms. Mind is a tool for the classification and modeling of objects. Its origin marks an evolutionary transition from protosemiotic agents, whose signs directly control actions, to eusemiotic agents, whose signs correspond to ideal objects. The hallmark of mind is a holistic perception of objects, which is not reducible to individual features or signals. Mind can support true intentionality of agents because goals become represented by classes or states of objects. Basic components of mind appear in the evolution of protosemiotic agents; thus, the emergence of mind was inevitable. The classification capacity of mind may have originated from the ability of organisms to classify states of their own body. Within primary modeling systems, ideal objects are not connected with each other and often tailored for specific functions, whereas in the secondary modeling system, ideal objects are independent from functions and become interconnected via arbitrarily established links. Testing of models can be described by commuting diagrams that integrate measurements, model predictions, object tracking, and actions. Language, which is the tertiary modeling system, supports efficient communication of models between individuals.

1 Introduction

Mind is traditionally considered as a human faculty responsible for conscious experience and intelligent thought. Components of mind include perception, memory, reason, logic, modeling of the world, motivation, emotion, and attention (Premack and Woodruff 1978). This list can be easily expanded to other kinds of human mental

A.A. Sharov (✉)
National Institute on Aging, Baltimore, MD, USA
e-mail: sharoval@mail.nih.gov

L. Swan (ed.), *Origins of Mind*, Biosemiotics 8, DOI 10.1007/978-94-007-5419-5_18,

activities. Defects in mental functions (e.g., in logic, attention, or communication) are considered as a loss of mind, partial or complete. In short, mind is a collection of mental functions in humans. However, this definition tells us nothing about the nature of mind. Human mental functions are so diverse that it is difficult to evaluate their relative importance. The only way to identify the most fundamental components of mind is to track its origin in animals, which inevitably leads us to the idea that mind exists beyond humans. Animal mental activities (i.e., "animal cognition") are definitely more primitive compared to those of the human mind, but they include many common components: perception, memory, modeling of the world, motivation, and attention (Griffin 1992; Sebeok 1972). The lack of abstract reasoning in animals indicates that reason is not the most fundamental element of mind but rather a late addition.

By accepting the existence of mind in animals, we commit ourselves to answer many difficult questions. For example, where is the lower evolutionary threshold for mind? Does mind require brain or at least some kind of nervous system? In other words, we enter the quest for the "minimal mind," which is the topic of this chapter. This evolutionary approach is opposite to Turing's criterion for machine intelligence, which is based on the ability of a human to distinguish between a computer and a human being based solely on communication with them (Turing 1952). To be indistinguishable from a human, a machine should have a "maximal mind" that is functionally equivalent to the human mind. Here, I propose that minimal mind is a tool for the classification and modeling of objects and that its origin marks an evolutionary transition from protosemiotic agents, whose signs directly control actions, to eusemiotic agents, whose signs correspond to ideal objects.

2 Agents

Mind is intrinsically related to life because it is a faculty of living systems. However, according to cybernetics, it can also exist in artificial devices (Nillson 1998). To present a unified approach to mind, we need first to discuss briefly the nature of life and artifacts. Machine metaphor is often perceived as a misleading simplification of the phenomena of life and mind (Deacon 2011; Emmeche and Hoffmeyer 1991). The motivation to separate life and mind from machines comes from the fact that simple machines are manufactured and programmed by humans, whereas organisms are self-produced and develop from eggs into their definite shape (Swan and Howard 2012). Also, machines change their state following deterministic rules rather than internal goals and values. But, despite these differences, the progress in understanding life and mind seems to lie in bridging the gap between life and artifacts rather than in building a wall between them. In particular, biological evolution can be seen as a sequence of inventions of various instruments that are needed to perform living functions (Dennett 1995). Cellular processes are based on molecular machines that copy sequences of nucleic acids, synthesize proteins, modify them, and assemble them into new molecular machines. Thus, components of organisms are manufactured, and living systems are indeed artifacts (Barbieri 2003). Although man-made machines lack some features of living organisms, this deficiency should be attributed to our insufficient knowledge and experience. Humans only just began learning how to make self-programmable and self-repairable mechanisms, whereas living cells mastered these skills billions of years ago.

One of the heuristics of systems methodology is "functionalism," which assumes that systems should be compared based solely on their functions rather than their material composition. This idea was initially proposed as a foundation for "relational biology" (Rashevsky 1938; Rosen 1970) and later was formulated as "functional isomorphism" (Putnam 1975). If an artificial system performs the same (or similar) functions as a living organism, then there is good reason to call it "alive." However, it would be confusing to apply the term "living organism" to artificial devices. Instead, it is better to use the term "agent" which fits equally well to living organisms and artificial devices. Agents should not be viewed only as externally programmed devices, as is commonly done in cybernetics. Although all agents carry external programs, the majority of agents, including all living organisms, also have self-generated programs. An agent is a system with spontaneous activity that selects actions to pursue its goals. Goals are considered in a broad sense, including both achievable events (e.g., capturing a resource, reproduction), and sustained values (e.g., energy balance). Some goals are externally programmed by parental agents or higher-level agents, and other goals emerge within agents. Note that mind is not necessarily present in agents. Simple agents can automatically perform goal-directed activities based on a program.

In the field of artificial intelligence, ideas of functionalism are often misinterpreted as a primacy of the digital program over the body/hardware and environment. Internet-based programs like the virtual world of "Second Life" may convince people that their functionality can be fully digitized in the future. However, programs are not universal but instead tailored for specific bodies and environments and therefore can be exchanged without loss of functionality only between similar agents in similar environments. Thus, "digital immortality" is a myth (Swan and Howard 2012). Self-producing agents have many body-specific functions associated with metabolism, assembly of subagents, growth, development, and reproduction. Obviously, these functions cannot be realized in a qualitatively different body. But functional methodology works even in this case because the body can support a large number of alternative activities, and it needs information to organize and control these activities. In summary, agents require *both* specific material organization (body) and functional information to control their actions.

Agents are always produced by other agents of comparable or higher functional complexity (Sharov 2006). This statement is an informational equivalent of the gradualism principle in the theory of evolution (Sharov 2009b). The reason why agents cannot self-assemble spontaneously is that they carry substantial functional complexity. Long evolutionary (or learning) timelines are required to develop each new function via trial and error; therefore, simultaneous and fast emergence of numerous novel functions is unlikely. The origin of life does not contradict the principle of gradualism because primordial agents were extremely simple and started from single functions (Sharov 2009a). The production of artificial agents by humans also satisfies the principle of gradualism because humans have a higher level of functional complexity than any human-made devices. Methods of agent manufacturing may include assembly from a set of parts as well as self-organization and self-development. Although the majority of human-made agents are assembled, some of them use elements of development. For example, satellites can unfold and reassemble in space after launch. Self-assembly is a common approach in nanotechnology and in synthetic organisms.

3 Functional Information

Agents are unusual material objects whose dynamics cannot be effectively described by physics, although they do not contradict physics. Instead, a semiotic description appears more meaningful: agents carry functional information, which is a collection of signs that encode and control their functions. The adjective "functional" helps to distinguish functional information from quantitative approaches developed by Shannon and Kolmogorov (Shannon 1948; Kolmogorov 1965). Although signs are material objects, they have functions within agents that are not directly associated with their physical properties.

Semiotics stems from the work of Charles Sanders Peirce, who defined a sign as a triadic relationship between a sign vehicle, object, and interpretant, which is a product of an interpretive process or a content of interpretation (Peirce 1998). However, not all agents can associate signs with content or meaning. Thus, I prefer a more generic definition of signs as objects that are used by agents to encode and control their functions (Sharov 2010). Most signaling processes that take place within the cells of living organisms do not invoke ideal representations, but they encode and/or control cellular functions and thus have a semiotic nature. Peirce deemphasized the role of agents in informational processes and did not consider the agent or organism as a component of the triadic sign relationship. He thought that meanings belonged to nature rather than to agents. For example, he wrote about nature's ability to acquire habits, which is consistent with his philosophy of objective idealism. Similar views were expressed by Jesper Hoffmeyer who assumed "minding nature" (Hoffmeyer 2010). In contrast, I view signs only in connection with agents who use them and see no reason to consider nature an agent. Although it may be hard to refute claims that the universe or Gaia are superorganisms (Lovelock 1979), I take a conservative approach and use the notion of "agent" only for those systems that clearly show a reproducible goal-directed activity and carry functional information to organize this activity (Sharov 2010).

Functional information is inseparable from agents who use it. Living organisms are products of their genome, which controls their development and growth. In contrast, cybernetics often distinguishes information (software) from computational devices (hardware). The distinction of software and hardware is meaningful only for slave agents like computers, which are produced and externally programmed by humans. A computer is similar to a ribosome in a living cell, because ribosomes are manufactured and externally programmed to make proteins. Programmed agents are often viewed as nonsemiotic systems (Barbieri 2008). However, this idea appears confusing because the execution of a program is a part of the semiotic activity of all agents, and agency is not possible without it. We humans are programmed genetically by our ancestors, behaviorally by our parents, and culturally by our society. These programs support our identity as a *Homo sapiens* species, as well as our race, sex, nationality, personality, and a whole range of physical and mental abilities. In addition to external programs, humans and most other organisms develop their own programs. When we learn new behaviors and skills, we convert them into programs that can be executed automatically or with minimal intervention from our consciousness. These self-generated programs comprise our personal identity. Our freedom comprises only a tiny fraction of our functional behavior. In fact, freedom would be destructive if it were not well balanced with programmed functions that can correct mistakes. But evolution would not be possible if all agents were 100% externally programmed, and nonevolving agents would perish in changing environments. Thus, the role of fully programmed agents is limited to supportive functions for other agents that are able to evolve and learn.

The meaning of functional information is grounded in a communication system, which is a set of compatible communicating agents (Sharov 2009c). For example, the genome alone does not mean anything; it has meaning only in relation to the organisms that use it. An egg can be viewed as a minimal interpreter of the genome (Hoffmeyer 1997). Although the structure of an egg is encoded by the genome, a real egg is needed to interpret the genome correctly. Thus, heredity is based on a combination of [genome+egg] rather than on the genome alone. This leads us to the idea that functional information is not universal but has its meaning only in relation to a certain communication system. Even a single agent is involved in a continuous self-communication through memory and therefore can be viewed as a communication system. Memory is a message sent by an agent to its own future state, and its purpose is to preserve the agent's ability to perform certain functions. Heredity is an extended self-communication or intergeneration memory (Sharov 2010). Other communication systems include multiple agents that exchange signals or messages. The most common example of such horizontal communication in living organisms is sexual reproduction, where the egg encounters an unfamiliar paternal genetic sequence. Agents from different communication systems do not exchange functional information on a regular basis because their interpretation modules are not fully compatible. For example, most interspecies hybrids in mammals are nonviable or sterile as a result of misinterpretation of the paternal genome. Communication systems often have a hierarchical structure. For example, species are partitioned into populations, which in turn are partitioned into colonies or families. Subagents within organisms (e.g., cells) make their own communication systems. Communication is often asymmetric when one kind of agent manipulates the functional information of another kind of agent. For example, agents can (re)program their subagents or offspring agents. Asymmetric communication often occurs between interacting organisms of different species (e.g., parasites reprogram their hosts, or preys mislead predators via mimicry and behavioral tricks). Because communication systems are multiscale and interdependent, evolution happens at multiple levels simultaneously.

4 Emergence of Mind from Elementary Signaling Processes

Mind is not a necessary component of agents. Bacteria are examples of mindless agents that operate via elementary signaling processes such as DNA replication, transcription, translation, and molecular sensing. They do not perceive or classify objects in the outside world as humans do; instead, they detect signals that directly control their actions. Direct control, however, may include multiple steps of signal transfer as well as logical gates. Following Prodi, I call this primitive level of semiosis "protosemiosis" (Prodi 1988). Protosemiosis does not include classification or modeling of objects; it is "know-how" without "know-what." Because molecular signaling is so different from higher levels of semiosis, Eco excluded it from consideration in semiotics (Eco 1976). However, the analysis of molecular signs in bacteria helps us to understand the origin and nature of signs in animals and humans; thus, protosemiosis should not be dismissed. Protosigns (i.e., signs used in protosemiosis) do not correspond to any object, which may seem confusing because our brains are trained to think in terms of objects. Although we associate a triplet of nucleotides in the mRNA with an amino acid as an object, a cell does not have a holistic internal representation of amino acid; thus, it is not an object for a cell. Instead, a triplet of nucleotides in the mRNA is associated with an action of tRNA and ribosome, which together append an amino acid to the growing protein chain.

Mind represents a higher level of information processing compared to protosemiosis because it includes classification and modeling of objects and situations (e.g., food items, partner agents, and enemies). These classifications and models represent the "knowledge" an agent has about itself and its environment, which are Innenwelt and Umwelt following the terminology of Uexküll (1982). I proposed calling this new level of semiosis "eusemiosis" (Sharov 2012). Information processing in eusemiosis can no longer be tracked as a sequence of signal exchanges between components. Instead, it goes through multiple semi-redundant pathways, whose involvement may change from one instance to another but invariantly converge on the same result. Thus, attractor domains are more important for understanding the dynamics of mind than individual signaling pathways. The classification of objects can be viewed as a three-step process. The first step is immediate perception, when various receptors send their signals to the mind, and these signals collectively reset the mind to a new state (or position in a phase space). The second step is the internal dynamics of mind which start with the new state of mind and then converge to one of the attractors. This process is equivalent to recognition or classification. Each attractor represents a discrete meaningful category (e.g., fruit or predator), which I call "ideal object." In contrast to real objects that are components of the outside world, ideal objects exist within the mind and serve as tools for classifying real objects. Finally, at the third step, the ideal object acts as a checkpoint to initiate some other function (physical or mental).

Ideal objects do not belong to a different parallel universe as claimed by Popper (1999). Instead, they are tools used by agents to perceive and manipulate the real world. Following the "law of the instrument" attributed to Mark Twain, to a man with a hammer, everything looks like a nail. Thus, ideal objects within mind determine how the outside world is perceived and changed. Ideal objects are implemented as functional subunits within complex material systems, for example, as specific patterns of neuronal activity or "brain-objects" (Swan and Goldberg 2010). But the material implementation of ideal objects is flexible whereas the function is stable. Similarly, computer programs are functionally stable despite the fact that they are loaded each time into a different portion of physical memory and executed by a different processor (if available).

"Object" is one of the most complex and abstract notions in human thought. However, we should not transfer all this complexity to simple agents like worms or shellfish. For example, we usually distinguish between objects and their attributes, where attributes are generic (e.g., whiteness) and can be applied to various classes of objects. Although we cannot directly assess the minds of simple agents, it is unlikely that they can contemplate generic attributes. Simple agents distinguish between classes of objects, but they do it unconsciously without considering attributes as independent entities. Humans can think of hypothetical ideal objects (e.g., unicorns), which include certain combinations of abstract attributes. Obviously, simple agents are not able to do that. Another difference is that humans can recognize individual objects, whereas simple agents cannot distinguish objects within the same functional category. Learning and modeling capacities of mind have progressed substantially in evolution (see below), and we should not expect that simple agents have the same flexibility in connecting and manipulating ideal objects as humans do.

Mind is a necessary tool for intentional behavior, which I consider a higher level of goal-directed activity. In contrast to protosemiotic agents, mind-equipped agents have holistic representations of their goals, which are perceived as ideal objects and integrate a large set of sensorial data. For example, immune cells of eukaryotic organisms can recognize a viral infection by the shape of the viral proteins as well as by specific features of viral nucleic acids and launch a defense response by producing interferon, antibodies, and cytokines. Memory T cells keep information on the properties of viral proteins acquired during the previous exposure to the same virus.

Goals may emerge internally within agents; however, they can also be programmed externally. For example, instinctive behaviors of organisms are programmed genetically by ancestors. In this case, ideal objects develop somehow together with the growing brain. External programming of goals is typical for artificial minds in robotic devices equipped with automated image processing modules (Cariani 1998, 2011). For example, a self-guided missile is programmed to classify objects into targets and nontargets and to follow the target.

Agents with an externally programmed mind can support a given static set of functions, but they lack adaptability and would not be able to keep a competitive advantage in changing environments. Thus, autonomous agents need adaptive minds capable of improving existing ideal objects and creating new ones via learning. Mind can generate new behaviors by creating novel attractors in the field of perception states and linking them with specific actions. If such behaviors prove useful, they can become habits and contribute to the success of agents. Requirement of learning does not imply that mind-carrying agents learn constantly. Minds may persist and function successfully in a nonlearning state for a long time. Most artificial minds are static replicas of some portion of the dynamic human mind. But minds cannot improve without learning.

The statement "minds cannot improve without learning" is correct if applied to individual agents; however, limited improvements of minds are possible in lineages of self-reproducing nonlearning agents via genetic selection. Mutations may cause the appearance of new attractors in the dynamic state of nonlearning minds or new links between ideal objects and actions. If these heritable representations help agents to perform some functions, the agents will reproduce and disseminate new behaviors within the population. This process, however, is slow and inefficient because of several problems. First, genetic selection can hardly produce any results in such highly redundant systems as minds because most changes of individual elements have no effects on the behavior. In other words, the fitness landscape is almost flat. Second, mind is a complex and well-tuned system; thus, any heritable change to individual elements that does have a phenotype is likely to be disruptive. Third, the functionality of mind has to be assessed in each situation separately because it may work in some cases but not in others. Genetic selection depends mostly on the worst outcome from a single life-threatening situation, and thus, it is ineffective for improving the performance of mind in individual situations. But despite these problems, it is conceivable that limited improvements of mind can be achieved by genetic selection. This helps us to explain how most primitive nonlearning minds appeared in the evolution of protosemiotic agents. Moreover, simple learning algorithms may emerge in the evolution of mind solely via genetic selection, making minds adaptable and partially independent from the genetic selection (see below). But genetic mechanisms are still important for the functionality of mind even in humans because the architecture of the brain is heritable.

5 Components of Minimal Mind Can Emerge Within Protosemiotic Agents

Because the emergence of mind is a qualitative change in organisms, it is difficult to understand the intermediate steps of this process. Here, I argue that all necessary components of mind, which include semi-redundant signaling pathways, stable attractors, and adaptive learning, can emerge at the protosemiotic level. Moreover, these components emerge not as parts of mind (which does not exist yet) but as tools that increase the efficiency of other simpler functions.

Redundancy of signaling pathways may seem to be a waste of valuable resources; however, it appears beneficial for agents in the long run. First, redundancy ensures the reliability of signaling. If one pathway is blocked (e.g., as a result of injury, stress, or infection), then normal functions can be restored via alternate pathways. Each cell has multiple copies of all kinds of membrane-bound receptors because cells cannot predict the direction of incoming signals and thus distribute receptors around the whole surface. Second, redundant signaling pathways may generate novel combinatorial signals. For example, one photoreceptor can only distinguish different intensities of light, but multiple photoreceptors can identify the direction of light and even distinguish shapes. Third, redundant signaling pathways increase the adaptability of agents because some of them may start controlling novel functions in subsequent evolution.

Stable attractors are common to most autoregulated systems, including simple devices with a negative feedback (e.g., centrifugal governor of the steam engine). Stability is necessary for all living organisms to maintain vital functions at optimal rates. Any function that escapes regulation may become harmful and lead to disease or death. However, simple stability in the form of steady states is usually not sufficient for living organisms. Reproduction, growth, and the development of organisms require more complex regulation pathways that combine stability with change in a form of limit cycles, branching trajectories, and even chaotic attractors (Waddington 1968).

Genetic mechanisms are not suitable for learning because the sequence of nucleotides in the DNA is not rewritable (although limited editing is possible). In contrast, simple autocatalytic networks can switch between two stable states ("on" and "off") and serve as a dynamic memory for the cell. Moreover, such networks can support primitive learning (e.g., sensitization and habituation) as well as associative learning as follows from a simple model of two interacting genes (Ginsburg and Jablonka 2009). In this model, genes *A* and *B* are activated by different signals S_a and S_b, and the product P_a of gene *A* has three functions: (1) it induces a specific phenotype or physiological response; (2) it stimulates temporarily the expression of gene *A* so that the gene remains active for some time after the initial signal S_a; and (3) it makes the expression of gene *A* dependent on the product P_b of gene *B*. If gene *A* is silent, then signal S_b activates gene *B*, but its activity does not produce any phenotype. However, if signal S_b comes shortly after signal S_a, then the product P_b will activate gene *A* and produce a phenotype. This network belongs to the protosemiotic level because it is based on fixed interactions between few components.

Because all components of minimal mind can appear within protosemiotic agents, the emergence of mind seems inevitable. But there is still a problem of how to combine these components. In particular, agents have to increase the depth of their hierarchical organization by making a set of partially independent subagents, whose state may switch between multiple attractors with adjustable topology. These subagents, which can be viewed as standard building blocks of mind, should then become connected via adjustable links. It appears that epigenetic mechanisms can convert DNA segments into a network of sub-agents with flexible control, as discussed in the following section.

6 Epigenetic Regulation May Have Supported the Emergence of Minimal Mind

It is difficult to pinpoint the emergence of mind on the evolutionary tree of life. However, it is certain that mind appeared in eukaryotic organisms with well-developed epigenetic regulation. Epigenetic mechanisms include various changes in cells that are long-lasting but do not involve alterations of the DNA sequence. I will consider only those epigenetic mechanisms that are mediated by chromatin structure because they are likely to have facilitated the emergence of mind. Chromatin consists of DNA assembled together with histones, which are specific proteins that support the stability of DNA and regulate its accessibility to transcription factors. Histones can be modified in many ways (e.g., acetylated, methylated, phosphorylated, or ubiquitinated) by molecular agents, and these modifications affect the way histones bind to each other and interact with DNA and other proteins. Some modifications convert chromatin to a highly condensed state (heterochromatin); other modifications support loose chromatin structure (euchromatin), which allows binding of transcription factors and subsequent activation of mRNA synthesis (Jeanteur 2005). Molecular agents can both read and edit histone marks. In particular, they can modify newly recruited histones after DNA replication in agreement with marks on the partially retained parental histones (Jeanteur 2005). As a result, chromatin states survive cell division and are transferred to both daughter cells. Thus, chromatin-based memory signs can reliably carry rewritable information through cell lineages and control differentiation of embryos (Markoš and Švorcová 2009). The chromatin state depends not only on histone marks but also on other proteins that establish links between distal DNA segments, as well as links between chromatin and nuclear envelopes. These proteins, which include insulators, mediators, cohesions, and lamins, create and maintain a complex 3-dimensional structure of the chromatin (Millau and Gaudreau 2011). Distal links create new neighborhoods and change the context for chromatin assembly.

Epigenetic mechanisms are important for the origin and function of mind because (1) they support a practically unlimited number of attractors that are spatially associated with different DNA segments, (2) these attractors can be utilized as rewritable memory signs, and (3) chromatin attractors can become interconnected via products of colocalized genes. Chromatin structure is repaired after mild perturbations by special molecular agents that edit histone marks. These repair mechanisms ensure the stability of attractors in the field of chromatin states. However, strong perturbations may cross the boundary between attractors, and chromatin would converge to another stable (or quasi-stable) state, which means overwriting the chromatin memory. Specific states of chromatin are spatially associated with certain genes, and these genes become activated or repressed depending on the chromatin state. Active genes produce proteins (e.g., transcription factors) which may regulate chromatin state at other genome locations. Association of chromatin with DNA is not sequence specific, which gives organisms the flexibility to establish regulatory links between any subsets of genes.

The combination of these three features of chromatin can support adaptive learning at the cellular level. As a toy model, consider a gene that can be activated via multiple regulatory modules in its promoter. Initially, the chromatin is loose at all regulatory modules, and therefore, DNA is accessible to transcription factors. Eventually, a successful action of a cell (e.g., capturing food) may become a "memory-triggering event," which forces the chromatin to condense at all regulatory modules except for the one that was functional at the time of the event. Then, as the cell encounters a similar pattern of signaling next time, only one regulatory module would become active – the one that previously mediated a successful action. Modification of chromatin (i.e., opening or closing) is controlled by the production of certain transcription factors that move from the cytoplasm to the nucleus and find specific DNA patterns where they bind. But how can transcription factors differentiate between active and nonactive regulatory modules so that only nonactive modules become closed? This kind of context-dependent activity is possible, thanks to the interaction between multiple transcription factors that are located close enough along the DNA sequence. For example, binding of the P300 protein to the regulatory module indicates ongoing activity of this module (Visel et al. 2009), and transcription factors may have opposite effects on the chromatin depending on whether they are bound to DNA alone or in combination with P300. This kind of mechanism may support associative learning at the initial steps of the emergence of mind. An important component of this mechanism is the ability of an agent to classify its own states as "success" or "failure," and activate memory in the case of success.

The importance of chromatin is supported by the fact that mechanisms of learning and memory in the nervous system include DNA methylation and histone acetylation (Levenson and Sweatt 2005; Miller and Sweatt 2007). However, it is plausible that mind appeared even before the emergence of the nervous system. For example, unicellular ciliates have elements of nonassociative learning (Wood 1992) and even associative learning (Armus et al. 2006). Plants, fungi, sponges, and other multicellular organisms without nervous systems are all likely to anticipate and learn, although their responses are much slower than in animals (Ginsburg and Jablonka 2009; Krampen 1981). It is reasonable to assume that mind functions were initially based on intracellular mechanisms, and only later, they were augmented via communication between cells. Then a multicellular brain should be viewed as a community of cellular "brains" represented by the nuclei of neurons. The idea that cellular semiosis is the basis for the functionality of the brain has been recently proposed by Baslow (2011). The human brain consists of 100 billion neurons, and each neuron has thousands of synaptic links with other neurons. Synapses of single neurons are all specialized in various functions; some of them are active, while others are repressed. Thus, a neuron has to "know its synapses" because otherwise signals coming in from different synapses would be mixed up. In addition, neurons have to distinguish temporal patterns of signals coming from each synapse (Baslow 2011). Individual neurons need at least minimal mind capacity to classify these complex inputs.

Baslow proposed that the "operating system" of neurons is based on metabolism (Baslow 2011). Although active metabolism is indeed required for the functioning of neurons, it does not seem to be specific for mind and cannot explain how cells learn to recognize and process new signaling patterns. The cellular level of mind is more likely to be controlled by epigenetic regulatory mechanisms in the nucleus. In multicellular organisms, however, many additional processes are involved in learning and memory, such as the establishment of synaptic connections between neurons and the specialization of neural subnetworks for controlling specific behaviors.

Mind appears as a new top-level regulator of organism functions, but it does not replace already existing hardwired protosemiotic networks. Many low-level functions do not require complex regulation; they are well controlled by direct signaling, and replacing them with a learning mechanism would be costly and inefficient. However, some hard-programmed processes like embryo development may acquire partial guidance from the minds of individual cells or from the brain. Neurons establish functional feedback regulation of growing organs, where nonfunctional cells or cell parts (e.g., synapses) are eliminated (Edelman 1988). In other words, cells attempt to find a "job" in the body that fits to an available functional niche and the cell's prehistory. If a job is not found, then the cell goes into apoptosis.

7 The First Object Classified by Minimal Mind Was the Body

The initial task of mind was to classify those objects that are most important for the life of an organism. Because an agent's body is most intimately linked with a large number of functions, we can hypothesize that the body was the first object to be classified by mind. The purpose of classifying body states is to assign priorities to various functions, such as the search for food, defense from enemies, and reproduction. Functions of protosemiotic agents are directly controlled by internal and external signs, and therefore, priorities are fixed by a heritable signaling network. In contrast, agents with mind can learn to distinguish body states and adjust the priority of functions based on previous experience.

Of the two components of mind, Innenwelt (classifications and models of self) and Umwelt (classifications and models of external objects), Innenwelt is primary and Umwelt is secondary. Simple agents do not distinguish between internal and external sensations. It requires additional complexity for agents to realize that there are external objects beyond signals that come from receptors. The main difference between "internal" and "external" worlds is a higher predictability of the internal world and a lower predictability of the external world. Thus, it is reasonable to presume that Umwelt emerged as a less predictable portion of a former Innenwelt. This evolutionary approach to the differentiation of "external" from "internal" is profoundly different from cybernetics, where the boundary between the system and environment is defined a priori.

The capacity of mind to classify and model objects is closely related to the ability of agents to track objects. In particular, agents can rely on the assumption that objects keep their properties over time. For example, a predator that is chasing an object identified previously as prey does not need to repeat identification over and over again. Similarly, modeling appears most beneficial if the agent keeps track of the predicted object. Thus, tracking of objects by agents augments the utility of classification and modeling. The advantage of body as the first classified and modeled object is that it is always accessible, and thus, agents do not need additional skills for object tracking.

8 Modeling Functions of Mind

Modeling, which can be defined as prediction or anticipation of something unperceived, is the second major function of mind after the classification of objects. Elements of modeling are present in any classification, because ideal objects are already models. Recognition of an object is based on the anticipated combination of traits, as follows from the extensively explored area of image recognition. Some of these models are fixed, whereas others include parameters that are adjusted to increase the likelihood of a match between the model and sensorial data (Perlovsky et al. 2011). For example, distance to the object can be used as a parameter which affects the size and resolution of the image, as well as its position relative to other objects. These simple models belong to the primary modeling system, where ideal objects are not connected and therefore not used for prediction or anticipation of something different than what is perceived. Some of them are pure sensations, and others are integral sensation-actions. As an example of sensation-action, consider a moth that by instinct starts laying eggs after recognizing its host plant.

Advanced models that establish relationships between ideal objects belong to the secondary modeling system (Sebeok 1987). For example, if a bird attempts to eat a wasp and gets stung, then it connects the ideal object of a wasp with pain. As a result, this bird will not attempt to eat anything that looks like a wasp because the image of a wasp reminds it of pain. It was suggested that the secondary modeling system is handled by the interpretive component of the brain, whereas cybernetic and instinctive components handle the primary modeling (Barbieri 2011). The secondary modeling system establishes links between various ideal objects and therefore allows agents to develop flexible relationships between signs and functions. The origin of the secondary modeling system can be associated with the emergence of powerful sense organs that provided animals with more information than was needed for immediate functions. As a result, the classification of objects became more detailed and partially independent from their utility. Using a combination of a large number of traits, animals are able to recognize individual objects, associate them with each other, and make a mental map of their living space. Individual objects are then united into functionally relevant classes. Animals also can use abstract ideal objects that correspond to individual traits (e.g., color, shape, or weight) of real objects. Dynamic models associate the current state of an object with future states of the same object. They are used by predators to predict the movement of their prey. Association models predict the presence of one object from the observation of another kind of object. For example, animals associate smoke with forest fires and attempt to escape to a safe location.

One of the recent approaches to model building is dynamic logic (Perlovsky et al. 2011). The idea is to maximize the likelihood of matching between the set of models with adjustable parameters to the set of empirical data. Each model corresponds to a potential object, which can be added or deleted in the process of optimization. The accuracy of comparison between object-models increases, and model parameters are adjusted as optimization progresses. This approach explains two important aspects of modeling. First, detection of objects is not possible without models because models specify what we are looking for. And second, objects can be measured using optimal parameters of object-models (although this is not the only way to measure objects). Because the data are referenced by space and time, models include motion equations and yield plausible trajectories of object-models. However, all object-models identified with this method are primary ideal objects (i.e., they belong to the primary modeling system). Connections between primary objects have to be established at a higher level of the hierarchy of objects (Perlovsky et al. 2011).

Models are the main subject of Peirce's semiotics, where the perceived object is a sign vehicle that brings into attention the interpretant or associated ideal object. The primary modeling system operates with icons, which are associated with isolated ideal objects (sensations or sensation-actions), whereas the secondary modeling system also includes indexes which are the links between ideal objects (Sebeok and Danesi 2000). Peirce, however, viewed sign relationships as components of the world rather than models developed by agents. He believed that models were embedded in the world. The danger of this philosophy (i.e., objective idealism) is that it easily leads to dogmatism as models become overly trusted. But how can we evaluate the relationship between a model and reality? Models can be used in two ways: they can be trusted and they can be tested. When a bird does not attempt to catch wasps after being stung, it trusts the model of a wasp. However, not all models generate reproducible results, and therefore, models need to be tested and modified if necessary.

9 Testing Models

Model testing is one of the most important activities in science, and it has direct implications for epistemology (Cariani 2011; Popper 1999; Rosen 1991; Turchin 1977). Animals also test models, but they do not run experiments for the sake of testing hypotheses as humans do. Instead, they evaluate the success rates of their behavioral strategies and establish preferences for more successful behaviors. In this way, predators learn how to chase and capture prey, and birds learn how to attract the attention of predators away from their nests.

Model testing is a complex procedure that determines if predictions generated by the model match the real world. In the simplest case, an agent measures the initial state of the object, and the obtained results are used as input for the model. Then the output of the model is compared to the measurement of the final state of the object, and if they match, the test is considered successful (Cariani 2011; Rosen 1991; Turchin 1977). To formalize model testing, we need to generalize our terms. First, the expression "initial state of the object" implies that agents have a method for

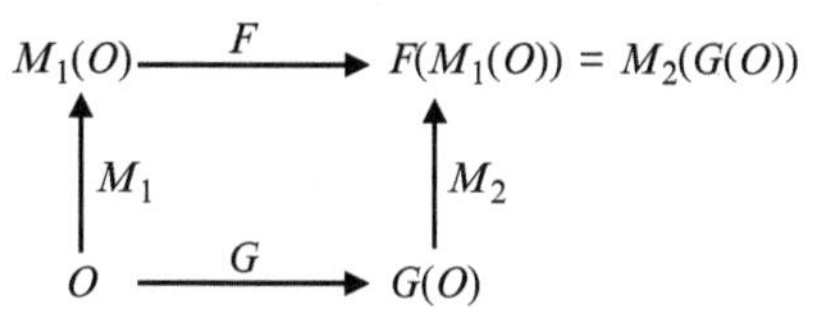

Fig. 1 Commuting diagram of model testing. M_1 and M_2 are measurement methods for the initial object O and final object $G(O)$, respectively; G is the object tracking function, and F is the map between ideal objects in the model

tracking objects. In particular, each object O is associated with the final object $G(O)$, where G is the tracking function. Second, objects are characterized either quantitatively by measurements or qualitatively by the identification of individual features or by classifying whole objects. In result, each object O becomes associated with some ideal object $M(O)$ in mind, which is interpreted as a measurement of that object. In general, agents use multiple measurement methods M_1, M_2, … $M_{n,}$ which are applicable in different situations. Similarly, in science, we use different measurement devices and sensors to characterize objects. Finally, the model is a map, F, between ideal objects in mind. For example, a dynamic model associates initial measurements of an object with measurements of its final state. Then successful model testing can be represented by a commuting diagram (Fig. 1), where measurement of the final state of the object, $M_2(G(O))$, matches to the model output from the measurement of the initial state of the object used as input, $F(M_1(O))$. Two measurement methods M_1 and M_2 may be the same, but in the general case, they are different. If the equation $M_2(G(O))=F(M_1(O))$ is true for all available objects, then the model F is universal relative to measurement methods M_1 and M_2 and tracking method G.

Commuting diagrams, similar to Fig. 1, were proposed previously (Cariani 2011), but function G was interpreted as objective natural dynamics of the world. In contrast, I associate function G with an agent's ability to track or manipulate objects. An example of nontrivial object tracking is the association of the "morning star" with the "evening star" (i.e., planet Venus) on the basis of the model of planetary movement. This example illustrates that all four components of the model relation (F, G, M_1, M_2) are interdependent epistemic tools, and one component may help us to improve another component.

Cariani suggested that the manipulation of an object is a mapping from the ideal representation to the object itself (Cariani 2011), which has the opposite direction compared to the measurement. This approach, however, implies that real objects are created from ideal objects without any matter. In contrast, I suggest associating the manipulation of objects with various tracking functions G. Some G functions may represent a passive experiment, where objects are mapped to their natural future state, whereas other G functions represent active experiments where objects are mapped into their products after specific manipulations. If we want to construct meta-models that describe multiple methods of object manipulation, then each method i should be linked with a corresponding model F_i and object tracking method G_i.

Commuting diagrams of model testing capture a very important aspect of epistemology: the equivalence is achieved in the domain of ideal objects rather than in the domain of real objects. Thus, different models may equally well capture the same process or relationship in the real world. The second conclusion is that models are always tested together with measurement methods and tracking methods, which are usually ignored in physics. As a result, agents from one communication system cannot take advantage of models developed within another communication system if measurement methods and tracking methods do not match.

According to the critical rationalism of Popper, a model, whose predictions are wrong, should be removed from the domain of science (Popper 1999). However, this rarely happens; instead, model components (F, G, M_1, M_2) are adjusted to make the diagram in Fig. 1 commuting. Popper condemned this practice because it makes hypotheses nonfalsifiable. However, Popper's argument does not make sense from the evolutionary point of view. If animals rejected any model that once had generated a wrong result, then they would soon run out of models and fail to perform their functions. Any model is a product of evolution and learning and integrates long-term experience of agents. It is better to have a nonaccurate or nonuniversal model than no model at all. This explains why models are so persistent both in biological evolution and in human culture.

10 Model Transfer Between Individuals

Most models used by animals are not communicated to other individuals. Thus, each animal has to develop its own models based on trial and error as well as heritable predispositions. However, social interactions may facilitate the development of models in young animals. For example, animals may copy the behavior of their parents and eventually acquire their models in a faster way than by pure trial and error. However, efficient communication of models is possible only by language, which corresponds to the cultural level of semiosis, following the terminology of Kull (2009). In language, signs do not only correspond to ideal objects, they also replicate the structure of relationships between ideal objects in the model. Thus, language itself becomes the modeling environment called the tertiary modeling system (Sebeok and Danesi 2000). Language is based on symbols which are signs whose meanings are established by convention within the communication system. Then, a message with two (or more) interconnected symbols is interpreted as a link between corresponding ideal objects within the model. Thus, the tertiary modeling system is based on symbols (Sebeok and Danesi 2000).

In conclusion, minimal mind is a tool used by agents to classify and model the objects. Classification ends up at the ideal object, which serves as a checkpoint to initiate certain physical or mental functions. Mind is projected to appear within eukaryotic cells with well-developed epigenetic regulation because these mechanisms can convert DNA segments into standard information-processing modules with multiple attractor domains and flexible control. Classification and modeling of objects started from the body of agent and then expanded to external objects. Modeling functions of mind progressed from primary models that simply support classification of objects to secondary models that interconnect ideal objects and finally to tertiary models that can be communicated to other agents.

References

Armus, H. L., Montgomery, A. R., & Gurney, R. L. (2006). Discrimination learning and extinction in paramecia (*P. caudatum*). *Psychological Reports, 98*(3), 705–711.

Barbieri, M. (2003). *The organic codes: An introduction to semantic biology*. Cambridge/New York: Cambridge University Press.

Barbieri, M. (2008). Biosemiotics: A new understanding of life. *Die Naturwissenschaften, 95*(7), 577–599.

Barbieri, M. (2011). Origin and evolution of the brain. *Biosemiotics, 4*(3), 369–399.

Baslow, M. H. (2011). Biosemiosis and the cellular basis of mind. How the oxidation of glucose by individual neurons in brain results in meaningful communications and in the emergence of "mind". *Biosemiotics, 4*(1), 39–53.

Cariani, P. (1998). Towards an evolutionary semiotics: The emergence of new sign-functions in organisms and devices. In G. V. de Vijver, S. Salthe, & M. Delpos (Eds.), *Evolutionary systems* (pp. 359–377). Dordrecht/Holland: Kluwer.

Cariani, P. (2011). The semiotics of cybernetic percept-action systems. *International Journal of Signs and Semiotic Systems, 1*(1), 1–17.

Deacon, T. W. (2011). *Incomplete nature: How mind emerged from matter*. New York: W. W. Norton and Company.

Dennett, D. C. (1995). *Darwin's dangerous idea: Evolution and the meanings of life*. New York: Simon & Schuster.

Eco, U. (1976). *A theory of semiotics*. Bloomington: Indiana University Press.

Edelman, G. M. (1988). *Topobiology: An introduction to molecular embryology*. New York: Basic Books.

Emmeche, C., & Hoffmeyer, J. (1991). From language to nature – The semiotic metaphor in biology. *Semiotica, 84*(1/2), 1–42.

Ginsburg, S., & Jablonka, E. (2009). Epigenetic learning in non-neural organisms. *Journal of Biosciences, 34*(4), 633–646.

Griffin, D. R. (1992). *Animal minds*. Chicago: University of Chicago Press.

Hoffmeyer, J. (1997). Biosemiotics: Towards a new synthesis in biology. *European Journal for Semiotic Studies, 9*(2), 355–376.

Hoffmeyer, J. (2010). Semiotics of nature. In P. Cobley (Ed.), *The Routledge companion to semiotics* (pp. 29–42). London/New York: Routledge.

Jeanteur, P. (2005). *Epigenetics and chromatin*. Berlin: Springer.

Kolmogorov, A. N. (1965). Three approaches to the quantitative definition of information. *Problems of Information Transmission, 1*(1), 1–7.

Krampen, M. (1981). Phytosemiotics. *Semiotica, 36*(3/4), 187–209.

Kull, K. (2009). Vegetative, animal, and cultural semiosis: The semiotic threshold zones. *Cognitive Semiotics, 4*, 8–27.

Levenson, J. M., & Sweatt, J. D. (2005). Epigenetic mechanisms in memory formation. *Nature Reviews Neuroscience, 6*(2), 108–118.

Lovelock, J. E. (1979). *Gaia. A new look at life on earth*. Oxford: Oxford University Press.

Markoš, A., & Švorcová, J. (2009). Recorded versus organic memory: Interaction of two worlds as demonstrated by the chromatin dynamics. *Biosemiotics, 2*(2), 131–149.

Millau, J. F., & Gaudreau, L. (2011). CTCF, cohesin, and histone variants: Connecting the genome. *Biochemistry and Cell Biology, 89*(5), 505–513.

Miller, C. A., & Sweatt, J. D. (2007). Covalent modification of DNA regulates memory formation. *Neuron, 53*(6), 857–869.

Nillson, N. J. (1998). *Artificial intelligence: A new synthesis*. San Francisco: Morgan Kaufmann Publishers.
Peirce, C. S. (1998). *The essential Peirce: Selected philosophical writings* (Vol. 2). Indiana: Indiana University Press.
Perlovsky, L., Deming, R., & Ilin, R. (2011). *Emotional cognitive neural algorithms with engineering applications. Dynamic logic: From vague to crisp* (Vol. 371). Warsaw: Polish Academy of Sciences.
Popper, K. (1999). *All life is problem solving*. London: Routledge.
Premack, D. G., & Woodruff, G. (1978). Does the chimpanzee have a theory of mind? *The Behavioral and Brain Sciences, 1*, 515–526.
Prodi, G. (1988). Material bases of signification. *Semiotica, 69*(3/4), 191–241.
Putnam, H. (1975). *Mind, language and reality* (Vol. 2). Cambridge: Cambridge University Press.
Rashevsky, N. (1938). *Mathematical biophysics*. Chicago: University of Chicago Press.
Rosen, R. (1970). *Dynamical system theory in biology*. New York: Wiley-Interscience.
Rosen, R. (1991). *Life itself: A comprehensive inquiry into the nature, origin, and fabrication of life*. New York: Columbia University Press.
Sebeok, T. A. (1972). *Perspectives in zoosemiotics*. The Hague: Mouton.
Sebeok, T. (1987). Language: How primary a modeling system? In J. Deely (Ed.), *Semiotics 1987* (pp. 15–27). Lanham: University Press of America.
Sebeok, T. A., & Danesi, M. (2000). *The forms of meaning. Modeling systems theory and semiotic analysis*. New York: Mouton de Gruyter.
Shannon, C. E. (1948). A mathematical theory of communication. *Bell System Technical Journal, 27*(379–423), 623–656.
Sharov, A. A. (2006). Genome increase as a clock for the origin and evolution of life. *Biology Direct, 1*, 17.
Sharov, A. A. (2009a). Coenzyme autocatalytic network on the surface of oil microspheres as a model for the origin of life. *International Journal of Molecular Sciences, 10*(4), 1838–1852.
Sharov, A. A. (2009b). Genetic gradualism and the extraterrestrial origin of life. *Journal of Cosmology, 5*, 833–842.
Sharov, A. A. (2009c). Role of utility and inference in the evolution of functional information. *Biosemiotics, 2*(1), 101–115.
Sharov, A. (2010). Functional information: Towards synthesis of biosemiotics and cybernetics. *Entropy, 12*(5), 1050–1070.
Sharov, A. A. (2012). The origin of mind. In T. Maran, K. Lindström, R. Magnus, & M. Tønnensen (Eds.), *Semiotics in the wild* (pp. 63–69). Tartu: University of Tartu.
Swan, L. S., & Goldberg, L. J. (2010). How is meaning grounded in the organism? *Biosemiotics, 3*(2), 131–146.
Swan, L. S., & Howard, J. (2012). Digital immortality: Self or 01001001? *International Journal of Machine Consciousness, 4*(1), 245–256.
Turchin, V. F. (1977). *The phenomenon of science*. New York: Columbia University Press.
Turing, A. (1952). Can automatic calculating machines be said to think? In B. J. Copeland (Ed.), *The essential Turing: The ideas that gave birth to the computer age* (pp. 487–506). Oxford: Oxford University Press.
Uexküll, J. (1982). The theory of meaning. *Semiotica, 42*(1), 25–82.
Visel, A., Blow, M. J., Li, Z., Zhang, T., Akiyama, J. A., Holt, A., et al. (2009). ChIP-seq accurately predicts tissue-specific activity of enhancers. *Nature, 457*(7231), 854–858.
Waddington, C. H. (1968). Towards a theoretical biology. *Nature, 218*(5141), 525–527.
Wood, D. C. (1992). Learning and adaptive plasticity in unicellular organisms. In L. R. Squire (Ed.), *Encyclopedia of learning and memory* (pp. 623–624). New York: Macmillan.

Concept Combination and the Origins of Complex Cognition

Liane Gabora and Kirsty Kitto

Abstract At the core of our uniquely human cognitive abilities is the capacity to see things from different perspectives or to place them in a new context. We propose that this was made possible by two cognitive transitions. First, the large brain of Homo erectus facilitated the onset of recursive recall: the ability to string thoughts together into a stream of potentially abstract or imaginative thought. This hypothesis is supported by a set of computational models where an artificial society of agents evolved to generate more diverse and valuable cultural outputs under conditions of recursive recall. We propose that the capacity to see things in context arose much later, following the appearance of anatomically modern humans. This second transition was brought about by the onset of contextual focus: the capacity to shift between a minimally contextual analytic mode of thought and a highly contextual associative mode of thought conducive to combining concepts in new ways and "breaking out of a rut." When contextual focus is implemented in an art-generating computer program, the resulting artworks are seen as more creative and appealing. We summarize how both transitions can be modeled using a theory of concepts which highlights how different contexts shift the interpretation of a single concept.

What is the essence of our humanness? We propose that what is at the core of our uniquely human cognitive abilities is the capacity to *place things in context* or *see things from different perspectives*. This enables us not just to be creative, but to put our own spin on the inventions of others, modifying them to suit our own needs and

L. Gabora (✉)
Department of Psychology, University of British Columbia,
Okanagan campus, Arts Building, 3333 University Way, Kelowna, BC V1V 1V7, Canada
e-mail: liane.gabora@ubc.ca

K. Kitto
Information Systems School, Queensland University of Technology,
2 George Street, Brisbane, 4000, Australia
e-mail: kirsty.kitto@qut.edu.au

L. Swan (ed.), *Origins of Mind*, Biosemiotics 8, DOI 10.1007/978-94-007-5419-5_19,

tastes, in turn, leading to new innovations that build cumulatively on previous ones (Gabora 2003, 2008a, b, c, d; Gabora and Russon 2011; Gabora and Kaufman 2010). It enables us to modify thoughts, impressions, and attitudes by thinking about them in the context of each other and thereby weave them into a more or less integrated structure that defines who we are in relation to the world. Our compunction to put our own spin on the ideas and inventions of others results in accumulative cultural change, referred to as the *ratchet effect* (Tomasello et al. 1993).

Understanding how this capacity evolved, and testing it against other theories about what is responsible for our humanness, is difficult. All that is left of our prehistoric ancestors are their bones and artifacts such as stone tools that resist the passage of time. Methods for analyzing these remains are becoming increasingly sophisticated, but they still leave many questions unanswered and are often compatible with several competing theories. Thus, in seeking to explain the evolution of the uniquely human cognitive capacities that have transformed our lives, and even the planet we live on, formal computational and mathematical models provide an extremely valuable set of reconstructive tools. Steps toward a mathematical model of the evolution of the cognitive mechanisms underlying the evolution of the capacity to "see things in context" have been put forward (Gabora and Aerts 2009), and computational models of this have also been developed (DiPaola and Gabora 2007, 2009; Gabora 1994, 1995, 2008a, b; Gabora and Leijnen 2009; Leijnen and Gabora 2009, 2010; Gabora et al. in press; Gabora and Firouzi 2012; Gabora and Saberi 2011). The goal of this chapter is to explain these efforts in layperson terms, which fill in some gaps, and show how they constitute an integrated effort to formally model the evolution of the cognitive mechanisms that underlie our humanness.

1 First Transition: The Earliest Signs of Creativity

The last common ancestor of humans and other great apes lived between four and eight million years ago. The minds of our earliest ancestors, *Homo habilis,* have been referred to as *episodic* because there is no evidence that their experience deviated from the present moment of concrete sensory perceptions (Donald 1991). They were able to encode perceptions of events in memory, and recall them in the presence of a reminder or cue, but had little voluntary access to memories without environmental cues. They would, for example, not think of a particular person or object unless something in their environment concretely triggered its recall. They were therefore unable to voluntarily shape, modify, or practice skills and actions, and neither could they invent nor refine complex gestures or means of communicating.

Homo habilis was eventually replaced by *Homo erectus,* which lived between approximately 1.8 and 0.3 million years ago. This period is widely referred to as the beginnings of human culture. The cranial capacity of the *Homo erectus* brain was around 1,000 cc, which is about 25 % larger than that of *Homo habilis*, at least twice as large as that of living great apes, and 75 % that of modern humans (Aiello 1996; Ruff et al. 1997). *Homo erectus* exhibited many indications of enhanced intelligence, creativity, and an ability to adapt to their environment. For example, they made use of sophisticated task-specific stone hand axes, complex stable seasonal home bases, long-distance hunting strategies involving large game, and migration out of Africa.

This period marks the onset of the archaeological record, and it is thought to be the beginnings of human culture. It is widely believed that this cultural transition reflects an underlying transition in cognitive or social abilities. Some have suggested that such abilities arose with the onset of a *theory of mind* (Mithen 1998) or the capacity to imitate (Dugatkin 2001). However, there is evidence that nonhuman primates also possess theory of mind and the capacity to imitate (Heyes 1998; Premack 1988; Premack and Woodruff 1978), and yet, they do not compare to modern humans in intelligence and cultural complexity.

Evolutionary psychologists have suggested that the intelligence and cultural complexity of the *Homo* line is due to the onset of *massive modularity* (Buss 1999/2004; Barkow et al. 1992). However, although the mind exhibits an intermediate degree of functional and anatomical modularity, neuroscience has not revealed vast numbers of hardwired, encapsulated, task-specific modules; indeed, the brain has been shown to be more highly subject to environmental influence than was previously believed (Buller 2005; Wexler 2006).

2 A Promising and Testable Hypothesis

Donald (1991) proposed that with the enlarged cranial capacity of *Homo erectus*, the human mind underwent the first of three transitions by which it—along with the cultural matrix in which it is embedded—evolved from the ancestral, prehuman condition. This transition is characterized by a shift from an *episodic* to a *mimetic mode* of cognitive functioning, made possible by onset of the capacity for voluntary retrieval of stored memories, independent of environmental cues. Donald refers to this as a *self-triggered recall and rehearsal loop*. Self-triggered recall enabled information to be processed recursively and reprocessed with respect to different contexts or perspectives. This allowed our ancestors to access memories voluntarily and thereby to act out[1] events that occurred in the past or that might occur in the future. Thus, not only could the mimetic mind temporarily escape the here and now, but by miming or gesture, it could communicate similar escapes to other minds. The capacity to mime thus brought forth what is referred to as a *mimetic* form of cognition, so ushering in a transition to the mimetic stage of human culture. The self-triggered recall and rehearsal loop also enabled our ancestors to engage in a stream of thought, where one thought or idea evokes another, revised version of it, which evokes yet another, and so forth recursively. In this way, attention is directed away from the external world toward one's internal model of it. Finally, self-triggered recall allowed for voluntary rehearsal and refinement of actions, enabling systematic evaluation and improvement of skills and motor acts.

[1] The term *mimetic* is derived from "mime," which means "to act out."

3 Computational Model of First Transition

The recursive recall hypothesis is difficult to test directly, for even if correct, the brain tissues of our ancestors are long disintegrated, so we cannot directly study how the neural mechanisms underlying recursive recall evolved. It is, however, possible to computationally model how the onset of the capacity for recursive recall would affect the effectiveness, diversity, and open-endedness of ideas generated in an artificial society. This section summarizes how we tested Donald's hypothesis using an agent-based computational model of culture referred to as "EVOlution of Culture," abbreviated EVOC. EVOC successfully models how 'descent with modification' occurs in a cultural context. The approach can thus be contrasted with computer models of how individual learning affects biological evolution (Best 1999, 2006; Higgs 2000; Hinton and Nowlan 1987; Hutchins and Hazelhurst 1991). Details of the modeling platform are provided elsewhere (Gabora 2008b, c; Gabora and Leijnen 2009; Leijnen and Gabora 2009).

3.1 The EVOC World

EVOC uses neural network-based agents that (1) invent new ideas, (2) imitate actions implemented by neighbors, (3) evaluate ideas, and (4) implement successful ideas as actions. Invention works by modifying a previously learned action using learned trends (such as that more overall movement tends to be good) to bias the invention process. The process of finding a neighbor to imitate works through a form of lazy (what computer scientists refer to as "nongreedy," by which they mean that solutions provided at each stage in an iteration are not necessarily optimal) search. An imitating agent randomly scans its neighbors and assesses the fitness of their actions using a predefined fitness function. It adopts the first action it comes across that is fitter than the action it is currently implementing. If it does not find a neighbor that is executing a fitter action than its own action (see below for a discussion of fitness), it continues to execute the current action. Over successive rounds of invention and imitation, agents' actions improve. EVOC thus models how descent with modification occurs in a purely cultural context. Agents do not evolve in a biological sense—they neither die nor have offspring—but do in a cultural sense, by generating and sharing ideas for future actions.[2]

Following Holland (1975), we refer to the success of an action in the artificial world as its *fitness,* with the caveat that unlike its usage in biology, here, the term is unrelated to number of offspring (or ideas derived from a given idea). The fitness function rewards head immobility and symmetrical limb movement. Fitness of actions starts out low because initially all agents are entirely immobile. However,

[2] The approach can thus be contrasted with computer models of how individual learning affects biological evolution (Hinton and Nowlan 1987; Hutchins and Hazelhurst 1991). For an explanation of why we do not adopt the framework of memetics see (Gabora 1999b, 2004, 2008d).

some agent quickly invents an action that has a higher fitness than doing nothing, and this action gets imitated, leading to an increase in fitness. Fitness increases further as other ideas get invented, assessed, implemented as actions, and spread through imitation. The diversity of actions initially increases due to the proliferation of new ideas and then decreases as agents hone in on the fittest actions.

The artificial society consists of a toroidal lattice with 100 nodes, each occupied by a single, stationary agent. We used a von Neumann neighborhood structure (agents only interacted with their four adjacent neighbors). During invention, the probability of changing the position of any body part involved in an action was 1/6. (Since there are 6 body parts, this averages out to one body part change per action.) On each run, creators and imitators were randomly dispersed.

3.2 Chaining

This gives agents the opportunity to execute multistep actions. For the experiments reported here with chaining turned on, if in the first step of an action an agent was moving at least one of its arms, it executes a second step, which again involves up to six body parts. If, in the first step, the agent moved one arm in one direction and in the second step it moved the same arm in the opposite direction, it has the opportunity to execute a three-step action, and so on. The agent is allowed to execute an arbitrarily long action so long as it continues to move the same arm in the direction opposite to the direction it moved previously. Once it does not do so, the chained action comes to an end. The longer it moves, the higher the fitness of this multistep chained action. This is admittedly a simple action, but we were not interested in the impact of this action *per se*. The goal here was simply to test hypotheses about how chaining at the individual level affects dynamics at the societal level by providing agents with a means of implementing multistep actions such that the optimal way of going about one step depends on how one went about the previous step. This seems to be a common feature of many useful actions such as the repetitive motions involved in toolmaking, sawing, carving, weaving, and so forth.

Where c is "with chaining," w is "without chaining," and n is the number of chained actions, the fitness, F_c, of a chained action is calculated as follows:

$$F_c = F_w(n-1)$$

The fitness function with chaining provides a simple means of simulating the capacity for recursive recall.

3.3 Results

As shown in Fig. 1, the capacity to chain together simple actions to form more complex ones increases the mean fitness of actions across the artificial society. This is most evident in the later phase of a run. Without chaining, agents converge

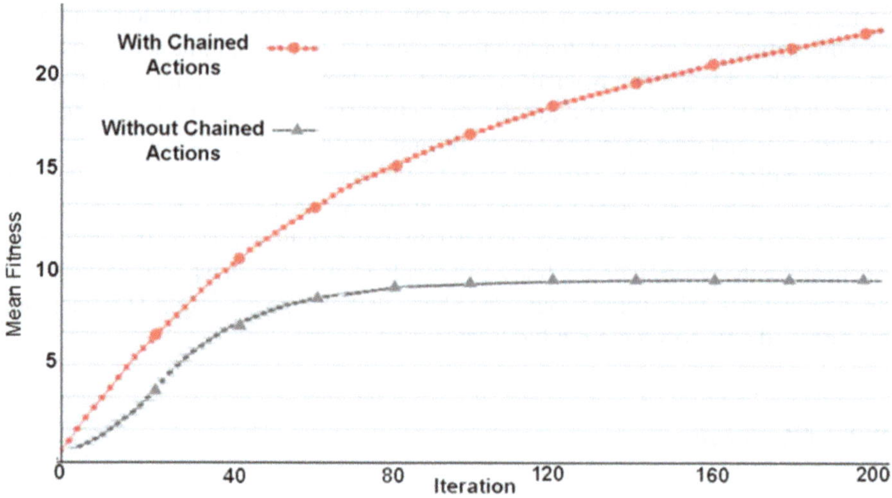

Fig. 1 Mean fitness of actions in the artificial society with chaining versus without chaining (From Gabora and Saberi 2011)

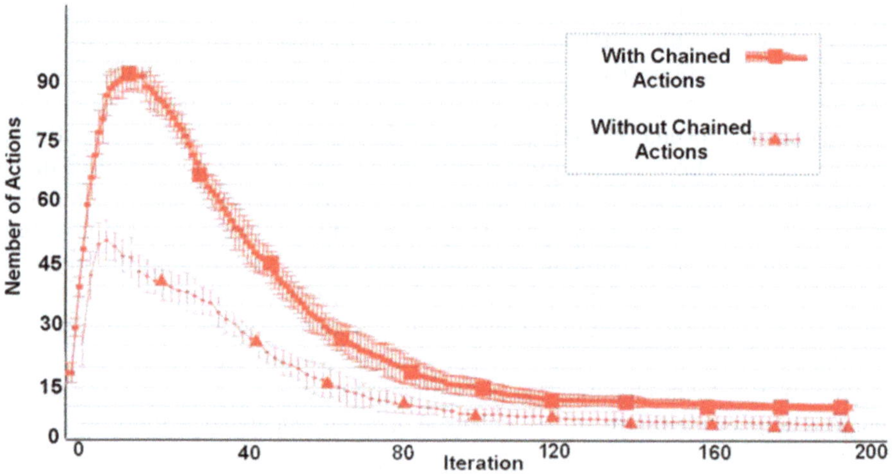

Fig. 2 Mean number of different actions in the artificial society with chaining (*continuous line*) versus without chaining (*dashed line*) (From Gabora and Saberi 2011)

on optimal actions, and the mean fitness of action reaches a plateau. With chaining, however, there is no ceiling on the mean fitness of actions. By the 200th iteration, the chaining process has led to more than double the maximum fitness attainable without chaining.

As shown in Fig. 2, chaining also increases the diversity of actions. This is most evident in the early phase of a run before agents begin to converge on optimal actions. Although in both cases there is convergence on optimal actions, without chained actions, this is a static set (thus mean fitness plateaus), whereas with chained actions, the set of optimal actions is always changing, as increasingly fit actions are found (thus mean fitness keeps increasing).

This shows that recursive recall increased the fitness of ideas while simultaneously increasing the number of different ideas across the artificial society. It thus supports the hypothesis that the onset of recursive recall was a critical step toward the kind of cognition we associate with humans.

We also tested the effect of chaining on the capacity to benefit from learning. Recall that agents have the capacity to learn trends from past experiences and thereby bias the generation of novelty in directions that have a greater than chance probability of being fruitful. Since chaining provides more opportunities to capitalize on the capacity to learn, we hypothesized that chaining would accentuate the impact of learning on the mean fitness of actions, and this too turned out to be the case (Gabora and Saberi 2011).

Note that in the chaining versus no chaining conditions, the size of the neural network is the same, but *how it is used* differs. This suggests that it was not larger brain size per se that initiated the onset of cumulative culture, but that larger brain size enabled episodes to be encoded in more detail, allowing more routes for reminding and recall, thereby facilitating the ability to recursively redescribe information stored in memory (Karmiloff-Smith 1992) and thereby to tailor it to the situation at hand.

4 Mathematical Modeling of Recursive Redescription: An Idea in Context

A limitation of this model is that the recursive recall does not work, as it does in humans, by considering an idea in light of one perspective, seeing how that perspective modifies the idea, recognizing in what respect this modification suggests a new perspective from which to consider the idea, and so on. Mathematical modeling of recursive redescription requires an approach that can incorporate the effect of context on the state of a concept. It is widely recognized that the standard analytical techniques of science are not up to the challenge of modeling these contextual effects, because when concepts appear in the context of each other, their meanings change in ways that are noncompositional, that is, they behave in ways that violate the rules of classical logic (Osherson and Smith 1981; Hampton 1987; Aerts and Gabora 2005a, b; Kitto 2006, 2008a; Aerts 2009; Kitto et al. 2011). Despite its potential impact, this challenge is not as insurmountable as it might at first seem, as there is one mathematical formalism which was invented precisely to describe such contextuality: quantum theory (QT). It is not the purpose of this chapter to describe either this theory[3] or applications of it to cognition in any detail. Rather, we seek here to explain why QT provides a viable formalism to describe the density of

[3] This is summarized nicely in Isham (1995).

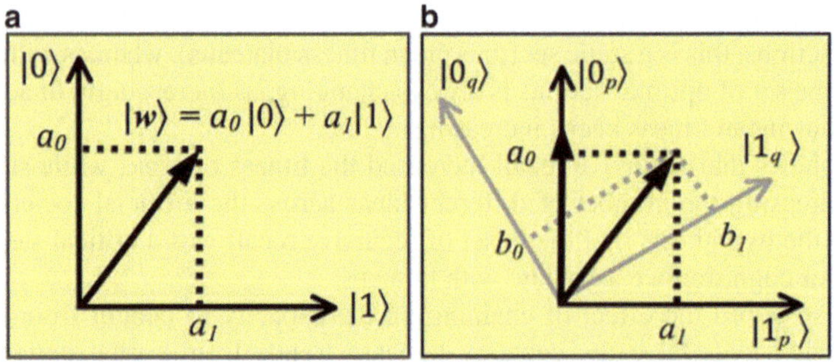

Fig. 3 Representingation of the idea $|w\rangle$ in two different contexts. (**a**) An idea has some probability of being interpreted with two different meanings, as represented by the projection of the idea onto two different basis states. (**b**) In the original context, p, the probability of collapse to basis state $|0\rangle$ and outcome a_0 was greater than collapse to basis state $|1\rangle$ with outcome a_1. In a different context q, this probability changes markedly, which can be seen by the different projections onto the new context

information storage that is required before the transition to recursive redescription can take place.

The quantum approach to concepts explicitly represents the context in which information occurs via a notion of measurement. Put simply, for quantum systems, a measurement does not simply record what is there but interacts with the system under consideration to reveal information about its state *in the context defined by the measurement setting* (Aerts et al. 2000; Kitto 2008b). In this theory, it is impossible to refer to the state of a system without reference to a measurement setting. Similarly, considering some concept w without reference to the context in which it occurs is implausible at best. FIRE, for example, might be a danger (in a FOREST FIRE), a tool (a COOKING FIRE), a light source, and community hub (a CAMP FIRE), and the meaning that we attribute to the concept FIRE will vary widely as a result. In Bruza et al. (2009), a simple model of this effect as it applies to the human mental lexicon was presented, and here, we shall briefly overview that model. In particular, we shall illustrate the manner in which the same idea can be attributed with more than one meaning, so contributing to the density of information storage.

In Fig. 3, we have drawn a *geometrical* representation of an idea in context. An idea, represented by $|w\rangle$, is represented by a vector which, depending on the context, can be interpreted two different ways, represented as $|1\rangle$ and $|0\rangle$. For example, a concept of FIRE represented in a particular context will have a certain potential of being interpreted as *dangerous* (e.g., FIRE is almost always interpreted as dangerous by the residents of Australia during summer.) Thus, unless the vector is perfectly aligned with one of the axes in the diagram, then a person who is asked about that concept will be genuinely undecided as to how they will interpret it. We represent this genuine indecision as a *superposition state*:

$$|w\rangle = a_0|0_p\rangle + a_1|1_p\rangle, \text{ where } |a_0|^2 + |a_1|^2 = 1.$$

However, in the different context represented by Fig. 3b, a different representation of the concept results:

$$|w\rangle = b_0 |0_q\rangle + b_1 |1_q\rangle, \text{ where } |b_0|^2 + |b_1|^2 = 1.$$

We posit that when concepts or ideas can be described as existing in a superposition state as in (2) and (3), they are experienced consciously as vague or "half-baked." Indeed, experimental evidence for such states has been obtained (Gabora and Saab 2011). By looking at ideas from different contexts, humans achieve a more well-rounded understanding of them. Indeed, humans frequently encounter situations where looking at a concept from one perspective brings to mind another perspective, and so on, until a detailed (and sometimes creative) understanding of the idea is achieved. Eventually, a particular interpretation upon. Each change from one version of the idea to another as it is viewed from a slightly different angle is described in this formalism as a sort of "measurement," which invokes an associated collapse of the state. The probability that a person will ascribe a particular interpretation to a given idea or concept is proportional to the length of the vector in that dimension (i.e., a projection onto the relevant basis state). This is represented in the formalism by taking the square of the length of the vector along the relevant axis (Isham 1995; Bruza et al. 2009; Kitto et al. 2011). More formally, the probability that an individual interprets idea $|w\rangle$ in the sense represented by the basis p is $P = |a_1|^2$. This is noticeably different from the interpretation that will be provided to the same idea in context q ($P = |b_1|^2$).

Through reference to Fig. 3b, we can immediately see that a different context results in a different probability value. In this formalism, a different context can result in a very different interpretation becoming likely. We can also extract the probability that a person will not associate a particular interpretation with a given concept ($P = |a_0|^2$ for context p and $P = |b_0|^2$ for context q).

Thus, returning to the idea of a FIRE, the probability that it will be interpreted as *dangerous* will be greater in the second context q than in the first. Perhaps this might be used to represent the likely danger that an early hominid attributed to the concept of FIRE in winter and summer, respectively. This allows for a dense representation of the concept FIRE. We do not need to encode each of the different meanings explicitly; they come from an interpretation associated with the idea that emerges at the moment of interpretation.

5 Second Transition: The "Big Bang" of Human Creativity

The European archaeological record indicates that a truly unparalleled cultural transition occurred between 60,000 and 30,000 years ago, at the onset of the Upper Paleolithic (Bar-Yosef 1994; Klein 1989; Mellars 1973, 1989a, b; Soffer 1994; Stringer and Gamble 1993). Considering it "evidence of the modern human mind at work," Richard Leakey (1984: 93–94) describes the Upper Paleolithic as "unlike previous eras, when stasis dominated, … [with] change being measured in millennia rather than hundreds of millennia." Similarly, Mithen (1996) refers to the Upper Paleolithic as the "big bang" of human culture, exhibiting more innovation than in the previous six million years of human evolution. This period exhibits the more or less simultaneous appearance of traits considered diagnostic of behavioral modernity. It marks the beginning of a more organized, strategic, season-specific style of hunting involving specific animals at specific sites; elaborate burial sites indicative of ritual and religion; evidence of dance, magic, and totemism; the colonization of Australia; and the replacement of Levallois tool technology by blade cores in the Near East. In Europe, complex hearths and many forms of art appeared, including cave paintings of animals, decorated tools and pottery, bone and antler tools with engraved designs, ivory statues of animals and sea shells, and personal decoration such as beads, pendants, and perforated animal teeth, many of which may have indicated social status (White 1989a, b).

Whether this period was a genuine revolution culminating in behavioral modernity is hotly debated because claims to this effect are based on the European Paleolithic record and largely exclude the African record (McBrearty and Brooks 2000; Henshilwood and Marean 2003). Indeed, most of the artifacts associated with a rapid transition to behavioral modernity at 40–50,000 years ago in Europe are found in the African Middle Stone Age tens of thousands of years earlier. However, the traditional and currently dominant view is that modern behavior appeared in Africa between 40,000 and 50,000 years ago due to biologically evolved cognitive advantages and spread, replacing existing species, including the Neanderthals in Europe (e.g., Ambrose 1998; Gamble 1994; Klein 2003; Stringer and Gamble 1993). Thus, from this point onward, there was only one hominid species, modern *Homo sapien,* and despite lack of overall increase in cranial capacity, their prefrontal cortex, and more particularly their orbitofrontal region, increased significantly in size (Deacon 1997; Dunbar 1993; Jerison 1973; Krasnegor et al. 1997; Rumbaugh 1997) in what was most likely a time of major neural reorganization (Klein 1999; Henshilwood et al. 2004; Pinker 2002).

Given that the Middle/Upper Paleolithic was a period of unprecedented creativity, what kind of cognitive processes were involved?

6 A Testable Hypothesis

Converging evidence suggests that creativity involves the capacity to shift between two forms of thought (Finke et al. 1992; Gabora 2003; Howard-Jones and Murray 2003; Martindale 1995; Smith et al. 1995): (1) *divergent* or *associative* processes are hypothesized to occur during idea generation, while (2) *convergent* or *analytic* processes predominate during the refinement, implementation, and testing of an idea. It has been proposed that the Paleolithic transition reflects a mutation to the genes involved in the fine-tuning of the biochemical mechanisms underlying the capacity to subconsciously shift between these modes depending on the situation, by varying the specificity of the activated cognitive receptive field. This is referred to as *contextual focus*[4] because it requires the ability to focus or defocus attention in response to the context or situation one is in. Defocused attention, by diffusely activating a broad region of memory, is conducive to divergent thought; it enables obscure (but potentially relevant) aspects of the situation to come into play. Focused attention is conducive to convergent thought; memory activation is constrained enough to hone in and perform logical mental operations on the most clearly relevant aspects. The theory is consistent with the notion that creativity involves both freedom and constraint; the generation of cultural novelty often starts with structural rules and frameworks (as in the templates of a haiku or a tragedy) as a basis to deviate from.

7 Support from the Computational Model

Again, because it would be difficult to empirically determine whether Paleolithic hominids became capable of contextual focus, we began by determining whether the hypothesis is at least computationally feasible. To do so, we used an evolutionary art system that generated progressively evolving sequences of artistic portraits. In this context, we sought to determine whether incorporating contextual focus into the computer program would enable it to generate art that people find aesthetically pleasing and "creative" on its own (*i.e.*, requiring no human intervention once initiated).

We implemented contextual focus in the evolutionary art algorithm by giving the program the capacity to vary its level of fluidity and control over different phases of the creative process in response to the output it generated. The creative domain of portrait painting was chosen because it requires both focused attention and analytical thought to accomplish the primary goal of creating a resemblance to the portrait sitter, as well as defocused attention and associative thought to deviate from the resemblance in a way that is uniquely interesting, that is, to meet the broad and often conflicting criteria of aesthetic art. Since the advent of photography (and earlier), portrait painting has not just been about accurate reproduction but also about achieving a creative or stylized representation of the sitter. Since judging creative art is subjective, a representative subset of the automatically produced artwork from this system was selected, output to high-quality framed images, and submitted to peer-reviewed and commissioned art shows, thereby allowing it to be judged positively or negatively as creative by human art curators, reviewers, and the gallery-going public.

[4] In neural net terms, contextual focus amounts to the capacity to spontaneously and subconsciously vary the shape of the activation function, flat for divergent thought and spiky for analytical.

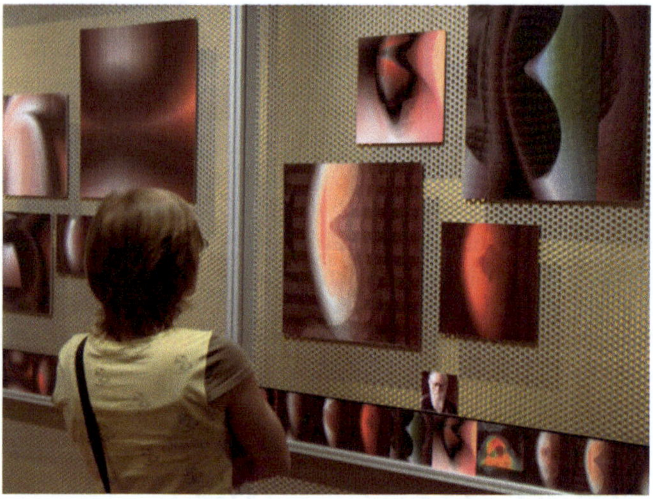

Fig. 4 Images produced by a computational art program that uses contextual focus at the MIT Museum in Cambridge, MA. These works have been seen by tens of thousands and perceived as creative art works on their own by the art public

The software incorporates several techniques that enable it to shift between different modes of thought, which are summarized here. (Implementation details are provided elsewhere; DiPaola 2009; DiPaola and Gabora 2007, 2009). Our goal was to incorporate the notion of contextual focus so that the software could shift between small ordered steps and large leaps through the landscape of artistic possibilities. This was carried out as follows: the system's default processing mode is an analytic mode, in which the primary aim is to achieve an accurate resemblance (similarity to the sitter image). Certain functional triggers (such as if the system is "stuck" and not improving) shift it to a more associative processing mode. This mode aims to achieve painterly aesthetic flair using principles of art creation (rules of composition, tonality, and color theory) as well as portrait knowledge space. Specifically, it takes into account (1) face versus background composition; (2) tonal similarity over exact color similarity, matched with a sophisticated artistic color space model that weighs for warm-cool color temperature relationships based on analogous and complementary color harmony rules; and (3) unequal dominant and subdominant tone and color rules, and other artistic rules based on a portrait painter knowledge domain as detailed in DiPaola (2009).

Incorporating contextual focus into the computer program not only improved its ability to generate a good resemblance but resulted in more abstract, aesthetically appealing portraits as well. Humans rated the portraits produced by this version of the portrait painting program with contextual focus as much more creative and interesting than a previous version that did not use contextual focus, and unlike its predecessor, the output of this program generated worldwide public attention. As shown in Fig. 4, sample pieces were exhibited at peer reviewed, juried, or commissioned shows in several major galleries and museums that typically only accept human art work, including the Tenderpixel Gallery in London, Emily Carr Galley in Vancouver, Kings Art Centre at Cambridge University, the MIT Museum, and the High Museum in Atlanta. The work was also selected for its aesthetic value to accompany a piece in *Nature* (Padian 2008). While these are subjective measures, they are standard in the art world. Thus, using contextual focus, the computer program automatically produced novel creative artifacts, both as single art pieces and as gallery collections of related art with interrelated creative themes, which provides compelling evidence of the effectiveness of contextual focus.

In sum, these results support the hypothesis that the impact of recursive recall was vastly accentuated by the capacity to shift between associative and analytic processing modes. This opened up a much greater variety of ways of seeing concepts from different contexts and examining ideas from different perspectives until one converges on an understanding that takes multiple facets into account. We suggest that a mechanism akin to contextual focus is what makes possible the cumulative creativity exhibited by successful computational models of language evolution (*e.g.*, Kirby 2001).

8 Modeling Contextual Focus: The Shifting Between Convergent and Divergent Thought

An even more compelling approach would result from developing a cognitive system that is capable of shifting between processing few features or properties of concepts and ideas (analytic or convergent thinking) to encoding many features or properties of concepts and ideas (associative or divergent thinking). The divergent mode would be highly conducive to the emergence of new concept combinations; since there are more properties encoded per concept, there are more potential connections, while the convergent mode would allow for focus and the honing in on useful ideas. Divergent thought is conducive to putting concepts together in new combinations. Using the quantum formalism discussed above, concept combination has been modeled using a tensor product and other more complex but accurate mathematical structures.

The details of this and related models have been discussed elsewhere (Aerts and Gabora 2005a, b; Gabora and Aerts 2002, 2009; Bruza et al. 2009; Kitto et al. 2011). However, the basic idea can be illustrated through a consideration of the two concepts FIRE and FOOD, and how they might have been combined in a creative manner by an early human. These two concepts are likely to have been thought of in a mutually exclusive manner by early humans, as FIRE would burn forests and fields so decreasing the expected yield of food. Thus, an increased experience of FIRE might have been expected to decrease the yield of food. However, at some point, FIRE was recognized as a tool; it could be used to create more food, by making inedible materials edible, rather than just being recognized as something that would decrease yields by burning food sources, etc. Representing FIRE as a superposition of useful ($|1_p\rangle$) and not useful ($|0_p\rangle$) and FOOD as a superposition of edible ($|1_q\rangle$) and inedible ($|0_q\rangle$), we can write the two combined concepts as

$$|\mathrm{FIRE}\rangle \otimes |\mathrm{FOOD}\rangle$$

$$= \left(a_0|0_p\rangle + a_1|1_p\rangle\right) \otimes \left(x_0|0_q\rangle + x_1|1_q\rangle\right)$$

$$= a_0x_0|0_p\rangle \otimes |0_q\rangle + a_0x_1|0_p\rangle \otimes |1_q\rangle + a_1x_0|1_q\rangle \otimes |0_p\rangle + a_1x_1|1_p\rangle \otimes |1_q\rangle$$

which is a superposition state that arises in the higher dimensional space represented by the four-dimensional basis states: $\left\{|0_p\rangle \otimes |0_q\rangle, |0_p\rangle \otimes |1_q\rangle, |1_p\rangle \otimes |0_q\rangle, |1_p\rangle \times |1_q\rangle\right\}$ (see Isham (1995) for more details about this kind of higher dimensional space). We immediately see that the combination of these two concepts has led to a combinatorial explosion of possibilities; in other words, this is a divergent process. If a person is now exposed to another concept, we can imagine a situation where their current cognitive state expands further still into a yet higher dimensional space. This process might go on for a number of steps; however, this increasingly more complex state is likely to be very difficult to maintain. Indeed, a potential downfall of processing in an associative mode and coming up with unusual combinations is that since effort is devoted to the reprocessing of previously acquired material, less effort may be devoted to being on the lookout for danger and simply carrying out practical tasks. Thus, associative thought was of little use until one could have a way of shifting back to a more analytical mode of thought. By reprocessing the new combination from increasingly constrained contexts or points of view, it would become clearer how to manifest it. Thus, while some associative thought is indisputably useful, it carries a high cognitive load, which increases as more and more concepts are combined. Eventually, there will be an adaptive advantage in settling upon one particular interpretation in a process of convergence.

The process of "measurement" discussed above performs this function, even in this scenario of rapidly expanding possibilities, and results in a convergent situation where one idea is finally settled upon, in turn lessening the load associated with maintaining a cognitive state. In the case above, an early human might have realized that FIRE when combined with FOOD could usefully render the inedible edible (as is represented by the state $a_1x_1|1_p\rangle \otimes |1_q\rangle$). The probabilities arising in this scenario might be very small, as the coefficients of Eq. (5) become smaller with each combination, thus indicating a situation where it is becoming more and more cognitively difficult to settle upon a particular meaning but also more likely that a highly improbable interpretation might be settled upon. Eventually, if enough humans experience this unusual cognitive state, there is a significant probability that one of them will start to cook inedible plants so rendering them edible. Initially, it may not be obvious how a new concept combination could make sense or materialize given the constraints of the world it is "born" into; for example, one does not know which features of each parent concept are inherited in the combination. Current work is being directed at finding natural representations of concepts, utilized naturally during the process of combination.

In summary, if early humans reached a stage where they could employ a divergent process of concept combination that was followed by a shift to a more constrained or convergent processing mode enabling them to actualize or manifest this new idea given the relevant practical and other considerations, then they would have found themselves at a significant adaptive advantage. They would have been capable of not just generating unusual new possibilities but also seeing them through, and so would have reached a new stage of cognitive activity.

9 Conclusions and Future Directions

Since concepts are the building blocks of human cognition, the explanation of how flexible, open-ended cognitive processes of thought arose will require a theory of concepts that can account for and model their contextual, noncompositional behavior. We showed how a quantum-inspired theory of concepts can be used to rigorously flesh out theories concerning the origins of modern cognition. Many species engage in acts that could be said to be creative, but humans are unique in that our creative ideas build on each other cumulatively. Indeed, it is for this reason that culture is widely construed as an evolutionary process (Bentley et al. 2011; Cavalli-Sforza and Feldman 1981; Gabora 1996, 1998, 2008a, b, c, d; Hartley 2009; Mesoudi et al. 2004, 2006; Whiten et al. 2011). Our unique cognitive capacities are revealed in all walks of life and have transformed the way we live and the planet we live on. We discussed two transitions in the evolution of human cognition: (1) its origins approximately two million years ago and (2) what has been referred to as the cultural explosion or "big bang" of human creativity approximately 50,000 years ago. We discussed cognitive mechanisms that have been proposed to underlie these transitions and summarized efforts to model them, both computationally and mathematically.

It has been hypothesized that the origins of complex human cognition can be attributed to the onset of *recursive recall,* in which one thought or stimulus evokes another in a string of associations (Donald 1991). This allowed for the chaining together of real or imagined episodes into a stream of thought or the chaining of movements into complex actions, such that feedback about one component affected performance of the next. This hypothesis has been shown to be compatible with likely changes in the architecture of human memory associated with the increase in cranial capacity at this time (Gabora 2003, 2008a). Moreover, in a test of this hypothesis using a computational model of cultural evolution in which neural network-based agents evolve ideas for actions through invention and imitation, chaining was shown to result in greater cultural diversity, open-ended generation of novelty, no ceiling on the mean fitness of cultural variants, and greater ability to make use of learning (Gabora and Saberi 2011). This shows that the hypothesis that recursive recall played an important role in the origins of complex cognition is computationally feasible. However, in the computational model, we simply compared runs in which agents were limited to single-step actions to runs in which they could chain simple actions into complex ones; chaining did not arise naturally through associative recall due to how items were encoded in memory. We suggest that it is not mere chaining that paved the way for complex cognition and cultural evolution, but chaining that involves the restructuring of concepts by viewing them from different contexts, and proposed that a formal model of this process will be required. We showed how a quantum-inspired theory of concepts can be used to model the transition to a state in which concepts and ideas are encoded in enough detail that associations among them are rich enough for a natural chaining through associative recall to occur, resulting in the capacity to progressively shape concepts, ideas, and actions by observing them from different contexts.

We discussed the hypothesis that the explosion of creativity in the Middle/Upper Paleolithic was due to onset of *contextual focus:* the capacity to shift between associative, conducive to forging new concept combinations, and analytic thought, conducive to manifesting them. Incorporating contextual focus (the capacity to shift between analytic and associative modes) into a computational model of portrait painting has resulted in faster convergence on portraits that human observers found preferable (DiPaola 2009; DiPaola and Gabora 2007, 2009). This supports the hypothesis that contextual focus provides a computationally plausible explanation for the cognitive capacities of modern humans.

A limitation of this work was that contextual focus was simply modeled as the capacity to shift between the competing goals of achieving an accurate resemblance of the sitter and deviating from the sitter's likeness by employing more abstract painterly techniques that exaggerate, minimize, or modify. This chapter also discussed a more sophisticated model of contextual focus using again the quantum-inspired model of concept combination. We show that if a cognitive system is capable of undergoing a transition from encoding few features or properties of concepts and ideas (analytic or convergent thinking) to encoding many features or properties of concepts and ideas (associative or divergent thinking), then new concept combinations are more likely to arise. The drawback is that such associative states are cognitively difficult to maintain, but we showed that if concept combination is followed by a shift to a more constrained or analytic processing mode, then an eventual interpretation can be settled upon, as the new idea or concept emerges from its previously "half-baked" state.

We are currently engaged in the move to more cognitively plausible computational implementations of creativity and its evolution. One of the projects that will soon be under way will implement contextual focus in the EVOC model of cultural evolution that was used for the "origin of creativity" experiments. This will be carried out as follows: The fitness function will change periodically so that agents find themselves no longer performing well. They will be able to detect that they are not performing well and, in response, increase the probability of change to any component of a given action. This temporarily makes them more likely to "jump out of a rut" resulting in a very different action, thereby simulating the capacity to shift to a more associative form of thinking. Once their performance starts to improve, the probability of change to any component of a given action will start to decrease to base level, making them less likely to shift to a dramatically different action. This is expected to help them perfect the action they have settled upon, thereby simulating the capacity to shift to a more associative form of thinking.

In short, we have developed several lines of investigation to formally test the feasibility of the hypothesis that human "mindedness" stems from onset of the capacity to see things in context or from multiple perspectives. We posit that this began with the onset of representational redescription at around the time of the appearance of *Homo erectus*, and that it was vastly enhanced by the onset of contextual focus, some time following the appearance of anatomically modern humans. Contextual focus enabled humans to shift between a minimally contextual analytic mode of thought, and a highly contextual associative mode of thought, conducive to "breaking out of a rut."

The hypotheses proposed here to underlie the evolution of our characteristically human ways of thinking and living in the world are speculative. However, we have shown that computational and mathematical models suggest that they are at least feasible. We believe they put us on our way toward modeling the mechanisms that could have made modern human cognition possible, along with the subsequent transformation of the planet we live on.

Acknowledgements This project was supported in part by the Australian Research Council Discovery grant DP1094974, the Natural Sciences and Engineering Research Council of Canada, and the Fund for Scientific Research of Flanders, Belgium.

References

Aerts, D. (2009). Quantum structure in cognition. *Journal of Mathematical Psychology, 53*, 314–348.

Aerts, D., Aerts, S., Broekaert, J., & Gabora, L. (2000). The violation of Bell inequalities in the macroworld. *Foundations of Physics, 30*(9), 1387–1414.

Aerts, D., & Gabora, L. (2005a). A state-context-property model of concepts and their combinations I: The structure of the sets of contexts and properties. *Kybernetes, 34*(1&2), 167–191.

Aerts, D., & Gabora, L. (2005b). A state-context-property model of concepts and their combinations II: A Hilbert space representation. *Kybernetes, 34*(1&2), 192–221.

Aiello, L. C. (1996). Hominine pre-adaptations for language and cognition. In P. Mellars & K. Gibson (Eds.), *Modeling the early human mind* (pp. 89–99). Cambridge: McDonald Institute Monographs.

Ambrose, S. H. (1998). Chronology of the later stone age and food production in East Africa. *Journal of Archaeological Science, 25*, 377–392.

Bar-Yosef, O. (1994). The contribution of southwest Asia to the study of the origin of modern humans. In M. Nitecki & D. Nitecki (Eds.), *Origins of anatomically modern humans*,Ý Plenum.

Barkow, J. H., Cosmides, L., & Tooby, J. (Eds.). (1992). *The adapted mind: Evolutionary psychology and the generation of culture*. New York: Oxford University Press.

Bentley, R. A., Ormerod, P., & Batty, M. (2011). Evolving social influence in large populations. *Behavioral Ecology and Sociobiology, 65*, 537–546.

Best, M. (1999). How culture can guide evolution: An inquiry into gene/meme enhancement and opposition. *Adaptive Behavior, 7*(3), 289–293.

Best, M. (2006). Adaptive value within natural language discourse. *Interaction Studies, 7*(1), 1–15.

Bruza, P. D., Kitto, K., Nelson, D., & McEvoy, C. (2009). Is there something quantum-like about the human mental lexicon? *Journal of Mathematical Psychology, 53*, 362–377.
Buller, D. J. (2005). *Adapting minds*. Cambridge: MIT Press.
Buss, D. M. (1999/2004). *Evolutionary psychology: The new science of the mind*. Boston: Pearson.
Cavalli-Sforza, L. L., & Feldman, M. W. (1981). *Cultural transmission and evolution: A quantitative approach*. Princeton: Princeton University Press.
Cloak, F. T., Jr. (1975). Is a cultural ethology possible? *Human Ecology, 3*, 161–182.
Deacon, T. W. (1997). *The symbolic species: The coevolution of language and the brain*. New York, NY: W.W. Norton.
DiPaola, S. (2009). Exploring a parameterized portrait painting space. *International Journal of Art and Technology, 2*(1–2), 82–93.
DiPaola, S., & Gabora, L. (2007, July 7–11). Incorporating characteristics of human creativity into an evolutionary art algorithm. In D. Thierens (Ed.), *Proceedings of the genetic and evolutionary computing conference* (pp. 2442–2449). University College London, England.
DiPaola, S., & Gabora, L. (2009). Incorporating characteristics of human creativity into an evolutionary art algorithm. *Genetic Programming and Evolvable Machines, 10*(2), 97–110.
Donald, M. (1991). *Origins of the modern mind: Three stages in the evolution of culture and cognition*. Cambridge, MA: Harvard University Press.
Donald, M. (1998). Hominid enculturation and cognitive evolution. In C. Renfrew & C. Scarre (Eds.), *Cognition and material culture: The archaeology of symbolic storage* (pp. 7–17). Cambridge: McDonald Institute Monographs.
Dugatkin, L. A. (2001). *Imitation factor: Imitation in animals and the origin of human culture*. New York: Free Press.
Dunbar, R. (1993). *Coevolution of neocortical size, group size, and language in humans. Behavioral and Brain Sciences, 16*(4), 681–735.
Finke, R. A., Ward, T. B., & Smith, S. M. (1992). *Creative cognition: Theory, research, and applications*. Cambridge, MA: MIT Press.
Gabora, L. (1994, July 4–6). A computer model of the evolution of culture. In R. Brooks & P. Maes (Eds.), *Proceedings of the 4th international conference on artificial life*, Boston, MA.
Gabora, L. (1995). Meme and variations: A computer model of cultural evolution. In L. Nadel & D. Stein (Eds.), *1993 lectures in complex systems* (pp. 471–486). Reading: Addison-Wesley.
Gabora, L. (1996). A day in the life of a meme. *Philosophica, 57*, 901–938.
Gabora, L. (1998). Autocatalytic closure in a cognitive system: A tentative scenario for the origin of culture. Psycoloquy, 9(67).
Gabora, L. (1999a, May 3–5). Conceptual closure: Weaving memories into an interconnected worldview. In G. Van de Vijver & J. Chandler (Eds.), *Proceedings of closure: An international conference on emergent organizations and their dynamics,* held by the Research Community on Evolution and Complexity and the Washington Evolutionary Systems Society, University of Gent, Belgium.
Gabora, L. (1999b). To imitate is human: A review of '*The Meme Machine*' by Susan Blackmore. *Journal of Artificial Societies and Social Systems* 2(2). Reprinted with permission in *Journal of Consciousness Studies, 6*(5), 77–81.
Gabora, L. (2003, July 31–August 2). Contextual focus: A cognitive explanation for the cultural transition of the Middle/Upper Paleolithic. In R. Alterman & D. Hirsch (Eds.), *Proceedings of the 25th annual meeting of the Cognitive Science Society* (pp. 432–437). Boston: Lawrence Erlbaum.
Gabora, L. (2004). Ideas are not replicators but minds are. *Biology and Philosophy, 19*(1), 127–143.
Gabora, L. (2008a). Mind. In R. A. Bentley, H. D. G. Maschner, & C. Chippindale (Eds.), *Handbook of theories and methods in archaeology* (pp. 283–296). Walnut Creek: Altamira Press.
Gabora, L. (2008b). EVOC: A computer model of cultural evolution. In V. Sloutsky, B. Love, & K. McRae (Eds.), *Proceedings of the 30th annual meeting of the Cognitive Science Society* (pp. 1466–1471). North Salt Lake: Sheridan Publishing.
Gabora, L. (2008c). Modeling cultural dynamics. In *Proceedings of the Association for the Advancement of Artificial Intelligence (AAAI) Fall symposium 1: Adaptive agents in a cultural context* (pp. 18–25). Menlo Park: AAAI Press.

Gabora, L. (2008d). The cultural evolution of socially situated cognition. *Cognitive Systems Research, 9*(1–2), 104–113.

Gabora, L., & Aerts, D. (2002). Contextualizing concepts using a mathematical generalization of the quantum formalism. *Journal of Experimental and Theoretical Artificial Intelligence, 14*(4), 327–358.

Gabora, L., & Aerts, D. (2009). A mathematical model of the emergence of an integrated worldview. *Journal of Mathematical Psychology, 53*, 434–451.

Gabora, L., & Saab, A. (2011). Creative interference and states of potentiality in analogy problem solving. *Proceedings of the Annual Meeting of the Cognitive Science Society* (pp. 3506–3511). July 20–23, 2011, Boston MA.

Gabora, L., & Leijnen, S. (2009). How creative should creators be to optimize the evolution of ideas? A computational model. *Electronic Proceedings in Theoretical Computer Science, 9*, 108–119.

Gabora, L., & Russon, A. (2011). The evolution of human intelligence. In R. Sternberg & S. Kaufman (Eds.), *The Cambridge handbook of intelligence* (pp. 328–350). Cambridge: Cambridge University Press.

Gabora, L., & Saberi, M. (2011). How did human creativity arise? An agent-based model of the origin of cumulative open-ended cultural evolution. In *Proceedings of the ACM conference on cognition & creativity* (pp. 299–306). Atlanta, GA.

Gabora, L., & Firouzi, H. (2012). Society functions best with an intermediate level of creativity. *Proceedings of the Annual Meeting of the Cognitive Science Society* (pp. 1578–1583). August 1-4, Sapporo Japan.

Gabora, L., Leijnen, S., & von Ghyczy, T. (in press). The relationship between creativity, imitation, and cultural diversity. *International Journal of Software and Informatics*.

Gamble, C. (1994). *Timewalkers: The prehistory of global colonization*. Cambridge, MA: Harvard University Press.

Hampton, J. (1987). Inheritance of attributes in natural concept conjunctions. *Memory & Cognition, 15*, 55–71.

Hartley, J. (2009). From cultural studies to cultural science. *Cultural Science, 2*, 1–16.

Henshilwood, C., d'Errico, F., Vanhaeren, M., van Niekerk, K., & Jacobs, Z. (2004). Middle stone age shell beads from South Africa, *Science, 304*, 404.

Henshilwood, C. S., & Marean, C. W. (2003). The origin of modern human behavior. *Current Anthropology, 44*, 627–651.

Heyes, C. M. (1998). Theory of mind in nonhuman primates. *The Behavioral and Brain Sciences, 211*, 104–134.

Higgs, P. (2000). The mimetic transition: A simulation study of the evolution of learning by imitation. *Proceedings: Royal Society B: Biological Sciences, 267*, 1355–1361.

Hinton, G. E., & Nowlan, S. J. (1987). How learning can guide evolution. *Complex Systems, 1*, 495–502.

Holland, J. K. (1975). *Adaptation in natural and artificial systems*. Ann Arbor: University of Michigan Press.

Howard-Jones, P.A., & Murray, S. (2003). Ideational productivity, focus of attention, and context. *Creativity Research Journal, 15*(2&3), 153–166.

Hutchins, E., & Hazelhurst, B. (1991). Learning in the cultural process. In C. Langton, J. Taylor, D. Farmer, & S. Rasmussen (Eds.), *Artificial life II*. Redwood City: Addison-Wesley.

Isham, C. (1995). *Lectures on quantum theory*. London: Imperial College Press.

Jerison, H. J. (1973). *Evolution of the brain and intelligence*. New York, NY: Academic Press.

Karmiloff-Smith, A. (1992). *Beyond modularity: A developmental perspective on cognitive science*, MIT Press. Boston, MA.

Kirby, S. (2001). Spontaneous evolution of linguistic structure: An iterated learning model of the emergence of regularity and irregularity. *IEEE Transactions on Evolutionary Computation, 5*(2), 102–110.

Kitto, K. (2006). *Modelling and generating complex emergent behaviour*. PhD thesis, School of Chemistry, Physics and Earth Sciences, The Flinders University of South Australia.

Kitto, K. (2008a). High end complexity. *International Journal of General Systems, 37*(6), 689–714.

Kitto, K. (2008b). Why quantum theory? In *Proceedings of the second quantum interaction symposium* (pp. 11–18). London: College Publications.
Kitto, K., Ramm, B., Sitbon, L., & Bruza, P. (2011). Quantum theory beyond the physical: Information in context. *Axiomathes, 21*(2), 331–345.
Kitto, K., Bruza, P., & Gabora, L. (2012, June 10–15). A quantum information retrieval approach to memory. In *Proceedings of the 2012 International Joint Conference on Neural Networks (IJCNN 2012), WCCI 2012 IEEE World Congress on Computational Intelligence, IEEE* (pp. 932–939). Brisbane: Brisbane Convention Centre.
Klein, R. (1989). Biological and behavioral perspectives on modern human origins in South Africa. In P. Mellars & C. Stringe (Eds.), *The human revolution*. Edinburgh: Edinburgh University.
Klein, R. G. (1999). *The human career: Human biological and cultural origins*. Chicago, IL: University of Chicago Press.
Klein, R. G. (2003). Whither the Neanderthals? *Science, 299*, 1525–1527.
Krasnegor, N., Lyon, G. R., & Goldman-Rakic, P. S. (1997). *Prefrontal cortex: Evolution, development, and behavioral neuroscience*. Baltimore, MD: Brooke.
Leakey, R. (1984). *The origins of humankind*. New York, NY: Science Masters Basic Books.
Leijnen, S., & Gabora, L. (2009). How creative should creators be to optimize the evolution of ideas? A computational model. *Electronic Proceedings in Theoretical Computer Science, 9*, 108–119.
Leijnen, S., & Gabora, L. (2010, August 11–14). An agent-based simulation of the effectiveness of creative leadership. In *Proceedings of the annual meeting of the Cognitive Science Society* (pp. 955–960). Portland, Oregon.
Martindale, C. (1995). Creativity and connectionism. In S. M. Smith, T. B. Ward, & R. A. Finke (Eds.), *The creative cognition approach* (pp. 249–268). Cambridge MA: MIT Press.
McBrearty, S., & Brooks, A. S. (2000). The revolution that wasn't: A new interpretation of the origin of modern human behavior. *Journal of Human Evolution, 39*, 453–563.
Mellars, P. (1973). The character of the middle-upper transition in south-west France. In C. Renfrew (Ed.) *The explanation of culture change*. London: Duckworth.
Mellars, P. (1989a). Technological changes in the Middle-Upper Paleolithic transition: Economic, social, and cognitive perspectives. In P. Mellars & C. Stringer (Eds.), *The human revolution*. Edinburgh: Edinburgh University Press.
Mellars, P. (1989b). Major issues in the emergence of modern humans. *Current Anthropology, 30*, 349–385.
Mesoudi, A., Whiten, A., & Laland, K. (2004). Toward a unified science of cultural evolution. *Evolution, 58*(1), 1–11.
Mesoudi, A., Whiten, A., & Laland, K. (2006). Toward a unified science of cultural evolution. *The Behavioral and Brain Sciences, 29*, 329–383.
Mithen, S. (1996). *The prehistory of the mind: A search for the origins of art, science, and religion*. London: Thames & Hudson.
Mithen, S. (1998). *Creativity in human evolution and prehistory*. London: Routledge.
Osherson, D., & Smith, E. (1981). On the adequacy of prototype theory as a theory of concepts. *Cognition, 9*, 35–58.
Padian, K. (2008). Darwin's enduring legacy. *Nature, 451*, 632–634.
Premack, D. (1988). "Does the chimpanzee have a theory of mind?" revisited. In R. W. Byrne & A. Whiten (Eds.), *Machiavellian intelligence: Social expertise and the evolution of intellect in monkeys, apes and humans* (pp. 160–179). Oxford: Oxford University Press.
Premack, D., & Woodruff, G. (1978). Does the chimpanzee have a theory of mind? *The Behavioral and Brain Sciences, 1*, 515–526.
Ruff, C., Trinkaus, E., & Holliday, T. (1997). Body mass and encephalization in Pleistocene Homo. *Nature, 387*, 173–176.
Rumbaugh, D. M. (1997). Competence, cortex, and primate models: A comparative primate perspective. In N. A. Krasnegor, G. R. Lyon, & P. S. Goldman-Rakic (Eds.), *Development of the prefrontal cortex: Evolution, neurobiology, and behavior* (pp. 117– 139). Baltimore, MD: Paul.

Smith, W. M., Ward, T. B., & Finke, R. A. (1995). *The creative cognition approach*. Cambridge, MA: MIT Press.

Soffer, O. (1994). Ancestral lifeways in Eurasia6The Middle and Upper Paleolithic records. In M. Nitecki & D. Nitecki, (Eds.), *Origins of anatomically modern humans*. New York: Plenum Press.

Stringer, C. & Gamble, C. (1993). *In search of the Neanderthals*. London: Thames & Hudson.

Tomasello, M., Kruger, A., & Ratner, H. (1993). Cultural learning. *The Behavioral and Brain Sciences, 16*, 495–552.

Wexler, B. (2006). *Brain and culture: Neurobiology, ideology and social change*. New York: Bradford Books.

White, R. (1989a). Production complexity and standardization in early Aurignacian bead and pendant manufacture: Evolutionary implications. In P. Mellars & C. Stringer (Eds.), *The human revolution: Behavioral and biological perspectives on the origins of modern humans* (pp. 366–90). Cambridge, UK: Cambridge University Press.

White, R. (1989b). Toward a contextual understanding of the earliest body ornaments. In E. Trinkhaus (Eds.), *The emergence of modern humans: Biocultural adaptations in the later Pleistocene*. Cambridge, UK: Cambridge University Press.

Whiten, A., Hinde, R., Laland, K., & Stringer, C. (2011). Culture evolves. *Philosophical Transactions of the Royal Society B, 366*, 938–948.

The Mind of the Noble Ape in Three Simulations

Tom Barbalet

Abstract The Noble Ape Simulation offers an account of the mind as something that can be observed, measured, and ultimately simulated through external effects. This version of the applied mind is not created through a single method but through layering three simulations relating to information chemistry, social constraints, and evolving narrative. As examples, additional simulation elements in Noble Ape are presented to offer the simulation methodology of Noble Ape. This chapter, rather than being a theoretical critique, is intended as a project report relating to three distinct yet interoperating simulated models of the mind. These are presented both as individual simulations and also the simulations' interactions. This produces a novel account of the applied mind. The methods used in creating such an applied mind provide an interesting insight into the possible origin of mind through pragmatic application rather than conjecture.

Keywords Artificial life • Simulation • Theories of mind • Robotics • Social robotics • Cognitive science • Cognitive simulation • Intelligent agents • Open source • Linguistics • Computational linguistics • Philosophy of mind • Philosophy of language

1 Background

I started Noble Ape at 19 years of age in 1996 in Australia. As Noble Ape is open source, there have been numerous contributing developers including engineers from Apple, Intel, and Cern. A substantial component of the Noble Ape Simulation discussed in this chapter has come from Bob Mottram, an industrial roboticist based in the United Kingdom. This chapter would not be possible without the dedicated

T. Barbalet (✉)
Noble Ape, Campbell, CA, USA
e-mail: tom@nobleape.com

L. Swan (ed.), *Origins of Mind*, Biosemiotics 8, DOI 10.1007/978-94-007-5419-5_20,

work Mottram has donated to Noble Ape. Through the work effort described in this chapter, Mottram worked remotely and often contributed source code independently. This is one of the charming idiosyncrasies of open source development. Multiple participants can work on the same piece of software for distinct purposes at the same time with minimal communication.

Noble Ape can be thought of as a series of different simulations:

- A landscape simulation that creates a large environment
- A biological simulation that models the underlying biology in the environment
- A weather simulation that creates the meteorological aspects of the environment
- Three independent but intertwined agent simulations:
  - A cognitive simulation
  - A social simulation
  - A narrative engine

The latter three are the primary topic of this chapter. The weather and biological simulations will also be discussed as they offer a connection with the cognitive simulation and the account of the broader methodological perspective of the project.

2 Artificial Life

Noble Ape is considered an artificial life project. Artificial life does not have an exact disciplinary definition. It covers a variety of different kinds of software, hardware, and chemistry that look to show *life as it could be* (Langton 1997). Artificial life is an idea that predates computation and exists in its most basic form in thought experiments about life – speculative life, if you will. Concepts in artificial life can be found as early as Hobbes' Leviathan (1651).

Computation has moved the field from thought experiments into a variety of different approaches including evolutionary computing, intelligent agents, genetic algorithms, applied genetic programming, and cellular automata. The field was broadly defined through a number of popular surveyings (Emmeche 1991; Levy 1992) and authors who developed their own early artificial life simulations (Dawkins 1987).

While the early artificial life simulations were relatively simple and similar to other kinds of software, artificial life software that has been in development for more than a decade is in a comparably advanced state. Modern computing, specifically continuous development adapting to modern multi-core processors, has advanced the capabilities of artificial life software. Noble Ape has been able to give back into this cycle too through its use by Apple and Intel to optimize processor power (Barbalet 2009).

It is important to recognize this chapter in this light. The work presented here relates to software that can be obtained both in source and executable form free of charge for additional scrutiny. The descriptions offered here are not speculative but relate to software that, although it may appear whimsical, has been of great practical benefit.

3 Motivations

Noble Ape was created with a basic hope: through sweat equity it would be possible to create a philosophically rich simulation of the mind. The problem was divided into two parts. An environment needed to be created that would have the depth and interest for these simulated minds to flourish. Also, perhaps far more difficult, the simulated minds would need to show a degree of tenacity to be a compelling representation of real-world cognitive dilemmas.

At the time of initial development, I knew of no peers in this kind of project. I later learned about the work of Larry Yaeger with Polyworld (1994). The distinction between Noble Ape and Polyworld was that Noble Ape did not have a neural network as the intelligence in silico. Initially Noble Ape relied on the cognitive simulation described in this chapter.

The early development of Noble Ape was a youthful opposition to dominant and failing ideas that went against my own experience. As a student of philosophy, I was frequently told that computer simulations offered no insight into the mind. I was presented with straw arguments relating to buggy software and failed robot experiments far from the work I read about at MIT in a similar time frame (Kirsh 1991).

The misguided view of software intelligence as failing and sub-utilitarian was in stark contrast to my own experiences in creating software (Barbalet 1997a). In my early teens, I developed computer games with compelling simulated comrades and enemies. Through my late teens, I wrote heuristic antiviral software that detected both known computer viruses and also predicted computer viruses from heuristic analysis of known symptoms and projected symptoms. Prior to Noble Ape, while I wrote antiviral software, I also wrote compiler software (software that took English language readable code and translated it into machine code) that was based on some of the dynamic and adaptive methods I saw used in computer viruses. My compiler software was intentionally non-malicious as it related to transforming abstract information without the infrastructure to be transmitted from machine to machine. These compiled models of adaptive intelligence seem distant from the poor accounts of software intelligence I was provided with in my philosophy studies.

My choice of study in mathematics, physics, and philosophy was a primary indication of my general level of distain for computer science with flawed neural networks and obsessive historical self-induced paradoxes (similar to the philosophy I found). As the early Noble Ape development showed (Barbalet 1997b), I was fixated on finding solutions to the origin of mind and a means of simulating the mind. Computer science and, as I found through my studies, philosophy were not going to provide the answers or even the direction for this insight.

I felt very strongly that trying to find a biological mirror of the mind in software failed to identify how little was known about the relating biology. In fact these attempts to simulate the mind through biologically inspired neural networks appeared to confirm the skeptical philosophical view that was omnipresent in my philosophy education. The early development of Noble Ape, in particular the biological simulation and the cognitive simulation, were intentionally developed in stark contrast to the failed but commonly accepted means of simulating both biology through representative biological models in software and attempts to simulate the mind through neural networks.

The energy and anger of youth tends to taper. The practical nature of maintaining a development like Noble Ape required progressive compromise. It is important to note the development moved from being distinctly radical to relatively mainstream not through a movement in the project but through a movement of the thinking on simulations that contained intelligent agents.

Part of the normalization of Noble Ape came through its utilitarian use. Within 7 years of the project starting, it was embraced by a generation of engineers at Apple and 2 years following another group of engineers at Intel (Barbalet 2005a). Through this period, limited additional work could be done on the simulation. As the primary maintainer, roughly 5 years were spent updating the project to the changes required by the Apple and Intel engineers.

It was this normative maintenance culture that appealed to Bob Mottram. The cognitive simulation (unique and original to the project), the social simulation (based in social robotics), and the narrative engine (based in early artificial life simulation) are combined in the Noble Ape development. This combination of simulations within a unified project represents my pluralist and utilitarian philosophical views on the origin of mind. The project also identifies the only productive way these models can be used is in concert: not in contrast or competition. It is also important to note that the latter two contributions for the simulations of mind in Noble Ape probably would not have been accepted in the early history of the development.

Moreover, it is perfectly feasible that additional simulations will be added through the continued developmental history of Noble Ape. It is also quite possible that the simulation models used could be unified. This should provide further philosophical insight as the method used to reduce these simulations should also provide finer conceptual structure to the origin of mind.

4 Biological Simulation

The biological simulation was the first new software developed for Noble Ape. Noble Ape was created rapidly as it was primarily a combination of existing projects I created. The landscape and visualization came from earlier landscape graphics environments I created (Barbalet 2004), and the cognitive simulation came from earlier agar (petri dish) simulations I created. The early development was undertaken on first-generation personal computing (PC-XT and PC-AT computers and 68000 Macintosh computers). For the scale of landscape being simulated, even macro population simulation (Volterra 1931) would have been too computationally intensive.

At the time, I was studying physics. It appeared the easiest way to minimize the exertion of processing power was to model the biological simulation on quantum mechanics (computationally, if not conceptually). The use of quantum mechanics in the biological simulation can be explained relatively simply. Take a point on the landscape and perform a summation of probabilities. These probabilities can be offered in a thought experiment. What would it take for a particular biological species to exist at that point? The landscape is a wave function. It is a continuous two-dimensional planar function. There are various properties of the landscape. The landscape at a particular point has an area associated with it. It has a height above some arbitrary level – a height above sea level, for example. It has a water value associated with it that relates to its proximity to saltwater or freshwater. There is a moving sunlight operator that represents how the simulated sun is hitting the point at a particular time. There's a total sunlight operator that is taken over all time. Also a salt operator that represents salt water or ground salt.

The height is the underlying quantum mechanics wave function, and these operators (height above sea level, area, moving sunlight, total sunlight, water, and salt) are applied to the wave function to give a value.

At any given point, there is a probability density that something will be there. This only becomes actuality when a noise map is put on the probability density. This cuts the probability density and shows where something actually is rather than a probability of its being there. Rather than creating a huge biological system including every part and a wide variety of other interactions, the biological simulation just interrogates the environment at a particular place and calculates the operators that are applicable. If the Noble Apes are foraging for food, the simulation can get the various operators that converge on whether the Noble Apes are interested in berries or whatever food is available and can interrogate the environment directly rather than having a large biological simulation.

Using a plant as an example, consider the surface area needed. Surface area is a point relative term based on a flat plane having little surface area, approaching a near infinite surface area as the landscape moves to a cliff. A tree can't grow well on a cliff, so the surface area has certain importance. There are various plants that thrive at particular heights. Water is also an important factor. Moving sunlight is less important, but total sunlight is critical, and depending on whether the plant likes or dislikes salt is a factor too. Insects may dislike being in direct sunlight so the moving sunlight indicates where some insects may not want to be.

A noise map is used to intersect the probability function coming from the operators acting on the wave function. The change on the noise map depends on whether the biology is a plant or an animal. If it is a plant, there needs to be reproducibility at a specific point, whereas if it is an animal, it needs to change over time. The plant noise map is static, whereas the animal noise map has periodic transitions.

The biological simulation provides a good example of the pragmatism that has been a defining factor in the creation of simulations for Noble Ape. A specific need for great detail and a limitation of processing power created a biological simulation that may not express all the components for a detailed biological understanding but produces enough biodiversity to provide a detailed simulation environment and simulated diet for the Noble Apes.

5 Weather Simulation

Added to Noble Ape in 2000, the weather simulation can be summarized as a water vapor simulation with a hard ceiling. The water vapor moves over the landscape. As there is increased pressure, the clouds form and rainfall occurs at the highest pressure points. The weather simulation is calculated at half the resolution of the landscape. This is due to the time to calculate the underlying weather. This calculation was heavily optimized to make it as fast as possible.

The weather simulation is less scalable than the biological simulation. It not only has been maintained through the functional purpose of providing accurate and diverse weather conditions to the simulation inhabitants but it also closely resembles the initial two-dimensional cognitive simulation. The weather simulation still has shared mathematics with the three-dimensional cognitive simulation.

There is a somewhat tongue-in-cheek grand unified simulation theory that the weather simulation and the cognitive simulation could have greater shared mathematical elements. The cognitive simulation was the subject of substantial optimization by engineers at both Apple and Intel for their respective processing hardware (Barbalet 2009). If it was possible to find connective mathematical elements, and have these elements optimized through modern processing hardware, both the weather and the cognitive simulations would see substantial speed improvements.

6 The Cognitive Simulation

The cognitive simulation predates a majority of the development of Noble Ape. It comes from my early simulation of agar (petri dish) bacterial growth. Through developing these simulations, I came to the idea that bacterial growth could represent information transfer. As the bacteria grew through the agar, the movement into corresponding cells was similar to information being transferred to the surrounding cells (Barbalet 2009). The mathematics for bacterial growth in agar and the final mathematics for the Noble Ape cognitive simulation were quite different, but they had mathematical similarities. Both were represented by competing equations: one associated with the movement through space and one associated with the movement through time. In the cognitive simulation, these two competing equations were labeled *desire* in terms of the traversing through space and *fear* for reacting through time (Barbalet 1997b). The original cognitive simulation was a two-dimensional simulation in a 128×128 cell space. The sensors (that pushed sense information into the cognitive simulation) were at one end, and the actuators (that took information from the simulation to produce action) were at the other end. The sensors' noise and excitement would ripple through to the actuators through the agar-like substrate accordingly.

In the two-dimensional simulation, the information flow has characteristics that were very similar to those of the weather simulation; however, it had a strong bias in linear movement providing just a single dimension of information transfer.

I moved to a three-dimensional model with the same underlying mathematics in a smaller area ($32 \times 32 \times 32$ cells). This added the ability for information to transfer in all three dimensions rather than the scanning two dimensions that ultimately led to a single productive dimension that related to the time transfer of the information.

In the current version of the cognitive simulation, Mottram changed the code slightly, so the sensors and actuators are once again equidistantly spaced. The addition of a third dimension gives a fixed processing length and an ability for the information to intermingle.

What the cognitive simulation presents is a description of the mind in a pre-language and pre-social state. It is the idea of the mind as survival organ. The mind must guide the agent to food and away from danger. Society, as it is represented in such a mind, is purely a fear negator and potentially also the guide toward feeding and procreating areas. The cognitive simulation provides a primitive survival model of the mind.

The cognitive simulation describes not only the process but also the information vessel where there are sensors and actuators that are passing information through the vessel. The properties of the vessel explain how the information is retarded and propagated. The sensors are firing information, and actuators are reacting to this information. The space between the sensors and actuators in the vessel is the mathematical space described by fear and desire. Conceptually the vessel description of the cognitive simulation has only one flaw. The space of the cognitive simulation wraps around. The x-axis wraps into itself as does the y-axis and the z-axis of the simulation space. This provides an additional property where the nearest sensor and actuator connection may be through the axis origin. The contribution of sensor information into the cognitive simulation may be maintained through multiple traverses through the cognitive simulation space. These rippling waves of information transfer are negated through both desire properties and also the ability for sensors to provide strobing feedback that can stabilize returning information signals.

Desire is reinforcing for actuator responses. Rather than reacting violently to information that is being put through the sensors, desire reinforces this information, slightly retarding it through the spatial mathematics it employs. The agent does not react so fearfully. In contrast, fear amplifies the sensor signals and causes more reactive movement when this information is received by the actuators. Both fear and desire coexist in the cognitive simulation to counterbalance these competing properties.

The cognitive simulation size for the Noble Apes has remained the same since it moved to three dimensions. Those size constraints should be expanded for some interesting effects. With the additional simulations of the mind described in this chapter, in particular the narrative engine, a $64 \times 64 \times 64$ cell to even a $256 \times 256 \times 256$ cell cognitive simulation would greatly benefit the broader agent model.

There are a number of other species that exist in the Noble Ape environment. The Noble Apes have a primary role because they are sentient human-like creatures. There are felines, birds, and smaller mammals. These species would benefit from having simple cognitive simulations that are similar to the Noble Apes. The weighting between fear and desire as well as the size of the cognitive simulation could be altered. Consider a feline having a cognitive simulation of $8 \times 8 \times 8$ cells. Rather than having a heavy fear weighting to the cognitive simulation, the simulated feline would have a stronger weighting to desire as they are the primary predator in the environment. They have little need for fear and are more governed by their general desires.

7 Social Simulation

Noble Apes with just the cognitive simulation were not particularly social. They were a reactive and fearful group of simulated agents. Mottram came to the Noble Ape Simulation with a background in social robotics, in particular a strong interest in the work of Cynthia Breazeal at MIT (2002). Mottram's initial feedback having reviewed the simulation was that there needed to be a set of social factors and constraints hardcoded into the simulation.

Mottram saw grooming as an important primate social behavior that was absent from the Noble Apes. He set about implementing something comparable to grooming as he realized that grooming served both a utilitarian function (the removal of parasites) and also a psychological function (of determining and reinforcing status and bonding relationships between individuals). In keeping with the theme of nobility, Mottram added an *honor* value that was indicated of the social status of each individual in the group. He also added a value indicating the number of parasites carried by each Noble Ape together with a simple mathematical model of parasite reproduction, energy cost to the Noble Ape, and their transmissibility between Apes.

Mottram hardcoded interactions that would create a simple economy based on social status. When one Noble Ape was groomed by another, they spent some of their honor value, while the groomer acquired a corresponding amount of honor for performing the service of removing parasites (and hence reducing energy depletion). The honor value might then be later used to bias mating decisions. Mottram also started adapting some of the genetic aspects of the simulation and created ideas of families, social groups, and clans.

Although in the initial implementation of this grooming-based economy of status, Noble Apes were not explicitly aware of their own honor value or that of others; the later addition of the narrative engine permitted them to become aware of this factor.

If the Noble Apes had a self-aware notion of their own honor, then it would change their interactions and the simulation would digress into an honor optimization algorithm. Honor was heavily muted in things that the Noble Apes could access. Primarily it just had unexplained effects when, for example, they were meeting other Noble Apes or they were squabbling. This simulated honor contained elements of luck based on probabilistic outcomes.

Mottram also explicitly hardcoded for social drives (Breazeal 2002). The hunger drive represented a biological quantity but also represented an interaction with food.

The social drive represented an interaction with other entities and had various feedbacks associated with social interaction. The fatigue drive related to tiredness, an overabundance of swimming, and a variety of other factors. The sex drive also contained elements of social interaction and genetic predetermined preferences. These drives, like honor, were represented as a single variable each.

8 Social Graph

In addition to social variables, Mottram and I worked together to produce a social graph. The social graph described a spatial map where the relationship of each Noble Ape in space is represented by their social connections and time is represented in simulated time. The social graph could be considered another simulation in and of itself. It is foreseeable in the future development that the social graph becomes a fully independent simulation.

The social graph interaction produced a very rich graphical view of Noble Ape society. Social groups of Noble Apes appear in cloud-like formations through the social graph. Each individual Noble Ape only has a social group of six other Noble Apes. Although six others may seem extremely small, the larger families and genetic groups maintain hardcoded connections. The Noble Ape will be able to implicitly recognize kin, but it may not have the same memory of this hardcoded kin as it has of an individual in its social group memory. The six Noble Apes in the social group memory of each Ape magnified over the population total produce a rich social environment that is represented as a rich graphical environment.

This graphical view illustrates dramatically the friends and enemies of each Noble Ape. Moreover, conditions of social ejection are shown graphically. Some conditions of Noble Ape squabbling eject one or two Noble Apes out of a family or clan group. There are choices the other Apes need to make about whether or not they want to interact socially with the socially ejected party.

The social graph tracks a variety of smaller things, but it can be used in a spatial graph setting as well. The difficulty in understanding simulations like Noble Ape is that they are just so rich. Vast numbers of interactions occur. Any additional abstraction that can convey meaning is greatly beneficial. The social graph provides this ability to see an aspect of the simulation that would have been very difficult to do through observing the simulation over time and interpolating through the information presented.

The social graph highlighted two properties of the social simulation that had been observed through simulation space interactions, but the profound effect on Noble Ape society was not properly understood until the social graph identified them explicitly.

The first property highlighted was that social relations can be asymmetric. This is identified in the social graph interaction where Noble Apes make mistakes. Information is forgotten by certain Apes at a faster rate and remembered by others. There are bitter Noble Apes who have had negative interactions that they haven't forgotten. Other Noble Apes forget these interactions and get on with their foraging. There is also implicit confusion the way the family groups are described. Some of the Noble Apes think that certain other Apes are in one family group, and some of the Apes think they're in another family group due to implicit mistakes in group meetings and information presented to the Noble Apes in conversation with other Noble Apes. The notion of primary truth is not there. It is relative and muddled. In code, the same event or idea is represented by something that is not referential to a single thing but in fact is completely uniquely represented per Noble Ape. As the Noble Ape replays these events through narrative either internal (in their own thinking) or external (telling any Noble Ape who will listen), it is possible for the Noble Ape's own description of the thing being discussed to change through the narrative process.

The second property highlighted through the social graph was the role that squabbling plays in the Noble Ape interactions. There is a wide variety of extremes associated with squabbling. Squabbling is a very broad description of anything from gesturing and shouting to noncontact swipes and aggressive posturing to violent blows and murder in some rare circumstances. As the Noble Apes get closer, more interaction can occur. Mottram hardcoded these interactions offering honor as the defining factor but also utilizing the level of social animosity the Noble Apes held to one another. As noted, Noble Apes implicitly have very small social groups in their recall social memory. For this reason, if a Noble Ape has a dispute with another Ape, this interaction may replace other Apes that they periodically meet and this replacement may make the Noble Apes more susceptible to creating a sometimes artificial nemesis.

9 Narrative Engine

The social simulation provides an underlying social structure that is relatively easy to understand both in short-term interactions and long-term trends primarily because it is heavily hardcoded. Each interaction has a specific condition and a coded response.

Through extended discussions with futurist linguist, Heron Stone, the challenge was made that Noble Ape should be able to simulate the linguistic phenomenon Stone advocated: every aspect of modern human existence appears to be based on an executed language program (Barbalet and Stone 2011). An internal narrative (thought) similar to the external narrative (speech) governs modern existence and should be able to be simulated through Noble Ape. While the idea of thought as language was not new, the ability to construct an internal and external narrative engine that literally drove the Noble Ape interactions was a challenge.

Up until this point, Noble Ape communications in the simulation were very basic. There was screaming and shouting and gesturing, but there was nothing that described the rich internal narrative that could capture things like belief or even things like social dance.

There is a variety of things captured by language both implicitly and explicitly. The challenge was to create a narrative engine where the Apes could have both internal dialogue (language-structured thought) and an external dialogue (language-structured speech).

Mottram and I came to this challenge at the same time. There was a shared interest in Corewar (Shock and Hupp 1982) and in the artificial life simulations like Tierra (Ray 1991). Corewar provided a thorough treatment of early stable byte-code languages. Byte-codes mean literally small atomic blocks of computer executable code. Stable byte-code languages had the benefit that although code could be modified (and the effects of these code changes could be dramatic for only a single change of the code), the actual code remained execution stable. The narrative engine for Noble Ape would have to be execution stable. Execution unstable in contrast would mean there would be byte-codes that could *crash* the Noble Ape's language, creating a fatal or irrecoverable error.

The narrative engine commands captured five kinds of things: data, sensors, actuators, operators, and conditionals. The data maintained data elements that were not executed but stored. Sensors captured a variety of simulated external senses of the Noble Apes. Actuators captured the abstracted movement of the Noble Apes. Operators covered both logical and arithmetic operators. Conditionals covered casual logic.

The original narrative engine implementation offered by Mottram had the limitation of just a single narrative. The Noble Apes had this narrative both internally and communicated this narrative externally and it existed as a single entity. I noted that this method did not capture radicalization or an ability to exist in a society and hold independent beliefs (Barbalet and Stone 2011). It was critical to have an internal and an external narrative. These two narratives needed to be quite distinct.

In the current narrative engine, each Noble Ape has an external and an internal narrative that is a stream of byte-codes. When Noble Apes meet and converse, they are running a shared program that alters their own byte-code. This is happening in parallel with their conversing companion. External narrative is exchanged and altered in parallel; this creates a conversation.

When the Noble Ape is not in conversation, the same process is going on but rather than the external narrative being run with another external narrative, the internal narrative of the Noble Ape converses with the external narrative and vice versa. The Noble Apes literally talk to themselves without uttering a simulated sound.

Mottram tied the movement or the physical action of the Noble Ape to the internal narrative. This is an ongoing point of development discourse as I contend the internal narrative should be totally private. At the same time, I concede that the spoken external narrative is not the best place to gather movement from. This mapping of movement from the internal narrative also lends a simulated weight to saying one thing but doing another.

Mottram and I had distinctly different views on the initial conditions of the narratives. My view was that the narrative byte-code should have an even and random probability of occurring in the initial internal and external narrative states. Mottram held the view that the byte-code should be genetically weighted and also contain a distinctly higher ratio of sensors to all other narrative engine types similar to the sensory wonder of a baby. The random case produced faster productive narratives both internally and externally. The genetically ordered with heavy sensor predetermined method produced more natural timescales in terms of productive and mature narrative creation.

10 Narrative Engine as Narrative Generator

The narrative engine-generated byte-code is alien when compared to the English language. It is relatively unintelligible to even those familiar with the byte-code syntax. As with the social graph to understand the social simulation, there is a need for an equivalent technology to turn the Noble Ape narrative byte-code into a human-readable form.

I wrote a scripting language to compliment Noble Ape called ApeScript (Barbalet 2005b). Rather than describing a piece of software, ApeScript creates a programming model for writing a single time-cycle of Noble Ape interaction. Nontrivially, ApeScript can cover more than just a single time-cycle of interaction, but the time-cycle (a simulated minute) is the unit of execution in simulation. ApeScript is created to cover a series of possible situations where the actual circumstances leading into the execution of the ApeScript code define which paths in the ApeScript code will be executed.

The same conditions are in place for the narrative engine byte-code. It is based in the same unit of time and has roughly the same possibilities of code paths.

At the time of writing, the initial work has been performed to translate the byte-code into ApeScript. Curiously the combined ApeScript and byte-code translation is a subset (or intersection) of both languages. It produces a robust syntax that translates both ways. ApeScript is not English, and this final translation is outside the time frame of this chapter; however, it is a direction the development needs to go to provide the following possibility.

The ability to provide a detailed description of the Noble Ape external and internal narratives would provide a compelling additional element. As with the social graph, it would give immediate feedback to a great level of detail on exactly what was happening with Noble Ape societies from an individual up to a community. If the ability to provide bidirectional translation is maintained (as the intersection of ApeScript and the byte-code narrative provides), the ability to inject English language programming back into the simulation environment is possible. Assuming that the English language programming is an intersecting set of wild English (Barbalet and Stone 2011), ApeScript and the narrative byte-code, it may not appear as fluidly readable as wild English but it would provide an ability to add a wide variety of concepts external to the simulation that would have to otherwise be grown organically through the simulation interactions or artificially hardcoded.

11 In Concert

The cognitive simulation, the social simulation, and the narrative engine are not independent simulations. Each takes from elements of the external simulation environment, and each has its own dependencies. All three simulations can be turned off allowing only one or two remaining simulations to run and interact or none of these simulations to run, to test other aspects of the Noble Ape Simulation environment. For clarity, the interactions that are explicitly hardcoded are nullified in this context.

The shared external simulation space should not be discounted in this analysis. It may appear that the cognitive simulation has the most ethereal connection to the external simulation environment. This is not the case. From the early origins of Noble Ape, the connection between movement and the forced feedback loop from the external simulation back into the cognitive simulation resulting in movement ensures the external simulation is the most important contributor to the cognitive simulation (Barbalet 1997b).

The narrative engine is the mediator between the cognitive simulation and the social simulation. Prior to the narrative engine, the Noble Apes existed as reactive agents with additional surprises through social interactions. The movement to hard-code more behaviors created a reinforcement of certain behaviors.

The narrative engine allows for the possibility of future undoing of this hardcoding. It should be possible for all the elements of the hardcoded social simulation to be removed and potentially suggested to the narrative engine. This would allow the Noble Apes to truly evolve their own social norms where concepts like honor are socially agreed upon and also open to individual and historical misinterpretation.

It is possible for the cognitive simulation to hybridize with the narrative engine as well. Consider if the narrative engine byte-codes were communicated through the cognitive simulation substrate. In this regard, the simulations discussed could all resolve to a single system and still maintain their functionality with the potential addition of new behaviors that could not have been explicitly hardcoded.

12 Noble Apes and Humans

This chapter offers a nontechnical surveying of the Noble Ape Simulation to show fundamentally that software can be a useful analytical tool for philosophy. Rather than discussing specific philosophical dilemmas posed by different philosophical models of the mind to determine the possible origin of mind, this chapter has offered a pragmatic surveying of the strengths and weaknesses of simulation methods used to model aspects of the mind as it is externally represented. This has been done intentionally to avoid implicit and oftentimes artificial paradoxes these philosophical models present. As should be clearly demonstrated through Noble Ape, three or more views of the mind can coexist in productive agents.

The connection to origin described here is relatively simple. From basic reactive chemistry through early social needs to language-dominated primates, the origin of the mind can be reduced to basic reactive chemistry; however, this is not a unique solution. There is a multiplicity of solutions.

The solution outside chemistry is equally plausible. It is perfectly credible that a mind could come from computation like the narrative engine, and that this mind would have distinct but valid origins. The narrative engine mind does not have to come through computation either. The origin of language could force the mind as an internal representation of external conversations.

Similarly the mind could come through arbitrary social constraints that force the need for a mind on the entity within the social environment. The mind would exist just as much from the society as it does from the individual.

For coherence, I will continue to write simulation software that coexists rather than finding apparent artificial paradoxes. An artificial mind, whatever its origin, is a terrible thing to waste.

References

Barbalet, T. S. (1997a). *The original manuals of Noble Ape*. Raleigh: Lulu.

Barbalet, T. S. (1997b). Noble Ape Philosophic. *Noble Ape Website*. Retrieved February 10, 2012, from http://www.nobleape.com/man/philosophic.html

Barbalet, T. S. (2004). Noble Ape simulation. *IEEE Computer Graphics and Applications, 24(2)* (pp. 6–12). Los Alamitos: IEEE Computer Society.

Barbalet, T. S. (2005a). Apple's CHUD tools, Intel and Noble Ape. *Noble Ape Website*. Retrieved February 10, 2012, from http://www.nobleape.com/docs/on_apple.html

Barbalet, T. S. (2005b). ApeScript notes. *Noble Ape Website*. Retrieved February 10, 2012, from http://www.nobleape.com/man/apescript_notes.html

Barbalet, T. S. (2009). Noble Ape's cognitive simulation: From agar to dreaming and beyond. In R. Chiong (Ed.), *Nature-inspired informatics for intelligent applications and knowledge discovery: Implications in business, science, and engineering*. Hershey: IGI Global Information Science Reference.

Barbalet, T.S., & Stone, H. (2011). *Stone Ape Podcast*. Retrieved February 10, 2012, from http://www.nobleape.com/stone/

Breazeal, C. L. (2002). *Designing sociable robots (Intelligent robotics and autonomous agents)*. Cambridge, MA: MIT Press.

Dawkins, R. (1987). *The blind watchmaker*. New York: Norton.

Emmeche, C. (1991). *The garden in the machine*. Princeton: Princeton University Press.

Kirsh, D. (1991). Today the earwig, tomorrow man? *Artificial Intelligence, 47*, 161–184.

Hobbes, T. (1651). *Leviathan*. Retrieved February 10, 2012, from http://archive.org/details/hobbessleviathan00hobbuoft

Langton, C. G. (1997). *Artificial life: An overview (Complex adaptive systems)*. Cambridge, MA: MIT Press.

Levy, S. (1992). *Artificial life: A report from the frontier where computers meet biology*. New York: Pantheon.

Ray, T. S. (1991). Evolution and optimization of digital organisms. In K. R. Billingsley et al. (Eds.), *Scientific excellence in supercomputing: The IBM 1990 contest prize papers* (pp. 489–531). Athens: The Baldwin Press.

Shock, J., & Hupp, J. (1982, March). The worms programs – Early experiences with a distributed computation. *Communications of the ACM, 25*(3), 172–180.

Volterra, V. (1931). Variations and fluctuations of the number of individuals in animal species living together. In R. N. Chapman (Ed.), *Animal ecology* (pp. 409–448). New York: McGraw-Hill.

Yaeger, L. S. (1994). Computational genetics, physiology, metabolism, neural systems, learning, vision, and behavior or PolyWorld: Life in a new context. In C. Langton (Ed.), *Proceedings of the artificial life III conference* (pp. 263–298). Reading: Addison-Wesley.

From the Natural Brain to the Artificial Mind

Massimo Negrotti

Abstract Discussing the mind, we face a clear asymmetry: While the brain can be scientifically observed, the mind cannot. However, in order to reproduce something, we need to observe it. The claim according to which the artificial reproduction of some mental activities would be helpful in understanding the mind is weak in principle. For instance, what any school of A.I. tries to reproduce is not the mind but a model of it coming from a specific psychological or ontological paradigm that assumes the existence of the mind as something given. Therefore, the "eradication" of the mind from the brain evolution and activity adds a further degree of arbitrariness to the unavoidable bias and transfiguration that characterizes every attempt to reproduce natural objects, that is, to design *naturoids*.

1 Introduction: The "Brain Shift"

To assert the existence of something is not the same as observing it. This is certainly the case with the human mind because nobody can affirm having observed it, while we must accept the idea that the brain exists, for this is empirically evident. We are all inclined to believe that our mental states or processes come from the brain, although many of us refuse to believe that this same organ is sufficient to explain our reasoning, feelings, and so on. Our cultures have been so deeply and, for such a long time, dominated by the certainty of the existence of the mind, that even on a purely linguistic level, we would find it strange to speak of a "tired brain" rather than a "tired mind," or to ask what is going through someone's brain rather than through his or her mind. But, for all intents and purposes, we would understand each

M. Negrotti (✉)
University of Urbino 'Carlo Bo', Urbino, Italy
e-mail: massimo.negrotti@libero.it

L. Swan (ed.), *Origins of Mind*, Biosemiotics 8, DOI 10.1007/978-94-007-5419-5_21,

other anyway because the two concepts—brain and mind—clearly converge on a unique reality, though one largely unknown.

In fact, dualism presents itself through two historical mainstreams. On the one hand, we have the metaphysical tradition according to which humans possess a twofold reality—namely, the body and the soul. This claim cannot be based on any empirical evidence, of course. Nevertheless, in the course of human history, and even right up to the present day, the existence of the soul has been believed and asserted by many, including many philosophers.

On the other hand, we have the modern dualistic approach, which starts with René Descartes, and progresses, in various forms, via Franz Brentano to contemporary thinkers such as Karl Popper and David Chalmers. The interesting point is that, in contemporary debates, the soul is no longer the issue at stake, and this is probably due to the widely diffused influence of our scientific cultural premises. As a consequence, more subtle or special concepts, such as *consciousness* and *intentionality*, are at the center of current debate among philosophers, neuroscientists, and psychologists. Such concepts are certainly linked to the mind, but, at the same time, they implicitly refer also to the traditional view of the soul. Nevertheless, the simple fact that the soul needs a definite metaphysical foundation induces most scholars to avoid any explicit reference to it.

However, in conceiving the mind as something clearly separated from the material structure of the brain, contemporary dualism traces back to traditional metaphysics, although it replaces philosophical certainties with a wide spectrum of theoretical views and models, and, in the end, with an overall uncertainty regarding what exactly is to be conceived in the concept of *mind*.

Undoubtedly, the underlying reason for the change sketched above is the unavoidable "discovery" of the central role of the brain—and of its regions—in many cognitive or emotional activities. This explains why, for instance, authors such as Popper and Eccles (1984) suggest that the main problem is not that of recognizing the existence of the mind—as certain as the existence of the brain—but, rather, that of describing the interfaces between them; for instance, Jerry Fodor assigns to mental representations the ability to set up a symbolic linkage between the mind and the body (Fodor 1983). Daniel Dennett is one of the most explicit philosophers in assigning, to the brain, functions with the power of triggering consciousness, thus bestowing upon it the role of cause, while consciousness becomes the effect (Dennett 1992). John Searle, following a more sophisticated strategy, speaks of the relationships between mind and brain as something deriving from a sort of "non-event causation" linking the brain and the mind, although it remains mysterious how a "non-event cause" might generate anything other than a "non-event effect," which would seem to be no effect at all (Searle 1999). Gerald Edelman (2004) and Francis Crick (1994) recognize, instead, that in order to understand consciousness, it is necessary to understand what is going on in the brain. Roger Penrose, taking the road of a fine-structure investigation of the brain processes, maintains that even the role of neurons is open to question because they are "too big," while the most interesting level of analysis—if we are ever to discover the roots of consciousness—concerns the cytoskeleton and the quantistic workings of its microlevel components (Penrose 1989).

Although only a minority of scholars explicitly embrace a monistic view (see, for instance, Rorty 1980), it seems clear that a growing power of attraction is being played by the brain and by its neurological functions. In a sense, therefore, even current dualism appears as if it were a residue of ancient metaphysics. In fact, it gives up the radical disjunction between the body and the soul—in terms of both origin and stuff—but simultaneously tries to keep alive the "existence" of something that, although it comes from the physical structure of the brain, cannot be understood as a regular physical function or effect.

In my opinion, such a position comes from a sort of "due respect" paid to the widely shared metaphysical tradition upon which our cultures are based. This takes the form of a die-hard view according to which any physical matter must be conceived as something brute, separated from the superior value of nonphysical reality. In this framework, even scientific observation of the world, while appreciated for its production of pragmatic knowledge, is widely conceived as a mere matter-based kind of activity, explicitly or implicitly classified by many scholars in the humanities, therefore, as belonging to a lower class with respect to the speculative and nonphysical realms. This intellectual standpoint derives from the Hellenistic culture, and its legacy induces many, even today, to think that the lack of empirical evidence doesn't matter at all because the essential truth of things shouldn't be reduced to their empirical phenomenology.

Significantly, in the past century in sociology and anthropology, we can find a meaningful analogy between the concept of mind and that of culture. Here, the structure of society plays the role of the brain, and culture is conceived as the mind of society to such an extent that, as Pitirim Sorokin says, "[t]he superorganic is equivalent to mind in all its clearly developed manifestations" (Sorokin 1947, p. 3).

Culture takes the physiognomy of something strictly similar to the human mind also in the definition given by Alfred Kroeber, when he says that,

> Superorganic does not mean nonorganic, or free of organic influence and causation; nor does it mean that culture is an entity independent of organic life in the sense that some theologians might assert that there is a soul which is or can become independent of living body. Superorganic means simply that when we consider culture, we are dealing with something that is organic but which must also be viewed as something more than organic. (Kroeber 1948)

Such views were strongly criticized because of their more or less conscious tendency toward metaphysics, and probably for this reason, Kroeber, in the last years of his life, modified his position, underlining the methodological role, rather than the substantive one, of the concept of *superorganic*, which should be assumed as an abstract instrument of intelligibility. That is to say, he "came to maintain that culture was *nothing but* an abstraction form" (Bidney 1996). In other words, the superorganic and the mind, in the best case, may play the role of hypothetical constructs that are provisionally useful for designing research on cultural or mental behavior as processes instantiated by physical structures and not as empirical and autonomous realities in themselves.

In all likelihood, the mind, too, will gradually disappear as a *sui generis* "substance," taking up, instead, the role of a more reasonable methodological tool—namely, an abstraction that is useful for expressing what the brain does and how it becomes externalized by communication, in turn generating cultural forms. Lastly, we should emphasize that monism is almost always based on the uniqueness of the brain and not on the uniqueness of the mind. Nobody, apart from a few neo-idealists, currently hopes to find a monistic view of the mind as the unique reality, given that the advancements of neuroscience also have an irresistible appeal, and, therefore, even philosophical speculation enters more and more into the discussion. Nevertheless, the majority of thinkers still refuse to accept the overlap between the working of the brain and the instantiation of thought and feelings. Rather, they are looking for the biophysical sources of something that eventually transcends both biology and physics.

2 The Reification of the Mind

While refusing to believe in the metaphysics of the soul, many contemporary scientists and philosophers of mind cannot but keep alive the traditional belief in the existence of an entity that is nevertheless recognized as a nonphysical one. This paradox is not based exclusively on the history of our civilization. Contemporary dualism also comes from the intriguing questions raised by cybernetics and by information theory as applied to biology. As is well known, cybernetic loops and recursivity—the so-called self-reference phenomena that underlie biological autopoiesis—are at the core of some biological schools of thought.

Although cognition is often outlined as a relational biological process, the insistence upon the self-referential ability of the human brain leads to a belief in the nonphysical nature of the mind and of the property of consciousness. Actually, if an observer is able to observe himself, then this happens *as if he were* outside himself. Therefore, according to this view, since the brain is the only physical entity at stake, the self-observer is a nonphysical external actor—namely, the *mind* that accounts for self-consciousness. This aspect has been discussed many times by relating it to Gödel's seminal theorems denying the possibility, in a consistent formal system, of proving the truth of all true statements within that system. On the other hand, other authors are inclined to think that the human mind is not a formal system and that its most striking property is precisely that of working *as if it were not* a "part" of the brain system. According to this position, the mind is able to, for instance, evaluate the truth of a sentence, as if the process of evaluation would occur outside a formal system, escaping, this way, from Gödel's theorem. The sentences uttered by Gödel himself could be taken as a good example of this ability of the human mind (Webb 1980). Anyway, as has been noted,

> Gödel's theorems do not prevent the construction of formal models of the mind, but support the conception of mind as something which has a special relation to itself, marked by specific limitations. (Bojadziev 1997)

Our ancient habit of thinking that an effect must be brought about by an external cause—while in cybernetic loops, a feedback comes from a part of the system itself—prevents us from accepting that the brain has this "special relation to itself." Consequently, we conjure up an "external" actor, called the mind, in the same way as we have constructed so many myths or metaphysical entities to account for this or that natural phenomenon, our very existence included. However, the construction of metaphysical doctrines or beliefs is a wholly legitimate activity of our brain, which, it seems, is so inclined for some mysterious intrinsic reasons that would seem to lie beyond scientific investigation.

Although recursive abilities are strategic for achieving consciousness, it is far from clear why a nonphysical actor, which we call "mind," should be required for obtaining such performance. Indeed, by introducing it, we are suddenly faced with three problems instead of one. In addition to dealing with the brain and its highly complex nature, we must also deal with the mind, endowed with its own supposed features, and finally, we must face up to the not-inconsiderable problem of connecting the nonphysical mind with the physical brain.

At this point, we should not neglect the rise of symbolic artificial intelligence (A.I.) that has powerfully influenced the debate regarding the mind-body problem, once more tending to privilege mind over brain. This has happened because the features of the mind are, on the face of it, rather less difficult to model than those of the brain, although some hope arises with the advent of so-called artificial neural networks, which mimic, at a superficial level, the way biological neurons create intelligent links. Unlike the brain, whose deep structures and inner workings are largely unknown, the mind and its properties have been described in many different ways. Apart from the numerous philosophical theories put forward over the centuries, we have models of the mind in each of the numerous psychological schools, and in linguistics, cognitive science, logic, anthropology and even philosophy.

After the classical debate on the feasibility of its most ambitious aims, which involved philosophers and engineers in the 1980s, A.I. researchers have, for the most part, chosen models and theories that are best suited for easy transference to computers, thus perhaps justifying the somewhat disparaging definition of A.I. as a technology oriented toward reproducing a theory or a model rather than discovering something new by adopting computer-based techniques. In fact, a computer is not a laboratory, but a "translator" of a model into a symbolic structure and process.

It is interesting to note that, today also, when A.I. researchers work on a "theory testing" level, they are constantly looking for some persuasive analogy that utilizes Ashby's principle of *functionally isomorphic* devices. The general idea is that, in order to better understand a natural object—for example, the human brain or mind—it may be useful to build concrete devices that, within certain limits, should behave in the same way as the natural object under study (Cordeschi 2002). Nevertheless, this strategy neglects the fact that, in doing so, researchers will encounter behaviors that will come not only from the tested theory or from the model as an abstract outline of the natural phenomenon but from the undesigned interplay among the features of the material components of the device.

In other cases, models very often derive from some widespread philosophical or sociological doctrine. For example, Marvin Minsky's theory of the mind as a society of simple and thoughtless agents comes from an old organicist philosophical and sociological tradition that assigns no special importance to the individual components of a society, holding instead that what matters is what "emerges" from the coexistence and interaction among the individual members. Thus, in contrast to Penrose—who supports the hypothesis of there being tracks of the mind at a quantum level in the deepest structures of neurons—Minsky says,

> I'll call "Society of Mind" this scheme in which each mind is made of many smaller processes. These we'll call agents. Each mental agent by itself can only do some simple thing that needs no mind or thought at all. Yet when we join these agents in societies – in certain very special ways – this leads to intelligence. (Minsky 1988)

Nevertheless, it should be noted that the actual successes or failures of A.I. projects have little to do with the widespread discussion of mind that A.I. has promoted and renewed. What the vast majority of A.I. programs actually do is not the reproduction of human *knowing* and *thinking* as such, but rather the logical or quantitative calculations that reproduce explicit rules shared by human beings, including A.I. researchers themselves. In this direction, recent proposals, like ontology engineering, try to set up large databases of linguistic terms defined at various levels of formalization and put together by means of semantic and functional relationships (Denicola et al. 2009). With such strategies, researchers are trying to emulate human common sense, but presumably they establish a quite different system since nobody knows the "rules" that common sense follows.

This same pragmatic attitude in studying the human mind, which privileges the search for successful outcomes instead of pure knowledge, seems to describe the neural network approach. Here, as is well known, despite the ambitious name that recalls the neural functioning of the human brain, the target is to get from the machine the recognition of an input pattern after having "trained" the network—be it hardware or software—to recognize it. This technique is widely adopted for many tasks—especially incomplete data sets—in many sciences and professional activities. Nevertheless, it is at least uncertain to what extent such devices could help us in understanding the human brain. As far as the human mind is concerned—conceived as an additional entity to the physical brain—it has been proposed that neural networks should work together with symbolic A.I. programs (Sun and Bookman. 1994) in order to effect a convergence of reasoning and recognizing that characterizes human mind. Anyway, in the cases of symbolic A.I., which are more inclined to model the human mind, and in the case of neural networks, which are more inclined to model the human brain, it seems clear that dualistic or monistic premises play a key role although each, in the end, has to deal with a unique reality.

3 Reproduction and Observation Levels

While we have an ever-growing knowledge of the brain as a physical organ, we still have many interchangeable models of what the mind is and does. The weakness of models of the mind, as compared to the ever more reliable scientific study of the brain, is not, in itself, a great danger, because what we have come to think of as functions of the mind can often be assigned directly to the brain without any practical consequence. Nevertheless, while we are free to assign to the mind a very wide spectrum of properties and functions—as, in so doing, we have no empirical and spatiotemporal criteria to fulfill—there arises the problem of establishing which of them can really be assigned to the natural brain.

Thus, for example, while our thought is surely generated by the brain—even if one attributes it to the mind—this does not mean that each result of our thought corresponds to a given preestablished brain structure. On the one hand, we can view the mind simply as the performance of the brain, but on the other, we must admit that each so-called mental activity of the brain is not necessarily traceable to some isomorphic brain structure, whereas, within certain limits, we can localize the brain structure involved in, say, the contraction of a given muscle. For example, we can use words or numbers in very different ways, or build and then change views and theories at will, exploiting the same basic biological structure. What changes is probably the specific architecture that each brain assumes. As water flows downhill, the physical forces remain the same, though the paths and the consequences may differ widely, depending on the constraints encountered by the water, or, in the case of the brain, depending on the networks activated at various levels in a given moment.

Recent advancements in neurology include so-called neuroimaging (functional magnetic resonance imaging, or fMRI) that provides visual evidence of the brain areas involved in a wide class of mental states, feelings, or decisions, thus vindicating, and expanding upon, the nineteenth-century hypotheses proposed by Franz Joseph Gall and by Pierre Paul Broca. In brief, these new, highly promising experimental developments

> ...can be defined as the class of techniques that provide volumetric, spatially localized measures of neural activity from across the brain and across time; in essence, a three-dimensional movie of the active brain. (Aguirre 2003)

While neuroimaging falls short of being a "movie" of the mind, it certainly makes it very problematic to reject the idea that the mind is nothing but the performance of a physical system whose activation *coincides* with what we call consciousness. In observing the activated areas of the brain in real time, as neuroimaging allows, we cannot (as yet, at least) identify actual thoughts or words as such. Nevertheless, one cannot plausibly imagine that the mind is something "surrounding" those areas—something "superior" to what these areas are and do. In fact, this would require some empirical evidence as it happens when we define a field exhibiting the measure of all its points. We know that the brain, due to its electrical activity, generates an electromagnetic field, but this cannot be taken as the proof of the existence of the mind, of course.

Just because a model of the mind cannot leave aside the brain as the engine of our consciousness, reasoning, decision-making, and so on, the attempt to reproduce mental behavior artificially is, if conceived as an enterprise that views the mind as a stand-alone system, without doubt destined to fail. Or rather, the reproduction of a mental process in a computer program—for example, via calculations or reasoning—can be successful not because it captures the complex way of reasoning of humans, but because it reproduces the final results of the brain's workings, that is to say, some established and expressible knowledge and logical or quantitative rules and their more or less complex combination.

An expert system, for instance, is a type of software that is able to provide consultancy, in terms of both explanation and prediction, to the user in a specific field of knowledge, such as medicine, law, or whatever. The system is able to do this with an acceptable success rate thanks to the "donation" from a human expert, who decants, as it were, his professional knowledge into a database. Then the software, through a set of inferential and statistical rules embedded in it, becomes able to deliver its consultancy as if it were, within certain limits, the human expert himself.

The key point is that what is modeled in an expert system is not a human brain, nor a supposed mind, but the final results—knowledge and rules—that humans have obtained after having worked for centuries on the best ways to reason with the facts within a given domain. This is why no A.I. program has yet been able to propose some new problem, although many such programs are undoubtedly useful in the problem-solving domain.

Within the long-running debate on the feasibility of A.I., John McCarthy maintained that a machine—even a simple thermostat—can think and have beliefs. He writes:

> [T]he thermostat can only be properly considered to have just three possible thoughts or beliefs. It may believe that the room is too hot, or that it is too cold, or that it is okay. It has no other beliefs; for example, it does not believe that it is a thermostat. (McCarthy 1990)

Almost a decade later, John Searle suggested that such a claim would imply the bad idea

> …that the hunk of metal on the wall that we use to regulate the temperature has beliefs in exactly the same sense that we, our spouses, and our children have beliefs. (Searle 1999, p. 410)

McCarthy and Searle were both right, because they were speaking of different things. McCarthy intended the operational logic embodied in the device, while Searle was referring specifically to *human* thought. The fact is that an algorithm incorporated in a thermostat is the explicit result of human reasoning—namely, that of the designer—and as such will demonstrate behavior reminiscent of artificially intelligent reasoning. Searle, by contrast, was concerned with the process of knowing which can, among other things, produce an algorithm. In the same way, humans generate knowledge that other humans are then taught in schools or universities. But these are very different processes.

Widening the well-known concept of "tacit knowledge" introduced by M. Polanyi in the 1960s (Polanyi 1966), and contrary to Maturana's thesis according to which the human brain "thinks in language" (Maturana et al. 1995), we may state that everything that happens in our brain is "silent" and possibly very well hidden in the microstructures and networks of interactions within the brain. We can utter sentences whose knowledge-content comes from the brain only after such content has been translated and transduced via several still unknown processes. Thought is to be understood as a truly preverbal process; only a small part of it can actually be externalized by language, of which an even smaller portion becomes shared knowledge in our culture.

This explains why advancements of our knowledge, both at individual and cultural levels, always take a huge amount of time as compared to the speed with which a new problem or a new strategy occurs in one's brain.

Therefore, models of mental processes, and their technological realizations, are successful when we implicitly define the "mind" as the name we give to our thought as already realized and communicated, such as inferential rules, mathematical or statistical ones, and common-sense based standards. It is quite unlikely that, instead of the established knowledge of a human expert in some discipline, we could exploit the "way of thinking" of Einstein or Mozart or anybody else. We can build an expert system based on Einstein's physics or Mozart's musical style, but we cannot enter into their way of relating to the world, or into the working of their brain when generating their theories or musical compositions. In Einstein's and Mozart's work, what is understandable and reproducible are the established and linguistically communicable results of their physical or musical thinking, and not the processes that have led to that.

Elsewhere (Negrotti 1999, 2010a, b) I have outlined a possible general theory of the technological reproduction of natural objects or processes—that is to say, the designing of *naturoids*. I wish here to make use of that theory in order to clarify the meaning of the foregoing discussion.

In order to design a naturoid, we should begin by observing the natural object or process we wish to reproduce. In fact, we may develop a model of a natural *exemplar*—to serve as the basis of a project—if and only if we can describe it after some empirical observation. For instance, if I wish to reproduce a kidney by means of current technological devices and techniques, I must be able to describe the natural exemplar—that is to say, the kidney—in the richest, most reliable, and objective way possible. All observation is, however, a process that is to be conceived as relating to some selected *observation level*: for instance, mechanical, chemical, electrical, biological, etc. To date, no project of naturoid production—beyond the level of chemically reconstructed molecules—can claim to have reproduced all the properties of a natural exemplar, and this depends, apart from other constraints, exactly upon the need to select one and only one observation level at any given moment and also upon the almost insurmountable difficulty of "joining" two or more such levels. Furthermore, even at a selected observation level, one has to decide what the *essential performance* of the natural object or process is that one wishes to reproduce.

The relativity of any observation level does not render impossible an objective description of a real object or process. It does, however, limit such a description to what is compatible with the particular level adopted, leaving, in the process, all else in the background. If, for example, I describe a certain exemplar from a chemical point of view, I am unable to capture its mechanical performances in my description. Nevertheless, the knowledge of its chemical properties may be sufficiently objective and useful for designing a naturoid that will be able to behave much like the natural exemplar in some respect, provided it is to work in a context characterized by the same observation level that I have selected—namely, the chemical one.

It often happens that a failure in the technological reproduction of a natural exemplar depends upon the wrong choice of observation level, as unfortunately happens not so rarely in bioengineering projects (Negrotti 2010a, b). But failure may also arise from the pretense of having the naturoid work according to an observation level that differs from the one at which it has been designed. Thus, as far as the problem of the reproducibility of the brain is concerned, much of the difficulty in the design depends on our rather poor knowledge of its possible observation levels. Furthermore, it would be even more difficult to decide which level should be considered the most indispensable in order to generate the brain's essential performance—namely, what we call mental states and processes.

However, we should also take into account that not all the designers of artificial objects remain faithful to the rule of the objective observability of the exemplars they wish to reproduce. Although this rule is widely accepted in the fields of naturoids designed within mature scientific disciplines, such as bioengineering, in several other fields, there is widespread use of arbitrary models constructed for describing equally arbitrary entities. This has happened many times in art, for instance, where painters have often represented metaphysical entities, such as God, assigning to them features imposed by a religious tradition acting as a "model" to be realized. But this happens regularly even in the field of A.I., since the model of the mind, or of one of its functions, is built up on the basis of this or that theory, even though none of the theories relates to an objectively established observation of the topic at issue. In other words, we cannot speak of a mental or mind-based observation level for the simple reason that we can observe only the brain and not the mind, the communicable results of the brain's workings and not the flow of mental processes.

From a methodological viewpoint, we cannot say to what extent the behavior of an A.I. program faithfully reproduces that which occurs in our brain, apart from the cases in which—as in the expert systems mentioned above and other computer science software—the project aims to reproduce only the results of our thinking and not the performance of thinking in itself.

A final question is: What would happen if we were able to reproduce a brain by means of technology, and, therefore, on the basis of some reliable model of this natural organ, though built according to only one observation level?

If its reproduction were to follow the ways of past and current methodology of the design of naturoids—whose limits are perhaps imposed by our very nature in observing the world—my opinion is that we would not see any "mind" emerging from it, even if we were successful in making the machine exhibit behavior that, if it were seen in human beings, would indicate some class of mental states and would be self-aware. In such a case, we would have surely built an artificial brain, and, in so doing, we would have discovered how useless, or unmanageable, the notion of mind is.

References

Aguirre, G. K. (2003). Functional imaging in behavioral neurology and cognitive neuropsychology. In T. E. Feinberg & M. J. Farah (Eds.), *Behavioral neurology and cognitive neuropsychology* (p. 85). New York: McGraw Hill.

Bidney, D. (1996). *Theoretical anthropology* (p. XXXI). New Brunswick: Transaction Publishers.

Bojadziev, D. (1997). Mind versus Gödel. In M. Gams, M. Paprzycki, & X. Wu (Eds.), *Mind versus computer* (p. 210). Amsterdam: Ios Press.

Brentano, F. (1874). *Psychologie vom empirischen Standpunkt*. Leipzig: Duncker & Humblot.

Chalmers, D. (1996). *The conscious mind: In search of a fundamental theory*. New York: Oxford University Press.

Cordeschi, R. (2002). *The discovery of the artificial: Behavior, mind and machines before and beyond cybernetics*. Dordrecht: Kluwer Academic Publishers.

Crick, F. (1994). *The astonishing hypothesis*. New York: Simon & Schuster.

Denicola, A., Missikoff, M., & Navigli, R. (2009). A software engineering approach to ontology building. *Information Systems, 34*, 258.

Dennett, D. (1992). *Consciousness explained*. New York: Back Bay Books.

Edelman, G. (2004). *Wider than the sky: The phenomenal gift of consciousness*. New Haven: Yale University Press.

Fodor, J. (1983). *The modularity of mind: An essay on faculty psychology*. Cambridge, MA: MIT Press.

Kroeber, A. (1948). *Anthropology: Race, language, culture, psychology, prehistory* (p. 253). New York: Harcourt Brace.

Maturana, H., Mpodozis, J., & Letelier, J. C. (1995). Brain, language and the origin of human mental functions. *Biological Research, 28*, 15–26.

McCarthy, J. (1990). The little thoughts of thinking machines. In V. Lifschitz (Ed.), *Formalizing common sense. Papers by John McCarthy* (p. 183). Nordwood: Ablex Publishing Corporation.

Minsky, M. (1988, March 15). *The society of mind* (p. 17). New York: Simon and Schuster.

Negrotti, M. (1999). *The theory of the artificial*. Exeter: Intellect Books.

Negrotti, M. (2010a). Designing the artificial: An interdisciplinary study. In R. Buchanan, D. Doordan, & V. Margolin (Eds.), *The designed world*. Oxford: Berg.

Negrotti, M. (2010b). Naturoids: From a dream to a paradox. *Futures, 42*(7), 759–768. Elsevier.

Negrotti, M. (2012). *From nature to naturoids. And back*. Heidelberg: Springer.

Penrose, R. (1989). *Shadows of the mind: A search for the missing science of consciousness*. Oxford: Oxford University Press.

Polanyi, M. (1966). *The tacit dimension*. New York: Doubleday & Co.

Popper, K., & Eccles, J. C. (1984). *The self and its brain: An argument for interactionism*. New York: Taylor & Francis.

Rorty, R. (1980). *Philosophy and the mirror of nature*. Princeton: Princeton University Press.

Searle, J. (1992). *The rediscovery of the mind*. Cambridge, MA: MIT Press.

Searle, J. (1999). Minds, brains, and programs. In N. Warburton (Ed.), *Philosophy. Basic readings*. New York: Routledge.

Sorokin, P. A. (1947). *Society, culture and personality: Their structure and dynamics*. New York: Cooper Square Publishers.

Sun, R., & Bookman, L. (Eds.). (1994). *Computational architectures integrating neural and symbolic processes*. Needham: Kluwer Academic Publishers.

Webb, J. (1980). *Mechanism, mentalism and metamathematics: An essay on finitism*. Dordrecht/Boston: D. Reidel Publ. Co.

References

Index

L. Swan (ed.), *Origins of Mind*, Biosemiotics 8, DOI 10.1007/978-94-007-5419-5,

D

N

O

MIX
Papier aus verantwortungsvollen Quellen
Paper from responsible sources
FSC® C105338

If you have any concerns about our products,
you can contact us on
ProductSafety@springernature.com

In case Publisher is established outside the EU,
the EU authorized representative is:
Springer Nature Customer Service Center GmbH
Europaplatz 3, 69115 Heidelberg, Germany

Printed by Libri Plureos GmbH
in Hamburg, Germany